SECOND EDITION

SLIP, TRIP, and FALL PREVENTION

A Practical Handbook

Reader Comments on the First Edition*

"This book presents a scientific description of slip resistance measurement, methods for ensuring accurate and repeatable quantification of slip resistance, and recommendations for maintaining slip-resistant workplaces. I highly recommend this text for anyone serious about controlling slip/fall hazards. The illustrations and photos are excellent, and the text is clear and direct. If your work requires the investigation, correction, or litigation of slip/fall accidents, this book belongs on your reference shelf."

Richard Sesek
University of Utah, SLC, UT

"Experienced safety professionals will quickly see the value in this book. But it is the corporate risk manager, HR generalist, or in-house facilities engineer charged with finding effective solutions who stands to gain the most. Steve Di Pilla ranks among the most respected voices in the world of slip and fall prevention. He goes out of his way to offer useful, unbiased information. This handbook offers practical real-world recommendations."

Steven Anderson
Zurich Risk Services

"This book is a comprehensive reference on slips and falls and "must have" for every safety professional reference library. Very useful hazard recognition and prevention guidelines are offered for indoor and outdoor slips and falls from floors, stairs, and ramps. In addition, this is one of the best references I've seen on slip-resistance measurement guidelines and tribology standards."

Wayne S. Maynard
Liberty Mutual Research Center

"Mr. DiPilla's book is delightfully clear, concise and comprehensive. It educates the reader from the scope of slips and falls issues, to common procedures for recognizing and mitigating walkway hazards, to management control methods and accident investigation/claims mitigation. The text and illustrations read in a simple, common-sense way, showing the truth of Voltaire's comment that 'common sense isn't so common.'

Another noteworthy feature of this book is the remarkably comprehensive, annotated references to slip resistance principles, flooring, footwear, and slip-resistance measurement techniques, both within the United States and internationally. The annotated listings of U.S. and international standards and guidelines in themselves provide the best collection of references I've seen on this complex subject. Remarkable."

David C. Underwood, Ph.D.

---

* Courtesy of Amazon.com, Inc. or its affiliates. All rights reserved.

SECOND EDITION

SLIP, TRIP, and FALL PREVENTION
A Practical Handbook

Steven Di Pilla

CRC Press is an imprint of the
Taylor & Francis Group, an **informa** business

CRC Press
Taylor & Francis Group
6000 Broken Sound Parkway NW, Suite 300
Boca Raton, FL 33487-2742

© 2010 by Taylor and Francis Group, LLC
CRC Press is an imprint of Taylor & Francis Group, an Informa business

No claim to original U.S. Government works

Printed in the United States of America on acid-free paper
10 9 8 7 6 5 4 3 2 1

International Standard Book Number: 978-1-4200-8234-0 (Hardback)

This book contains information obtained from authentic and highly regarded sources. Reasonable efforts have been made to publish reliable data and information, but the author and publisher cannot assume responsibility for the validity of all materials or the consequences of their use. The authors and publishers have attempted to trace the copyright holders of all material reproduced in this publication and apologize to copyright holders if permission to publish in this form has not been obtained. If any copyright material has not been acknowledged please write and let us know so we may rectify in any future reprint.

Except as permitted under U.S. Copyright Law, no part of this book may be reprinted, reproduced, transmitted, or utilized in any form by any electronic, mechanical, or other means, now known or hereafter invented, including photocopying, microfilming, and recording, or in any information storage or retrieval system, without written permission from the publishers.

For permission to photocopy or use material electronically from this work, please access www.copyright.com (http://www.copyright.com/) or contact the Copyright Clearance Center, Inc. (CCC), 222 Rosewood Drive, Danvers, MA 01923, 978-750-8400. CCC is a not-for-profit organization that provides licenses and registration for a variety of users. For organizations that have been granted a photocopy license by the CCC, a separate system of payment has been arranged.

Trademark Notice: Product or corporate names may be trademarks or registered trademarks, and are used only for identification and explanation without intent to infringe.

Library of Congress Cataloging-in-Publication Data

Di Pilla, Steven.
  Slip, trip, and fall prevention : a practical handbook / Steven Di Pilla. -- 2nd ed.
    p. cm.
  Previous ed. published under title: Slip and fall prevention, 2003.
  Includes bibliographical references and index.
  ISBN 978-1-4200-8234-0 (hardcover : alk. paper)
  1. Falls (Accidents)--Prevention. I. Di Pilla, Steven. Slip and fall prevention. II. Title.

T55.D565 2009
620.8'6--dc22   2009018494

Visit the Taylor & Francis Web site at
http://www.taylorandfrancis.com

and the CRC Press Web site at
http://www.crcpress.com

Contents

Foreword xix
Preface...xxi
Acknowledgments xxiii
Author xxv
Introduction...xxvii

Chapter 1 Physical Evaluation 1

    1.1   Introduction 1
    1.2   Expectation 1
            1.2.1   Missteps...1
    1.3   Distractions..2
    1.4   Level Walkway Surfaces 2
            1.4.1   Sidewalks..3
                     1.4.1.1   Rubberized Sidewalks 5
            1.4.2   Curbing..6
                     1.4.2.1   Curb Construction and Design 6
                     1.4.2.2   Curb Marking 8
            1.4.3   Curb Cutouts 8
            1.4.4   Other Walkway Impediments 10
                     1.4.4.1   Access Covers..10
                     1.4.4.2   Drainage Grates.....................................10
                     1.4.4.3   Posts 12
                     1.4.4.4   Bicycle Racks 12
                     1.4.4.5   Planters, Trash Receptacles, and
                                    Similar Furnishings 12
                     1.4.4.6   Architectural Designs.............................12
                     1.4.4.7   Doorstops and Other Small Trip
                                    Hazards 13
                     1.4.4.8   Temporary Fixes....................................13
    1.5   Level Walkway Surfaces and Water 16
    1.6   Parking Areas 19
            1.6.1   Tire Stops 19
            1.6.2   Speed Bumps..22
    1.7   Changes in Levels...24
            1.7.1 Air Steps 24
            1.7.2 Design...25
    1.8   Stairs..26
            1.8.1   Stair Missteps—Overstep and Heel Scuffs...........26
            1.8.2   Stair Statistics...27

		1.8.3	Stair Design 29
			1.8.3.1 Stair Nosings..................................30
			1.8.3.2 Stair Visibility..............................31
			1.8.3.3 Stair Landings..............................32

1.9 Handrails 33
    1.9.1 Graspability 35
    1.9.2 Supports..36
1.10 Guards (or Guardrails)..39
1.11 Ramps...40
    1.11.1 Ramp Design 40
    1.11.2 Ramp Landings 43
1.12 Floor Mats/Entrances and Exits 43
    1.12.1 General Precautions 44
    1.12.2 Minimizing Contaminants 45
    1.12.3 Mat Design 46
    1.12.4 Mat Design and Arrangement....................................46
    1.12.5 Mat Storage 47
    1.12.6 Mat Size..48
    1.12.7 Protection of Hard Flooring Surfaces 51
    1.12.8 Mat Cleaning Guidelines 51
1.13 Bathrooms 52
1.14 Elevators 52
1.15 Escalators..54
1.16 Carpet 56
1.17 Accessibility 57
1.18 Signage 57
1.19 Reference Standards 59
    1.19.1 NFPA International (Formerly National Fire
            Protection Association) 59
            1.19.1.1 NFPA 101 Life Safety Code (2006)............59
    1.19.2 ASTM International (Formerly American
            Society of Testing and Materials)...............................59
            1.19.2.1 ASTM F1637 Practice for Safe
                    Walking Surfaces...59
    1.19.3 American National Standards Institute (ANSI) 59
            1.19.3.1 ANSI A1264.1..59
            1.19.3.2 ANSI A1264.2 60
            1.19.3.3 ICC/ANSI A117.1 60
    1.19.4 Occupational Safety and Health Administration
            (OSHA)...60
            1.19.4.1 1910, Subpart D—Workplace Walking
                    and Working Surfaces.................................60
    1.19.5 Model Building Codes 60
    1.19.6 ASME International (Formerly American
            Society of Mechanical Engineers)..............................61

Contents

Chapter 2 Management Controls 67
    2.1 Introduction 67
    2.2 Engineering Precept 68
        2.2.1 No "Silver Bullet" 68
    2.3 Integrate Controls 68
        2.3.1 Program Component Design 69
        2.3.2 Hierarchy of Controls 69
            2.3.2.1 Elimination 69
            2.3.2.2 Engineering Controls 69
            2.3.2.3 Administrative Controls 70
            2.3.2.4 Personal Protective Equipment (PPE) 70
    2.4 Behavioral Safety and Pedestrian Traffic Flow 70
        2.4.1 Natural Paths/Observation 70
        2.4.2 Adapt or Adopt 70
    2.5 Building Inspection and Maintenance Programs 71
    2.6 Slip/Fall Hazard Self-Inspection Programs 72
    2.7 Spill and Wet Program 74
        2.7.1 Electronic Inspection System 75
    2.8 Recommended Practices for Snow Removal 78
        2.8.1 Objectives 78
        2.8.2 Personnel and Responsibilities 79
            2.8.2.1 Facility Manager 79
            2.8.2.2 Custodians 79
            2.8.2.3 Grounds Maintenance Staff 79
            2.8.2.4 Contracted Snow Removal 79
        2.8.3 Priorities for Removal 79
        2.8.4 Guidelines for Removal 80
        2.8.5 Snow Storage 81
        2.8.6 Application of Anti-Icing, Deicing, and Sand 81
            2.8.6.1 Temperature and Ice Melting 82
            2.8.6.2 Before the Storm 83
            2.8.6.3 During the Storm 84
            2.8.6.4 After the Storm 86
    2.9 Lighting 87
        2.9.1 Light Sources 87
            2.9.1.1 Mercury Vapor 87
            2.9.1.2 Metal Halide 88
            2.9.1.3 High-Pressure Sodium (HPS) 88
        2.9.2 Lighting Levels—Safety Only 88
        2.9.3 Lighting Levels—Categories 89
        2.9.4 Lighting Transitions 89
        2.9.5 Maintenance 90
    2.10 Contractual Risk Transfer 91
        2.10.1 General Administrative Measures 92

		2.10.2	Specific Control Measures	92
		2.10.3	Fundamental Guidelines	93
	2.11	Construction, Renovation, and Special Event Planning		93
	2.12	Loss Analysis		94
		2.12.1	Tracking Exposure	94
		2.12.2	Gathering the Data	94
		2.12.3	Developing Solutions	95

Chapter 3 Principles of Slip Resistance ... 105

	3.1	Introduction		105
	3.2	Principles of Friction		105
	3.3	Slip Resistance Defined		106
	3.4	Slip Resistance Factors		107
	3.5	The Mechanics of Walking		107
	3.6	Causation between Incident and Injury		108
	3.7	Slip Resistance Scale		109
	3.8	Surface Roughness		110
		3.8.1	Height or Sharpness	111
		3.8.2	Liquid Dispersion	111
		3.8.3	Measuring Roughness	111
		3.8.4	Roughness Thresholds	112
			3.8.4.1 HSE Roughness Thresholds	112
		3.8.5	Roughness Measurement Standards	114
	3.9	Wet Surfaces		115
		3.9.1	Hydroplaning	115
		3.9.2	Sticktion	115
	3.10	Human Perception of Slipperiness		116
	3.11	Classes of Tribometers		117
		3.11.1	Horizontal Pull (Dragsled)	117
		3.11.2	Pendulum	117
		3.11.3	Articulated Strut	118
	3.12	Hunter Machine		118

Chapter 4 U.S. Tribometers ... 121

	4.1	Introduction		121
	4.2	James Machine		121
		4.2.1	Operation	121
		4.2.2	Subsequent Versions	122
		4.2.3	Operational Issues	123
		4.2.4	Standards	125
			4.2.4.1 Committee D21 Polishes	125
	4.3	Horizontal Pull Slipmeter (HPS)		126

Contents ix

- 4.4 Horizontal Dynamometer Pull-Meter (C21 Ceramic Whitewares and Related Products) 128
  - 4.4.1 Issues 128
    - 4.4.1.1 Scope 129
    - 4.4.1.2 Significance and Use 129
    - 4.4.1.3 Apparatus Issues 130
- 4.5 Brungraber Mark I Portable Articulated Strut Slip Tester (PAST) 131
  - 4.5.1 Operation 132
  - 4.5.2 Availability and Standards 132
  - 4.5.3 F15 Consumer Products 133
    - 4.5.3.1 F462 133
- 4.6 Brungraber Mark II Portable Inclinable Articulated Strut Slip Tester (PIAST) 134
- 4.7 English XL Variable Incidence Tribometer (VIT) 137
  - 4.7.1 D01 Paint and Related Coatings, Materials, and Applications 141
    - 4.7.1.2 D5859 141
  - 4.7.2 WK11411 141
- 4.8 Test Pad Material 141
  - 4.8.1 Leather 141
  - 4.8.2 Neolite® Test Liner 142
  - 4.8.3 Rubbers and Other Footwear Materials 143
    - 4.8.3.1 Neoprene 143
    - 4.8.3.2 Four S 143
    - 4.8.3.3 TRRL or TRL Rubber 144
  - 4.8.4 Actual Footwear Bottoms 144
- 4.9 Uses for Tribometers 144
  - 4.9.1 Problem Identification—Prevention 144
  - 4.9.2 Little or No Prior History 144
  - 4.9.3 Claims Defense/Documentation 145
  - 4.9.4 Accident Investigation 145
- 4.10 Operator Qualifications of Competency 145
- 4.11 Equipment Calibration and Maintenance 145
- 4.12 Progress on the ASTM "Gold" Standard 146
  - 4.12.1 Validation and Calibration of Walkway Tribometers 146
  - 4.12.2 Field Testing with Walkway Tribometers 147
  - 4.12.3 Thresholds 147
  - 4.12.4 Benefits 148
- 4.13 Groundbreaking Research 148

Chapter 5 U.S. Standards and Guidelines 161

- 5.1 Introduction 161

5.2		Occupational Safety and Health Administration (OSHA)	161
	5.2.1	Section 1910.22 General Requirements	161
	5.2.2	Manlifts 1910.68(c)(3)(v)	162
	5.2.3	Fire Brigades 1910.156(e)(2)(ii)	162
	5.2.4	Appendix B to 1926 Subpart R—Steel Erection Regulatory (3) [Withdrawn]	163
5.3		Americans with Disabilities Act (ADA)	164
	5.3.1	ADAAG 4.5 Ground and Floor Surfaces/A4.5.1	165
5.4		Access Board Recommendations	166
5.5		Federal Specifications	166
	5.5.1	RR-G-1602D	166
5.6		U.S. Military Specifications (Navy)	166
	5.6.1	MIL-D-23003A(SH)	167
	5.6.2	MIL-D-24483A	167
	5.6.3	MIL-D-0016680C (Ships) and MIL-D-18873B	167
	5.6.4	MIL-D-3134J	167
	5.6.5	MIL-D-17951C (Ships)	167
	5.6.6	MIL-W-5044C	167
5.7		ASTM International (Formerly American Society for Testing and Materials)	167
	5.7.1	Technical Committee ASTM F-13	168
		5.7.1.1 ASTM F695	169
		5.7.1.2 ASTM F1240	169
		5.7.1.3 ASTM F1637	169
		5.7.1.4 ASTM F1646	170
		5.7.1.5 ASTM F1694	170
		5.7.1.6 ASTM F802	170
		5.7.1.7 ASTM F2048	170
5.8		NFPA International (Formerly National Fire Protection Association)	170
	5.8.1	NFPA 1901	171
	5.8.2	NFPA 101/5000	171
5.9		American National Standards Institute (ANSI)	171
	5.9.1	ANSI A1264.2-2006	172
	5.9.2	ANSI A1264.3-2007	173
	5.9.3	ANSI A137.1-1988	173
5.10		Underwriters Laboratories	173
	5.10.1	UL 410	174
5.11		Model Building Codes	176
5.12		Obsolete Standards	176
	5.12.1	Federal Test Method Standard 501a, Method 7121	176
	5.12.2	U.S. General Services Administration Specification PF-430C(1)	176
	5.12.3	ASTM D4518-91	177
	5.12.4	ASTM D-21 Gray Pages	177

Contents

5.13 U.S.-Based Industry Associations and Organizations Involved with Slip Resistance 177
    5.13.1 Ceramic Tile Institute of America (CTIOA) 177
    5.13.2 Tile Council of North America (TCNA) 178
    5.13.3 National Floor Safety Institute (NFSI) 179
        5.13.3.1 NFSI Position on Slip Resistance Testing 180
    5.13.4 Consumer Specialty Products Association (CSPA) 181
    5.13.5 Resilient Floor Covering Institute (RFCI) 182
    5.13.6 Footwear Industries of America (FIA) 182
    5.13.7 Contact Group on Slips, Trips, and Falls (CGSTF) 183
    5.13.8 International Ergonomics Association (IEA) 183
    5.13.9 National Safety Council (NSC) 183
    5.13.10 American Academy of Forensic Sciences (AAFS) 184

Chapter 6 Flooring and Floor Maintenance 195

6.1 Introduction 195
6.2 The Threshold of Safety 195
6.3 Identifying Types of Flooring and Their Properties 196
    6.3.1 Resilient Flooring 197
    6.3.2 Nonresilient Flooring 198
    6.3.3 Other Types of Flooring 199
6.4 Floor Finishes and Their Properties 200
    6.4.1 The Relationship of Shine to Slip 200
    6.4.2 Conventional Floor Finishes 201
    6.4.3 "Slip-Resistant" Floor Treatments 202
        6.4.3.1 Particle Embedding 202
        6.4.3.2 Surface Grooving and Texturing 202
        6.4.3.3 Etching 203
        6.4.3.4 Other Slip-Resistant Floor Treatments 203
6.5 The Impact of Wear 204
6.6 Floor Cleaning 204
    6.6.1 Floor Cleaning Products 205
        6.6.1.1 Basic Types of Floor Cleaners 205
        6.6.1.2 Six Categories of Floor Cleaners 206
    6.6.2 Floor Cleaning Issues and Methods 206
        6.6.2.1 Floor Stripping 207
        6.6.2.2 Wet Mopping Floors 208
        6.6.2.3 Floor Buffing/Polishing 208
        6.6.2.4 Floor Care Equipment Maintenance 209

		6.6.3	Floor Finish Indicators and Maintenance Issues 212

- 6.6.3 Floor Finish Indicators and Maintenance Issues 212
- 6.6.4 Outsourcing Floor Care 212
  - 6.6.4.1 Outsourcing Labor and Equipment 213
  - 6.6.4.2 Owning Equipment and Outsourcing Labor 213
  - 6.6.4.3 Owning the Equipment and Using In-House Labor 214
- 6.7 Carpet Maintenance 216
- 6.8 Assessment of Floor Treatment/Cleaning Products and Methods 216
- 6.9 Floor Maintenance Certification 217
  - 6.9.1 IICRC Hard Surface Floor Maintenance Specialist (FCT) 217
  - 6.9.2 Certified Floor Safety Technician (CFST) 217
  - 6.9.3 Rochester Midland Corporation 217
- 6.10 Floor Treatment Study 217

Chapter 7 Overseas Standards 237

- 7.1 Introduction 237
- 7.2 Slip and Fall Statistics Overseas 237
- 7.3 Overseas Standard Development 238
- 7.4 Ramp Tests 239
  - 7.4.1 Operational Issues 240
- 7.5 Pendulum Testers 243
  - 7.5.1 Operation 243
  - 7.5.2 ASTM E-303 243
  - 7.5.3 Issues 245
- 7.6 Digitized Dragsleds 247
  - 7.6.1 Issues 249
- 7.7 Other Dragsleds 251
- 7.8 Roller-Coaster Tests 252
  - 7.8.1 SlipAlert Operation 252
  - 7.8.2 Operational Issues 253
  - 7.8.3 Results 253
- 7.9 Portable Friction Tester 254
- 7.10 International Standards 255
  - 7.10.1 European Standards 255
    - 7.10.1.1 CEN 256
    - 7.10.1.2 CEN Standards Process 256
    - 7.10.1.3 CEN Slip Resistance Standards and Drafts 256
  - 7.10.2 German Standards 258

Contents   xiii

                7.10.2.1 DIN 18032 P2 DIN V 18032-2 Sport Halls—Halls for Gymnastics, Games and Multi-Purpose Use—Part 2: Floors for Sporting Activities; Requirements, Testing 259
                7.10.2.2 DIN 51097 Testing of Floor Coverings; Determination of the Anti-Slip Properties; Wet-Loaded Barefoot Areas; Walking Method; *Ramp* Test 259
                7.10.2.3 DIN 51130 Testing of Floor Coverings; Determination of the Anti-Slip Properties; Workrooms and Fields of Activities with Raised Slip Danger; Walking Method; Ramp Test 259
                7.10.2.4 Draft Standard DIN 51131 Testing of Floor Coverings—Determination of the Anti-Slip Properties— Measurement of Sliding Friction Coefficient 260
        7.10.3 British Standards 260
                7.10.3.1 Committee B/556 261
                7.10.3.2 Committee B/208 261
                7.10.3.3 Committee B/545 261
                7.10.3.4 Committee B/539 262
                7.10.3.5 Committee PRI/60 262
        7.10.4 Swedish Standards 262
                7.10.4.1 SS-EN 1893 Resilient, Laminate and Textile Floor Coverings— Measurement of Dynamic Coefficient of Friction on Dry Floor Surfaces 262
        7.10.5 Australia/New Zealand Standards 262
                7.10.5.1 The Standards Process 263
                7.10.5.2 Australian Standards 263
        7.10.6 Italian Standards 265
                7.10.6.1 DM 14 Guigno 1989 n. 236 265
        7.10.7 International Organization for Standardization (ISO) 266
                7.10.7.1 ISO Standards Process 266
                7.10.7.2 ISO Concerns 267
                7.10.7.3 ISO Flooring Committees 267
7.11 Overseas Organizations Involved in Slip Resistance 268
        7.11.1 The U.K. Slip Resistance Group 268
        7.11.2 Health and Safety Executive (HSE) 269
        7.11.3 SATRA Footwear Technology Centre 269

	7.11.4	Commonwealth Scientific and Industrial Research Organisation (CSIRO)	270
	7.11.5	INRS National Research and Safety Institute	271
	7.11.6	Berufsgenossenschaftliches Institut fur Arbeitssicherheit (BIA)	272
	7.11.7	Finnish Institute of Occupational Health (FIOH)	272
	7.11.8	International Association of Athletics Federations (IAAF)	272

Chapter 8 Footwear 295

8.1 Introduction 295
    8.1.1 Industry Conditions 295
    8.1.2 Potential Impact of Footwear 296
8.2 Footwear Design for Slip Resistance 296
    8.2.1 Sole Compounds 297
    8.2.2 Outsole Tread Patterns 297
    8.2.3 General Guidelines for Shoe Design and Selection 298
8.3 Labeling 298
    8.3.1 Labeling for Usage 298
    8.3.2 Labeling for Slip Resistance Testing 299
8.4 Advertising 300
8.5 Other Selection Guidelines 300
8.6 Other Protective Features 301
8.7 Maintenance 302
    8.7.1 Keeping Clean 302
    8.7.2 Wear and Inspection 302
    8.7.3 Replacement 303
8.8 Footwear Programs 304
    8.8.1 Mandate or Recommend 304
    8.8.2 Specifications 304
    8.8.3 Purchase Options 304
    8.8.4 Enforcement 305
8.9 Federal Specification—USPS No. 89C 305
8.10 Consensus Standards 305
    8.10.1 ASTM F08 Sports Equipment and Facilities ... 305
        8.10.1.1 ASTM F2333 Standard Test Method for Traction Characteristics of the Athletic Shoe–Sports Surface Interface 306
8.11 International Footwear Standards for Slip Resistance ... 306
    8.11.1 ISO 306
        8.11.1.1 ISO TC 94 Personal Safety— Protective Clothing and Equipment ... 306

Contents   xv

| | | 8.11.1.2 ISO TC 216 Footwear .... 309 |
| | 8.12 | GERMAN—DIN 4843–100 .... 309 |

Chapter 9 Food Service Operations 311

- 9.1 Introduction 311
- 9.2 Exposure Overview 312
- 9.3 Pedestrian Flow and Slips, Trips, and Falls 313
  - 9.3.1 Making the Undesirable Path Safer......................... 313
  - 9.3.2 Redirect Traffic onto Preferred, Safer Paths 314
  - 9.3.3 Customer Falls... 314
- 9.4 Floor Surfaces/Housekeeping 315
  - 9.4.1 Keep Floors Dry.. 315
  - 9.4.2 Keep Floors Clean... 316
    - 9.4.2.1 Common Cleaning Scenario..................... 316
    - 9.4.2.2 The Chemistry of Fat and Flooring 320
    - 9.4.2.3 Self-Washing Floors................................. 322
  - 9.4.3 Keep Walkways Clear 323
  - 9.4.4 Perceptions of Food Service Workers 323
  - 9.4.5 Spill Cleanup 323
- 9.5 Floor Mats 324
  - 9.5.1 Rubber Mats 324
  - 9.5.2 Olefin Fiber Mats... 326
  - 9.5.3 Wiper/Scraper Mats 326
- 9.6 Footwear.. 326
- 9.7 Multiple Intervention Study... 328

Chapter 10 Healthcare Operations 335

- 10.1 Introduction 335
  - 10.1.1 Impact of Age 336
- 10.2 Known Parameters of Patient Falls 336
  - 10.2.1 Type of Units 336
  - 10.2.2 Location of Falls.. 337
  - 10.2.3 Time of Day of Falls.. 337
  - 10.2.4 Activity at Time of Fall 337
  - 10.2.5 Length of Stay 338
- 10.3 Causes of Patient Falls.. 338
  - 10.3.1 Categorizing Causes of Patient Falls....................... 338
    - 10.3.1.1 Morse Fall Scale 338
    - 10.3.1.2 Tideiksaar Classification Method 340
  - 10.3.2 Personal Risk Factors of Patient Falls..................... 341
    - 10.3.2.1 Medical Conditions.................................. 341
    - 10.3.2.2 Physical Conditions 341
    - 10.3.2.3 Medication 342

10.3.2.4 Other Factors 343
10.3.2.5 Specialty Units.. 343
10.4 Calculating Fall Rates 344
10.4.1 Number of Patients at Risk Rate (commonly used in long-term care facilities)....................344
10.4.2 Number of Patients Who Fell Rate 344
10.4.3 Number of Falls per Bed 344
10.5 Interventions/Controls 344
10.5.1 Flooring...345
10.5.2 Lighting 345
10.5.3 Beds/Bedside..345
10.5.4 Bathrooms 347
10.5.5 Hallways 347
10.5.6 Footwear..347
10.5.7 Other Policies and Procedures 348
10.5.8 Assistive Devices..349
10.5.8.1 Restraints 349
10.5.8.2 Bed Alarm Systems 350
10.5.8.3 Bed Side Rails...351
10.5.8.4 Identification Bracelets 351
10.5.8.5 Hip Protectors...351
10.5.9 Coordination and Strength Training 352
10.6 Reducing Employee Falls 353
10.6.1 All Employees 354
10.6.2 Dietary..354
10.6.3 Housekeeping 355
10.6.4 Laundry 355
10.6.5 Crocs™ Footwear..355

Chapter 11 Profiles of Other High-Risk Industries 361

11.1 Introduction 361
11.2 All Occupancies 361
11.2.1 Exterior Controls 361
11.2.2 Interior Controls 362
11.3 Hospitality (Lodging)..362
11.3.1 Outside Hazards 362
11.3.2 Inside Hazards..362
11.3.3 Employees 363
11.3.4 Guest Room Bathrooms 363
11.3.5 Swimming Pool/Whirlpool Areas 364
11.3.6 Playgrounds 365
11.4 Mercantile (Retail) 367
11.5 Theaters 369
11.6 Trucking 371
11.6.1 Falls from Cabs 372

Contents   xvii

            11.6.2 Falls from Loading Docks...373
                      11.6.2.1 Dock Plates and Dock Boards 373

Chapter 12 Accident Investigation and Mitigation 381

       12.1 Introduction 381
       12.2 Pitfalls of Accident Reporting and Investigation 381
       12.3 Theories of Liability..382
       12.4 Accident Investigation 382
             12.4.1 Claimant and Witness Information 383
             12.4.2 General Occurrence Information 383
             12.4.3 Detailed Occurrence Information 384
             12.4.4 Location Information 385
             12.4.5 Stairs or Ramps 385
             12.4.6 Handrails 387
             12.4.7 Landings...387
             12.4.8 Lighting 387
             12.4.9 Management/Operational Control
                      Information 388
       12.5 Incident Reporting 388
       12.6 Occurrence Analysis 388
       12.7 Claim Mitigation 389
       12.8 Fraud Control Indicators...389
             12.8.1 Fraud—Manner of Claimant....................................390
             12.8.2 General Indicators 390
             12.8.3 Medical or Dental Fraud Indicators 390
             12.8.4 Lost Earnings Fraud Indicators................................390
       12.9 Fraud Control..390
       12.10 Admissibility of Expert Testimony 391
             12.10.1 The Frye Test..391
             12.10.2 The Daubert Ruling et al. 391
             12.10.3 *Kumho Tire Co. v. Carmichael* (97-1709)................392
             12.10.4 Practical Suggestions for Meeting Daubert/
                       Kumho...393
             12.10.5 Federal Rules of Evidence.......................................394
       12.11 Documentation 395
       12.12 Staff Issues 396

Bibliography 413

Index..429

Foreword

> The one permanent emotion of the inferior man is fear—fear of the unknown, the complex, the inexplicable. What he wants above everything else is safety.
>
> **Henry Louis Mencken**
> *Journalist (1880–1956)*

The evolution of tribology (the science of interacting surfaces in relative motion) has come a long way in the past half millennia. Slip resistance, as we know it, has evolved from the laws of friction first proposed by Leonardo da Vinci (1452–1519), which were rediscovered by Guillaume Amontons and published in 1699. As research in the field of pedestrian safety marches on, we gain more knowledge in kinesiology, tribology, and biomechanics. But the gains lead us to the realization that the things that we do know are outweighed by the things that we do not know. Such is true in almost all areas of science, where the gains in understanding lead to new questions. In the past 50 years, concerns about the safety of walking surfaces spawned committees in several national organizations. The safety of polish-coated floors became an issue in the second half of the twentieth century, followed closely by consumer concerns about bathtub safety.

Everyone, at one time or another, has probably experienced a trip and/or a slip and the subsequent fall. The embarrassment of the experience and what we perceive as looking like a fool tends to make us get up as quickly as possible and flee the scene, hoping that no one saw the buffoonery we just committed. Fortunately, there are those among us who suffer little from such experiences. And in those cases no report of the fall is ever made. The occurrence is somewhat akin to the proverbial tree in the forest that crashes to the ground (i.e., if no one saw the fall, then was there really a fall?). Falls are notoriously underreported, but the falls that do occur, and that are reported, have a significant impact on our society in terms of its safety, health, and welfare. The statistics attributed to fall injuries leave no question that falls are a significant problem and concern.

It has been almost six years since the last edition of this publication became available. In that time Mr. Di Pilla has continued to be a spearhead in the pedestrian safety arena. His leadership in the ASTM F-13 committee, the F-15 committee, the ANSI A1264 committee, and others is something that most of us do not see, but all of us benefit from. His resourcefulness, vigor, ethics, effort, and vision continue to make walking surfaces safer for everyone.

Steven Di Pilla has gathered the most current and relevant information available and compiled this guide as an information resource that was written to provide the reader with simple, proven, and effective knowledge and means to reduce fall occurrences. Understanding why falls occur serves as a foundation for mitigating falls. There are many books available on this subject, with most focusing on "slips and falls." This tome covers the gamut, from slips, to stumbles, to trips. From a safety engineering perspective, it is invaluable because it covers the four engineering priorities of dealing with hazards; elimination of the hazard through design; guarding the hazard; warning of the hazard; and training to minimize the risk of the hazard.

Having been fortunate enough to work closely with Mr. Di Pilla, I can attest to his competence and to what I believe is his overall objective: the reduction of injuries due to slips, trips, stumbles, and falls. Mr. Di Pilla looks at risk from a proactive perspective, rather than the typical reactive one. Risk management is a tool that many in the industry unfortunately use to fend off exposure and responsibility by passing the risk off to others through indemnity and assumed risk. Steve Di Pilla is of a different breed of professional, with certain gifts most of us do not possess.

This invaluable guide is a tool to be used and referenced by safety professionals, engineers, architects, designers, housekeeping organizations, property managers/owners, and anyone else who is interested in reducing the occurrence of falls. The citations, references, links, forms, etc. can assist in the development of a fall prevention program, or to improve an already existing one. Mr. Di Pilla has done us all a great favor in providing the second edition of this worthy guide.

Keith Vidal, P.E.
Vidal Engineering, LLC

Preface

This publication is intended as a reference for safety practitioners to use in assessing the exposure of slips and falls, reducing the potential for falls, and minimizing the severity of fall accidents. The following areas are covered:

- **Introduction:** To provide a framework for scope of the problem, statistics regarding fall occurrences and populations at risk are cited and discussed.
- **Chapter 1—Physical Evaluation:** This chapter deals with standards and guidelines relating to the design and layout aspects of a facility such as traffic flow, level walking surfaces, parking areas, changes in levels, stairways, handrails, ramps, entrances and exits, and floor mats.
- **Chapter 2—Management Controls:** Facility design and the related operational controls developed and implemented to maintain and complement good layout are equally important. This chapter addresses the less tangible features of slip and fall prevention and mitigation, such as spill and wet programs, self-inspection, construction/special event control, lighting, deicing and snow removal, and contractual arrangements.
- **Chapter 3—Principles of Slip Resistance:** To effectively assess, improve, and maintain good slip resistance, it is necessary to understand the fundamentals and special considerations in traction measurement and control. Included in this chapter is information on slip resistance factors, a history of the often-cited 0.5 threshold, an overview of key factors such as surface roughness and wet testing, and classes of tribometers.
- **Chapter 4—U.S. Tribometers:** There are only a few slip meters commonly used in the U.S. This chapter discusses each, including their history, operation, use, advantages, and operational issues. Also discussed are the various test pad materials used on tribometers as footwear surrogates, uses for tribometers, and related issues such as operator competency and equipment calibration and maintenance. Included are comparison charts of U.S. and non-U.S. tribometers.
- **Chapter 5—U.S. Standards and Guidelines:** There are federal and consensus standards-making organizations that have developed standards relating to slip resistance. Understanding the background, intent, and practical use of these documents can be an important reference tool. This chapter reviews the slip resistance specifications from UL, OSHA, ADA, Access Board, NFPA, Federal and Military Specifications, and ANSI. A review of U.S.-based industry associations that provide public information on slip resistance is also included.
- **Chapter 6—Flooring and Floor Maintenance:** A practical look at flooring is covered in this chapter, including identification and relative performance of flooring materials. Conventional and slip-resistant floor treatments are reviewed in detail, as well as floor cleaning products, equipment, and

outsourcing floor care. The importance of maintaining clean, dry floors cannot be overemphasized. Covered are optimal methods of floor cleaning, types of cleaning products, and floor maintenance equipment.
- **Chapter 7—Overseas Standards:** As businesses expand to include operations around the world, it becomes more relevant to gain an understanding of slip resistance protocols and standards outside of the United States. This chapter discusses the different approaches and prevalent test methods used including ramp tests, pendulum testers, and digitized dragsled devices. In addition, this chapter provides information about organizations involved in slip resistance research abroad.
- **Chapter 8—Footwear:** An additional area of slip and fall control is footwear in the workplace. Design, selection, maintenance, and footwear programs are essential considerations for employers. Also covered in this chapter are U.S. and international footwear slip resistance standards.
- **Chapter 9—Food Service Operations:** The food service industry is subject to high exposure and accident experience involving pedestrian falls, for employees and patrons. This chapter discusses the unique slip/fall issues and solutions associated with customer and employee falls in the food service environment. This includes a basic approach to proper floor maintenance and housekeeping, proper use of floor mats, and the importance of footwear.
- **Chapter 10—Healthcare Operations:** Due to the physical and sometimes mental condition of patients/residents, the healthcare industry has a population susceptible to high frequency and severity of falls. This chapter provides a detailed view of the special hazards and recommended controls, addressing the unique needs of these operations, with a focus on patients. Covered are known parameters of patient falls, risk factors, calculating fall rates, and an extensive discussion of options for interventions.
- **Chapter 11—Profiles of Other High-Risk Industries:** While many types of businesses have a lower exposure to pedestrian falls, others have higher-than-average loss experience. This chapter provides a snapshot of the special hazards and recommended controls of hospitality (lodging), mercantile, theater, and truck operations.
- **Chapter 12—Accident Investigation and Mitigation:** Prevention measures cannot eliminate all losses. To handle fall accidents effectively, proper investigation and mitigation controls must be in place. This includes gathering the needed information, investigating the facts, providing prompt and courteous aid to the injured, and combating fraudulent cases. A discussion of the admissibility of expert testimony is also covered.

Acknowledgments

The author thanks the following individuals for their valuable contributions:

John S. Ingram, *vice president of risk control services at ESIS, Inc.,* for his faith, support, and commitment to complete this book.

Keith Vidal, *president of Vidal Engineering, LLC,* for his continued guidance and counsel, and for being a great sounding board.

Tim Fisher, *vice president of Council on Standards and Practices for the American Society of Safety Engineers*, for his boundless energy and enthusiasm in the pursuit of the development of sound consensus standards.

Author

Steven Di Pilla is the director of research and development for global risk control services at ESIS, Inc., an ACE USA* risk management service company. He is responsible for identifying and filling the needs of staff consultants, researching technical issues, and developing references, training programs, standards, loss analyses, and external publications. He began his career at the former CIGNA Property & Casualty, now part of ACE USA, in 1980.

A professional member of the American Society of Safety Engineers (ASSE), he is former vice-chairman of the ASSE Standards Development Committee, immediate past chairman of the ASTM International Technical Committee F13 Pedestrian/Walkway Safety and Footwear, and former chairman of subcommittee F13.10 Traction. Other ASTM committee membership includes C21 Ceramics, D01 Paints, D21 Polishes, F06 Resilient Flooring, F08 Sports Equipment and Facilities, F15 Consumer Products, and E34 Safety and Health. He currently serves as chairman of the American National Standards Institute (ANSI) 1264.2, which deals with the slip resistance of working/walking surfaces, and is a member of Underwriters Laboratories STP 410, dealing with the measurement of slip resistance of walkway surfaces.

Di Pilla is an active member of numerous other professional organizations including the American Society for Industrial Security and the NFPA International Means of Egress Technical Committee for Safety to Life. He is the author of numerous studies and articles and has presented extensively.

Di Pilla holds a B.B.A. in property and casualty insurance from St. John's University–College of Insurance. He also holds associate designations in risk management (ARM), claims (AIC), and marine insurance management (AMIM) from the Insurance Institute of America. Certifications include Lightning Safety Professional (CLSP), English XL VIT Tribometrist (CXLT), and Global Environment of Insurance (GEI).

---

* ESIS, Inc. (ESIS©) is a risk management services company of ACE USA. "ACE USA" refers to the insurance, reinsurance, and risk management companies comprising the U.S.-based operating division of the ACE Group of Companies, headed by ACE Limited.

PLEASE READ CAREFULLY: The information contained in this publication is not intended as a substitute for advice from a safety expert or legal counsel you may retain for your own purposes. It is not intended to supplant any legal duty you may have to provide a safe premises, workplace, product or operation.

Introduction

Our greatest glory is not in never falling, but in getting up every time we do.

Confucius (551 bc–479 bc)

Each person takes an average of 5000–7000 steps a day. This represents a tremendous exposure.

Falls in the workplace are the number one preventable loss type; in public places, falls are far and away the leading cause of injury. More than one million people suffer from a slip, trip, or fall injury each year, and 17,700 died as a result of falls in 2005 (National Safety Council, 2007). This is a 1% increase from 2004, and represents a death rate of 6.0 per 100,000 of the population.

National Safety Council (NSC) statistics indicate that 25,000 slip and fall accidents occur daily in the United States. The NSC estimates that compensation and medical costs associated with employee slip and fall accidents alone are approximately $70 billion. Falls are estimated to cause 17% of occupational (work-related) injuries and 18% of public sector injuries (NSC); however, these figures are understated. Falls are notoriously underreported because accident rates are normally classified by injury type instead of cause of injury in workers' compensation and National Electronic Injury Surveillance System (NEISS) statistics.

In 2000, there were almost 10,300 fatal and 2.6 million medically treated nonfatal fall-related injuries. Direct medical costs totaled $200 million for fatal and $19 billion for nonfatal injuries. Of the nonfatal injury costs, 63% ($12 billion) were for hospitalizations, 21% ($4 billion) were for emergency department (ED) visits, and 16% ($3 billion) were for outpatient treatment. Medical expenditures for women, 58% of the older adult population, were two to three times higher than for men for all medical treatment settings. Fractures accounted for 35% of nonfatal injuries but 61% of costs (Stevens et al., 2006).

WORKPLACE FALLS

An estimated 300,000 disabling injuries occur each year in the U.S. workforce.

Of the 5,764 deaths at work in 2004, falls accounted for the second most frequent cause (behind contact with object or equipment), with 822 deaths (or 14%). In terms of cases with days away from work, falls were again ranked second, accounting for 255,600 days (or 20% of all lost work days).

Also in 2004, the total number of falls to lower level (79,800), falls on same level (167,010), and slips/trips (37,500) again rank number two, with 284,310 in nonfatal cases.

According to the NSC, the average cost of a fall/slip was $22,802 in 2003 and 2004, the second costliest type of loss (behind motor vehicle accidents), and compared to an average cost for all losses of $19,382. In fact, only fall/slip and motor vehicle accidents exceed this average cost.

PUBLIC SECTOR FALLS

It should be no surprise that falls represent as much as half of all liability loss frequency and severity for insurance companies. Studies show that this category of loss ranges from 7% (for manufacturing) to 52% (for care providers), and higher still for restaurants. The situation is comparable overseas.

There is, understandably, great interest in methods to prevent falls. Contrary to popular belief, most slip/falls are not due to carelessness. Many options are available in the design and maintenance of facilities to reduce or eliminate the potential for slip/falls.

A review of slip/fall losses reveals that, in addition to contributory negligence of the accident victim, there is often something the property owner/management could have done to reduce the severity or prevent the incident. That "something" is often repairing a defect in the environment or a lack of management controls that contributed to the likelihood or severity of the event.

CHANGING DEMOGRAPHICS

U.S. STATISTICS

The incidence of falls increases exponentially with age: an incidence rate of 30% in persons age 65 and over increases to 50% in persons age 80 and over. Twenty to 30% of older persons who fall suffer serious injury, such as hip and other fractures, dislocations, subdural hematoma, head injury, and other soft tissue injuries. More than 60% of people who die from falls are age 75 and over. Those who survive a fall suffer significant morbidity, with greater functional decline in activities of daily living (ADLs) and physical and social activities, and are at a greater risk of institutionalization than those at 65 to 74 years of age. Falls that do not result in serious injury may still have serious consequences for an older person, who may fear falling again, which can lead to reduced mobility and increased dependence through loss of confidence (Kalula, A WHO global report on falls among older persons. Prevention of falls in older persons: Africa case study).

As baby boomers mature, the U.S. population is aging at a rapid rate. Individuals 55 years and older are the fastest-growing segment of the population, currently at around 31%. More than 6,000 adults turn 65 every day. By the year 2030, the number of people over 65 is expected to double. By 2050, those over 65 will reach 80 million.

The increase in people age 85 and older is also substantial. By 2010, this segment will grow by 33%, from 4 million to 5.6 million.

In addition, the life expectancy of males and females continues to increase. By 2050, life expectancy for males is projected to go from 73 to 86, and from 79 to 92 for females.

These individuals are the country's workers and members of the general public. This older portion of U.S. citizens is most at risk for frequency and severity of slips and falls. Fall deaths increase with age, from a low of 40 (5 to 14 years old) to 11,900 (75

Introduction

and over). As the number of older people increases, so does their potential for fall injuries. This increasing exposure to falls requires increased attention to controls associated with reducing and preventing fall injuries for workers and for the general public.

According to the Centers for Disease Control (CDC):

- One of every three people 65 years and older falls each year. This amounts to more than 11 million people.
- Older adults are hospitalized for fall-related injuries five times more often than they are for injuries from other causes. About 40% of all nursing facility admissions are the result of falls.
- Of those who fall, 20 to 30% suffer moderate to severe injury, which reduces mobility and independence and increases the risk of premature death.
- In 1994, the total direct cost of all fall injuries for people age 65 and older was $20.2 billion. By 2020, this figure is expected to reach $32.4 billion.

Nearly 8,000 older adults died, and another 56,000 were hospitalized in 2005 after falls resulted in traumatic brain injury (TBI). TBI accounted for half of all unintentional fall deaths that year. TBIs are caused by a blow to the head, often resulting in long-term cognitive, emotional, or functional impairments. Such injuries can often be misdiagnosed among older patients (CDC, 2008).

In addition, more than 10 million Americans have osteoporosis, and 18 million have low bone density. Both of these factors contribute to the severity of a fall.

OVERSEAS STATISTICS

- The average rate of fall-related hospital admission among all those aged 60 and over in the United Kingdom is 169 per 10,000 population.
- The rate of fall-related hospital admission among all those aged 65 years and older in British Columbia (BC) Canada is 155 per 10,000 standard population.
- The rate of all those aged 65 years and over is 297 per 10,000 population in Western Australia.
- Among those aged 65 years and older in the United States and Canada, the main cause for fall-related hospital admission is hip fracture, accounting for more than 40% of all fall-related injuries among older persons treated in hospitals. The average length of stay for an older person due to a fall-related hospitalization is 4 to 15 days, and extends to 20 days if it is for a fall-related hip fracture.
- Sources from Australia and the United Kingdom show that the annual rate of fall-related ED visits (overlap with those admitted to hospital) among the older persons aged 65 years and over is 535 to 892 per 10,000 population.
- Falls account for 53.7% of all ED visits due to unintentional injuries among people aged 65 years and over in Canada. The rate of fall-related hospital admission or ED visits for females is more than twice the rate for males.
- In Finland and Australia, the average direct health system cost per fall injury episode (including those not requiring hospitalization) among those aged 65 years and over is between $1,049 and $3,611. The average cost of

hospitalization for a fall-related injury among those hospitalized is between US$6,646 and $17,483. Fracture of the hip is a main cause of costs to fall-related inpatient hospital treatment among those aged 65 years and older, accounting for 70% of inpatient costs (Fu, Health Service Impacts).

See Chapter 10, Healthcare Operations, for a more in-depth treatment of the impact of aging on slip/fall risk.

More Statistics

Slips and falls are a multifaceted problem because they are a major cause of loss, not only to the general public (i.e., invitees, guests, patrons, customers, clients) but also to employees. Slips and falls are the second most frequent cause of worker injuries and personal injury incidents.

Note: These statistics should be considered low estimates. Falls are notoriously underreported because accident rates are normally classified by injury type instead of cause of injury in workers' compensation and NEISS statistics. Many accidents that are otherwise classified began with a slip, trip, or fall.

Deaths*

Due to unintentional injury
- Falls are second only to motor vehicle accidents as causes of death, accounting for 17,700 deaths in 2005. Deaths from falls have steadily increased: 13,450 (1994); 13,986 (1995); 14,986 (1996); 15,447 (1997); 16,600 (1998); 17,100 (1999); and 17,500 (2004). Falls also account for the second highest death rate (6.0 per 100,000 people). The death rate has not fallen during any year since 1986.
- Deaths from falls are fairly evenly divided by month, with a high of 1473 in January and a low of 1204 in June, indicating a small impact of weather-related incidents.

Due to unintentional public injury (1999)
- Falls are the leading cause of unintentional public injury deaths, accounting for 6,800 deaths (up 17% from 1998, which was 8% higher than 1997).
- Unintentional Injury Deaths by Age and Type, 2003

Age Group	Number	Percentage
0–4	68	0.4%
5–14	45	0.3%
15–24	230	1.3%
25–44	921	5.3%

---

* Unless otherwise attributed, excerpted from *Injury Facts*, National Safety Council, 2007.

Introduction

Age Group	Number	Percentage
45–64	2,263	13.1%
65–74	2,048	11.9%
75+	11,654	67.6%
ALL	17,229	

- The rates of slip/fall deaths per 100,000 of the population vary by gender. Males have a rate of 6.2, while the rate for females is 5.6.

INJURIES

Nonfatal cases by industry sector (2004)

Sector	Nonfatal Cases			Fatalities		
	All	STF	%	All	STF	%
Agriculture, Forestry, Fishing, Hunting	19,750	4,770	24.2%	669	30	4.5%
Mining	9,350	2,030	21.7%	152	13	8.6%
Construction	**153,200**	**38,140**	**24.9%**	**1,234**	**437**	**35.4%**
Manufacturing	226,090	34,880	15.4%	463	5	1.1%
Wholesale Trade	81,140	16,770	20.7%	205	11	5.4%
Retail Trade	178,760	39,260	22.0%	377	13	3.4%
Transportation and Warehousing	120,010	23,510	19.6%	840	30	3.6%
Utilities	7,740	1,950	25.2%	51	8	15.7%
Information	21,150	5,950	28.1%	55	0	0.0%
Financial Activities	**34,930**	**11,370**	**32.6%**	**116**	**10**	**8.6%**
Professional and Business Services	90,500	22,670	25.0%	452	5	1.1%
Educational and Health Services	189,980	45,470	23.9%	157	6	3.8%
Leisure and Hospitality	**99,380**	**30,280**	**30.5%**	**247**	**18**	**7.3%**
Other Services	331,350	7,270	2.2%	207	5	2.4%

Injury-related hospital emergency department visits (1996, 1998, and 2004)
- Falls were the leading cause of injury-related emergency room visits, accounting for almost 8.5 million cases, or 21% of the total. Motor-vehicle accidents were a distant second at 4.7 million cases (or 11%).
- Hospital visits due to stair/step falls alone have been steadily increasing throughout the past two decades, from a low of 695,968 (1980) to 989,826 (1998) (CPSC/NEISS).
- In 2004, falls were the most frequent cause of ER visits for 9 of the 10 age groups (only in the 15–24 age group, where it was third behind struck by/against and motor vehicle occupant).

Nonfatal injury costs (1995–1996)
- Falls on stairs/steps represent the highest cost of nonfatal injuries for five of the 12 age groups (falls on floors account for three more age groups). Falls are in the top 10 of all 12 age groups. Overall, falls from

stairs/steps and floors rank number one and number two, respectively, accounting for 17% of nonfatal injury costs overall. The cost of stair-related injuries alone is estimated at $49.9 billion (Public Services Research Institute, from CPSC/NEISS data).

Nonfatal occupational injuries involving days away from work (1997)
- Same-level falls are the third most frequent type of nonfatal occupational injury, representing almost 14% of all losses (behind overexertion and contact with objects/equipment). Trade and services have the highest number of falls at 16.6% and 17.6%, respectively.
- Excluding motor vehicle accidents, the average cost of a worker fall ($12,470) is second only to burns ($12,792).

OTHER STATISTICS

- The annual Workplace Safety Index review by Liberty Mutual reported that employee slip and fall accidents in all industries hit an all-time high in 2002, the latest year for which figures are available, of $6.2 billion, up from $5.7 billion in 2001.
- Sidewalk slip-and-fall cases cost New York City taxpayers around $70 million a year. Sidewalk cases constitute the most frequently filed personal injury claims against the city. Payouts averaged $30,637 for the 2,226 cases resolved in 2004. In 2004, one case involved a woman who fell near a park in 1996 and then had a slew of medical complications, cost $1.3 million. (Lombardi, 2005)
- A study by the Food Marketing Institute (FMI) found that 53% of all workers' compensation claims and public liability suits against supermarkets resulted from injuries sustained as a result of slip-and-fall accidents.

For more information on U.S. injury statistics, see the current edition of the NSC's Injury Facts publication (http://www.nsc.org). For more slip/fall injury and death statistics on Food Service and Healthcare sectors, see these respective chapters.

1 Physical Evaluation

Do not look where you fell but where you slipped.

Proverb

1.1 INTRODUCTION

In the United States, there is substantial agreement regarding appropriate design dimensions for most walkway surfaces and associated components. The most recognized codes in this area are the model building codes and the life safety code. These codes have been in use for some time and consequently are well tested.

Physical evaluation related to slips and falls is one of the fundamentals to prevention. The importance of proper design cannot be overstated. Effective safety begins with good design. Ramps that are too steep, uneven steps, unmarked changes in levels, and missing handrails are major contributors to falls. Implementing effective management controls will have a limited impact if good design is not in place.

1.2 EXPECTATION

When walkway surface conditions encountered are different from what is expected, the potential for an accident increases.

"Expectation" is an underlying principle of effective slip/trip/fall evaluations. For example, if we are aware of the ice or water ahead, we can either attempt to avoid it *or* adjust our gait (e.g., the way we walk) to compensate for the differing surface. We will slow down, take smaller steps, and walk flatter. These subtle adjustments we make in the way we walk are likely to allow us to safely cross the hazardous area.

If we are not aware of the ice, however, we can neither avoid nor compensate our gait for the hazard. Numerous studies have consistently shown that individuals can cross a slippery area if alert to the condition.

Thus, the most effective option for preventing falls is to either eliminate unexpected conditions that constitute slip and fall hazards, or (less desirable) assure that pedestrians receive clear notice of the presence of such conditions so they can be adequately prepared to deal with them.

1.2.1 MISSTEPS

Slips and trips are only two of the immediate causes of falls. Missteps, including the potential for air steps, oversteps, unstable footing, and heel scuffs must also be considered when assessing physical hazards.

Missteps can be defined as "an unintentional departure from pedestrian gait appropriate for the walkway surface." Many falls are characterized as slips, even if they are not, because "slip" has become a broad term to describe missteps. As a result, they are less likely to be correctly coded or investigated (Pauls, 2007a).

1.3 DISTRACTIONS

It is important to recognize the impact that distractions can have on pedestrians. Avoid extensive signage and eye-catching images and designs in areas where pedestrians need to be aware of where they are walking. Unfortunately, this advice may be at odds with other objectives, such as in retail establishments where great effort is expended to capture the attention of customers in order to sell products. In such cases, the immediate area surrounding placement of displays should be otherwise free of slip and trip hazards.

Other areas of significant concern include stairs, escalators, floor surface transitions, changes in level (such as short flights of three steps or less), entryways where contaminants and moisture can accumulate, and congested areas.

1.4 LEVEL WALKWAY SURFACES

Based on several references, including the Americans with Disabilities Act (ADA) and ASTM International Standard F-1637 (see "Reference Standards" at the end of this chapter), a trip hazard is defined as a change in elevation in a walkway that is not a proper ramp or stairway, with a vertical face ¼ in. (6.4 mm) or higher, or a change in elevation of more than ¼ in. (6.4 mm) with an inclined face steeper than two horizontal on one vertical. Stride studies have shown that subjects with low-heeled shoes clear the ground by a mere ¼ in. (6.4 mm), and those with higher-heeled shoes by even less. Thus, a seemingly minimal but sudden increase in the walkway surface can readily result in a trip.

ADA Accessibility Guidelines for Buildings and Facilities (ADAAG, 4.5.1) state that "Ground and floor surfaces along accessible routes and in accessible rooms and spaces including floors, walks, ramps, stairs, and curb ramps, shall be stable, firm, slip-resistant, and shall comply with 4.5."

The following outlines some guidelines in evaluating the degree of hazard of level walking surfaces.

Online Resource: Federal Highway Administration—Sidewalk Assessment

Using this objective method of assessment, sidewalk professionals can evaluate conditions experienced by pedestrians in the public right-of-way and identify sites requiring accessibility and maintenance improvements. The information can also be used continue to improving sidewalk conditions for all users.
http://www.fhwa.dot.gov/environment/sidewalk2/sidewalks211.htm.

Physical Evaluation

1.4.1 Sidewalks

Conventional concrete sidewalks present a host of maintenance issues, most prominently the heaving (Figure 1.1) and breakage of panels due to tree roots, weather, and temperature extremes. Patching is at best a temporary solution.

Identify surfaces with cracks, potholes, or other conditions that could contribute to a fall. Settlement of asphalt and concrete surfaces can often create these conditions, as can concrete spalling. Walkway surfaces should be free of debris and other slippery material (e.g., gravel, mud, sand, food spills and other similar materials).

In practice, cracks in sidewalks are not normally a problem if the slabs are not heaved, but a vertical surface discontinuity of ¼ in. (6.4 mm) can trip many pedestrians, because it may be unexpected.

A simple and practical way to determine such small differences is to use pennies (Figure 1.2). With a thickness of $1/16$ in. (1.6 mm), stacking two pennies gives you a rule for $1/8$ in. (3.2 mm), and four pennies a rule for ¼ in. (6.4 mm) (Figure 1.3 and Figure 1.4).

FIGURE 1.1 Heaving of concrete block pavement.

FIGURE 1.2 An easy way of measuring small differences is with pennies.

FIGURE 1.3 The "before" picture—it is not difficult to see the tripping hazard in this picture, spanning the entire walkway area. Probably due to settling, the difference in level ranges from about ¼ in. to almost 2 in. (Photograph courtesy of K. Vidal.)

FIGURE 1.4 The "after" picture shows a good job of correcting the problem. (Photograph courtesy of K. Vidal.)

CASE STUDY: Uneven Exterior Walkway

Situation. The concrete block pavements at the front entrance are uneven, with many blocks lifted ½ in. (12.7 mm) or more above the level of the remainder of the walkway. This occurs due to settling and weather conditions.

Discussion. Based on several references, including the ADA and ASTM F-1637, a trip hazard is defined as a change in elevation in a walkway that is not a proper ramp or stairway, with a vertical face ¼ in. (6.4 mm) or higher, or a change in elevation of more than ¼ in. (6.4 mm) with an inclined face steeper than two to one. Thus, a seemingly minimal but sudden increase in the walkway surface can readily result in a trip.

Physical Evaluation

Recommendation. There are several options for addressing this concern:
- Replace the block pavement with stamped concrete, which is in essence a single slab that appears to be textured blocks. Many designs can be used. This option will eliminate the variation in the height of different parts of the walkway surface.
- Replace the block pavement with a standard concrete sidewalk design. This would provide more stability than the current design, but based on the amount/frequency of sidewalk replacement currently being performed on the campus, this option would still require some inspection and maintenance oversight.
- Reset/repair the existing blocks and initiate a self-inspection/maintenance program to assure frequent identification and correction of raised/damaged blocks. This is the least desirable option, because it is labor intensive and requires regular intervention.

1.4.1.1 Rubberized Sidewalks

A relatively new innovation, rubberized sidewalks can be an effective way to reduce the frequency and severity of trips. Typically made from recycled tire rubber with a polyurethane binder, these modular panels have an average cost per square foot (including break out and installation) of $15.

Rubberized sidewalks are durable, absorb shock, and are more resilient than concrete (Figure 1.5). These properties reduce the hazards of cracked and uneven sidewalks, while providing a "softer" surface if a fall does occur. Reversible pavers are available, which allows for at least 14 years of average use. Rubberized sidewalks have proven to be significantly more durable than concrete, resisting cracking, indentation, chipping, breakage, or other damage in freezing temperatures (Table 1.1).

Unlike concrete, rubber sidewalks can be lifted for tree root maintenance and replaced. Crews can trim tree roots every 2 or 3 years while roots are still in the offshoot stage. Usage has shown that the growth rate of roots is slower and roots grow in small, tender offshoots that are more easily trimmed. Because rubber is lighter and

FIGURE 1.5 Side-by-side comparison showing the original concrete sidewalk, and a Rubbersidewalks replacement, an interlocking modular recycled rubber paving system. (Courtesy of Rubbersidewalks, Inc. www.rubbersidewalks.com.)

TABLE 1.1
Comparison of Rubber, Concrete, and Asphalt Sidewalks

	Rubber	Concrete	Asphalt
Life cycle near tree roots	20+ years	5 years	2 years
Material cost (<1000 sf)	$10/sf	$5/sf	$3/sf
Labor	2 person crew	4 person crew	4 person crew
Weight	10.8 lbs /sf (@ $1^7/_8$") [47.6 mm]	47 lbs/sf (@ 2") [51 mm]	24 lbs per sq ft (@ 2") [51 mm]
Surface changes	Moderate fading, discoloration	Cracking, staining, discoloration	Chipping and holes
Mass changes	Possible setting requiring adjustment	Lifting, breaking	Lifting, deterioration
Shock	< 200 g @ 5 ft [1.524 m]	Over 200 g	Over 200 g

Source: Adapted from rubbersidewalks.com.

more resilient than concrete, tree roots receive sufficient water and oxygen through paver seams.

Although slip resistance is reported to be high, the test method used was the ASTM C1028 apparatus, which is not a reliable measure for wet surfaces (see Chapter 4, Tribometers).

Other benefits include:

- Modular system allows pavers to be periodically opened for inspection and easily reinstalled.
- Comfortable and healthy for walking or jogging.
- By absorbing sound, they reduce decibel level of foot and wheeled traffic.
- Increased ability to meet ADA accessibility requirements.

(Rubbersidewalks, Inc.)

1.4.2 CURBING

The standard height for curbs in the United States is 6 in. (15.24 cm). Curbs of different heights due to design or settling can create an unexpected condition for pedestrians, increasing the potential for falls (see Figure 1.6).

1.4.2.1 Curb Construction and Design

Curbs should be smooth and flat, and constructed to minimize movement that could result in vertical discontinuities to the walkway surface. Curbing is constructed in several ways. Less preferred is a separate extruded piece of concrete, independent of the sidewalk. Depending on soil conditions and weather, this construction carries a high risk of settling and results in uneven surfaces between the curb and the sidewalk. Extruded curbing is sometimes evident by the concrete gutter that extends

Physical Evaluation

FIGURE 1.6 Aside from the deteriorated condition of this curb, the height is excessive at 9 in. (Photograph courtesy of K. Vidal.)

out into the roadway. Construction methods such as monolithic pour or doweling of adjoining slabs are preferred. These are more stable because the curbing and sidewalk are a single piece of concrete, thus minimizing the potential for gaps and unevenness between them as well as heaving as settlement and weather take their tolls (see Figure 1.7).

FIGURE 1.7 In addition to the cracking and unevenness, a major concern of this sidewalk is the extreme settlement. The curbing (a separate piece) remains as it was constructed, which presents a significant trip hazard. The addition of drainage grates, which can be slippery and can accumulate water/ice when blocked, makes it a high-potential area for falls.

Curved or rolled curb forms that do not provide a distinct edge should be avoided. These designs offer mixed visual cues and may result in footing instability to susceptible populations, such as the elderly and those with a visual impairment.

1.4.2.2 Curb Marking

Curb markings in the Manual for Uniform Traffic Control Devices (MUTCD) 3A.04 regarding the color for curb markings gives leeway regarding the use of yellow for No Parking. This is another instance in which "uniformity" is ambiguous. The lack of prescriptiveness of MUTCD 3B.21 regarding colors to be used for outside curb markings reflects the lack of a single, nationally recognized standard.

According to http://www.caldrive.com/parking.html, California has a consistent statewide scheme, and the same meanings of red and blue are probably recognized by most drivers, but yellow and white curb markings have variable meanings.

Per a Federal Highway Administration (FHWA) FAQ, colored marking of outside curbs is more effectively used to emphasize the applicability of parking, stopping, and loading regulations stated on signs.

In terms of best practice, curbing leading to sidewalks and entrances should be painted a contrasting color. Unless curbs are painted yellow to denote "no parking" areas, good choices are white or red.

1.4.3 CURB CUTOUTS

The primary purpose of curb cutouts is to provide a means of making exterior walkways accessible to persons with disabilities, particularly those in wheelchairs; however, there are proper and improper designs for curb cutouts (Figure 1.8). If not done properly, cutouts can become substantial trip-and-fall hazards to those individuals as well as other pedestrians.

Whenever possible, however, building grading should be designed to avoid the need for cutouts, making the sidewalk and parking lot vertically seamless.

FIGURE 1.8 Although not well marked, the overall design with flared sides shown here is a good one.

Physical Evaluation

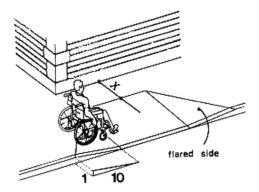

FIGURE 1.9 This figure shows a typical curb ramp (cut into a walkway perpendicular to the curb face) with flared sides having a maximum slope of 1:10. The landing at the top, measured from the top of the ramp to the edge of the walkway or closest obstruction is denoted as "x." If x (the landing depth at the top of a curb ramp) is less than 48 in. (1.2 m), the slope of the flared side shall not exceed 1:12. (from ADA Figure 12A).

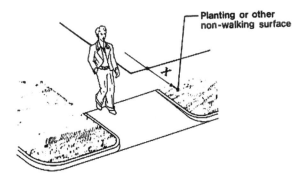

FIGURE 1.10 If the curb ramp is completely contained within a planting strip or other non-walking surface, so that pedestrians would not normally cross the sides, the curb ramp sides can be steep and include vertical returned curbs (from ADA Figure 12B).

The American National Standards Institute (ANSI) A117.1 standard, Accessible and Usable Buildings and Facilities from the International Code Council, provides curb cutout design specifications in section 406 (Figure 1.9 and Figure 1.10).

- Similar to all ramps, curb cutouts should not exceed 1:12.
- Do not permit curb edging to extend above the cutout slope. Instead, the ramp should have flared sides (a maximum slope of 1:10). If edging is present, a landscaped area or other barrier should be provided to prevent individuals from crossing over the raised curb area.

If cutout ramps are painted, use a slip-resistant paint. This usually means grit has been added to the mix, creating surface roughness. Although there is no requirement that curb ramps be marked, the issue of pedestrian expectation suggests there should be some visual cue indicating a change in walkway surfaces.

SIDEBAR: Walkway Surface Paint Study

A 2007 study tested the slip resistance of marking paint applied to three common outdoor walkway surfaces: brushed concrete (e.g., sidewalk), floated (smooth) concrete, and asphalt (e.g., parking lots or roadways). Three commercially available blue handicapped marking paints were used: a water-based latex emulsion paint, an acrylic fast-dry water-based paint, and a solvent-based alkyd paint.

In each case, the use of paint increased slip resistance under wet and dry conditions. In other than the asphalt surface, the difference was significant. All surfaces (as expected) yielded reduced slip resistance under wet conditions, though significantly less so with the addition of the marking paint.

The study showed the largest increase in slip resistance resulted from painting a smooth concrete surface. Since the dry film of paint is comprised of mostly hard pigment particles held together by the binder, it is possible the increase was due to silica and other pigments providing a more uniform surface with relatively greater roughness. The paint coating would thus provide a greater contact area between the walking surface and the shoe bottom, while the paint additives increased the relative roughness.

The results of this study testing suggest that, even without a surface roughening additive (e.g., sand, grit, and glass beads), the application of purpose-designed marking paint to exterior walkway surfaces areas has a positive effect on slip resistance (Curry et al., 2007).

However, because this was a single study and results have not been duplicated by other researchers, a conservative approach is recommended. Embedding an aggregate material to increase roughness is an inexpensive and prudent precaution to take.

1.4.4   Other Walkway Impediments

1.4.4.1   Access Covers

Access covers take many forms and may be present for many reasons. Usually, they provide access to valves or other connections located underground for a utility (e.g., water, gas, cable).

No practical reason exists to have a protrusion above the surface. Repaves should not result in sunken areas. Gradual transitions of pavement should be designed per ASTM and ADA guidelines, whether a surface protrusion or depression is created or is preexisting. Thus, junction boxes and any similar access covers (such as gas and sewers) in parking lots or in walkway paths should be designed to be flush with the surface (Figure 1.11).

Optimally, underground lines should be located so that access is not required in an obvious pedestrian path.

1.4.4.2   Drainage Grates

Commonly encountered in parking lots, drain gratings often have excessively wide openings, some as wide as 3 in. (76 mm). Some are designed with convex covers, making them even more hazardous to traverse, especially for those wearing high-heeled shoes and similar footwear (Figure 1.12).

Physical Evaluation

FIGURE 1.11 The condition, stability, and protrusions of this access cover clearly make it a poor example.

FIGURE 1.12 Note the large openings and uneven surface of this typical design of drainage grate commonly found in parking lots.

Drain grates are usually recessed to some degree to permit the channeling of water from other areas to the drain. Some are excessively graded, however, and settlement of the surface makes that slope even more extreme or creates depressions that result in pooling.

Visually, drainage grates tend to fade into parking lots since they are often as black as the asphalt. For purposes of pedestrian safety, consideration should be given to painting drainage grates a bright color like yellow. This will serve to increase the awareness of pedestrians to an increased walking hazard. It can also be of benefit to vehicles and cyclists traversing the area.

Optimally, when designing drainage, grates should be situated out of natural and expected pedestrian paths. ASTM F-1637 recommends that grates should have openings no wider than ½ in. (13 mm) in the predominant direction of travel.

FIGURE 1.13 This "pole farm" consists of seven poles in various states of disrepair, protecting nothing.

ADA Accessibility Guidelines for Buildings and Facilities (ADAAG, 4.5.4) agrees:

> If gratings are located in walking surfaces, then they shall have spaces no greater than ½ in. (13 mm) wide in one direction. If gratings have elongated openings, then they shall be placed so that the long dimension is perpendicular to the dominant direction of travel.

1.4.4.3 Posts

In many instances, posts are unnecessary evils. They can cause repeated damage while protecting little of consequence. Damaged posts can become greater hazards to pedestrians when broken off close to the ground and exposing sharp edges. Often, posts remain in place long after the object they were intended to protect has been removed (Figure 1.13).

If posts are used, they must be substantial enough to withstand the punishment anticipated. They should also be painted a high-contrast color so they can be readily seen.

1.4.4.4 Bicycle Racks

Bicycle racks should be placed away from vehicle and pedestrian traffic and arranged so walkway-level supports do not impinge on the walkway path and create a trip hazard.

In low-light or low-contrast condition, it is helpful if they are painted a bright color such as yellow or white (Figure 1.14).

1.4.4.5 Planters, Trash Receptacles, and Similar Furnishings

The configuration, visibility, and location of these furnishings may contribute to falls (Figure 1.15).

Planters that have sharp bases or that extend out into the walkway present a greater risk, as do other protrusions designed to fade into the scenery.

1.4.4.6 Architectural Designs

Architectural designs can unintentionally create fall hazards because they may not be easily seen and because they may be unexpected by the pedestrian.

Physical Evaluation

FIGURE 1.14 In this typical bike rack arrangement, the rack is set inside an untraveled area but bikes placed here protrude out into the walkway.

FIGURE 1.15 This blockage was moved into the walkway path to cover a hole in the sidewalk—a modest improvement at best.

1.4.4.7 Doorstops and Other Small Trip Hazards

The harder they are to see, the greater the hazard they pose. The cost benefits of small trip hazards need to be carefully evaluated to determine if the component is really necessary. If so, placement and marking should be considered (Figure 1.16).

1.4.4.8 Temporary Fixes

Temporary fixes are all too common, placing obstructions in the walkway. Some are designed to protect an opening from water infiltration or damage. Others are a misguided effort to prevent pedestrian falls. These halfhearted repairs often remain in

FIGURE 1.16 This is an example of a well-camouflaged doorstop.

FIGURE 1.17 Here is a substantial protrusion, intended to protect an opening left by removal of a light pole.

place for a period substantially longer than originally intended and frequently create or increase the hazard (Figure 1.17).

CASE STUDY: Curb, Curb Ramp, and Sidewalk (See Sections 1.4.2 and 1.4.3)

CURBS

The design and construction of exterior curbing along the walkway is of concern. Standard curb height in the United States is 6 in. (15 cm). Deviating from that standard presents a condition that pedestrians do not expect, especially in older individuals whose visual acuity (in terms of depth and color) is reduced.

Currently, facilities are built with 8-in. (20-cm) curbs. In some areas, curbs are even higher. A rash of falls at curbs is routinely reported upon opening a

FIGURE 1.18

facility. **It is recommended that construction of future facilities be provided with 6-in. curbs.**

Facilities that have been recently constructed use concrete instead of asphalt around the walkway/curb. This further reduces the visibility of the curb, because it readily blends visually with the color and texture of the parking lot and increases the likelihood of trips, missteps, and falls.

To mitigate the hazard of existing 8-in. (20-cm) curbs, it is recommended that the riser and at least 3 in. (76 mm) of the top (leading edge of the tread) of curbing be painted a contrasting color. This will call attention to the subtle change in level, and reduce the potential for pedestrians missing or misjudging their step up or down. Paint should be slip resistant, which usually involves adding grit to the mix, creating surface roughness.

CURB RAMPS

The curb ramps now in place are impractical and may not meet the requirements of the ADA. Regarding the built-up ramp (see Figure 1.18), the flat portion meeting the walkway must be at least 36 in. (91 cm) wide, and the flared sides of the ramp cannot exceed a slope of 1:10. In addition, this ramp is visually disruptive to the storefront, and takes up significant space. The curb ramp is likewise an awkward design primarily because the sidewalk slopes dramatically down (from the building toward the parking lot), making it difficult to navigate for wheelchair-bound individuals and pedestrians.

Unfortunately, the walkway is of insufficient depth to permit a curb cutout arrangement. However, the walkway is significantly wider as it approaches the main retail entrance. **It is recommended that a curb cutout be installed on either side of the main overhang.** This would provide several benefits:

- Allow full compliance with the ADA
- Allow the removal of the built-up ramp which is not aesthetically pleasing
- Allow the displays along the walkway to remain in place (currently they reduce the path below the minimum 36-in. (91-cm) width required to make it accessible—see Front Walkways, below)

FIGURE 1.19

- Reduce the amount of parking area needed to provide this access
- Provide a cleaner, neater, and more balanced appearance at the storefront

Front Walkways

The curb ramps in Figure 1.19 lead to the front walkway. However, an ice box, propane tank storage, and newspaper displays narrow the available path, impeding wheelchair access.

- On the right side, the ice box alone reduces the walkway to 20 in. (51 cm), far below the 36-in. (91-cm) minimum width required by ADA and needed for passage of a wheelchair. Were this to be removed, the propane storage next to it would still limit the width to 27 in. (53 cm).
- On the left side, the newspaper displays reduce the width to just 36 in. (91 cm), requiring the wheelchair to ride the edge of the curb.

It is recommended that these displays be moved to an area of the walkway where placement would permit a minimum of 38- to 40-in. (97- to 102-cm) widths, or remove them entirely from the walkway. As mentioned above, a more desirable alternative would be to relocate these curb ramps closer to the center, on either side. This would allow these displays to remain in place.

1.5 LEVEL WALKWAY SURFACES AND WATER

Facilities should be designed and maintained to prohibit (or at least minimize) the accumulation of water in walkway areas and around building areas.

Physical Evaluation 17

FIGURE 1.20 Illustrated here is a good example of grading away from the entrance area, thus avoiding water accumulation.

FIGURE 1.21 Here is a poor example of grading in which a parking lot is graded down toward the building, and toward the handicapped parking spots. This allows water to accumulate, which turns to ice in cold weather, creating hazardous conditions in a high-traffic area that is also designated for handicap use. The nearest drainage was halfway up the incline, too far away from the base to do much good.

Aside from the slip hazard posed by the water, cold weather increases the likelihood that such accumulations will turn into ice, thereby compounding the hazard (Figure 1.20 and Figure 1.21).

A variety of conditions can contribute to water accumulation:

- Improper grading of land—land should be graded away from the building, so that low ground is not at the building

- Depressions, holes, and other concave areas as a result of settling or repaving that cause water pooling
- Inadequate storm drainage design capacity, blocked drains, or clogged piping
- Insufficient gutter capacity, or drainage of gutter onto a pedestrian path
- Air-conditioning system condensation
- Lawn sprinkler overspray

CASE STUDY: Level Walkway Surfaces and Water

Problem: When it rains, the covered concrete walkways become slippery from windblown rain, local flooding, and drips from umbrellas and rain gear.

Issues:
- Smooth concrete
- Impermeable surface near significant water accumulation (no run-off, absorption)
- Standing water (puddles, longer time to dry, mold)
- Contamination from dirt and foliage nearby
- Mat type, design, maintenance

Questions:
- What is the age of the concrete?
- What are the discolorations?
- Is there a grading (angle)? What is it and how is it oriented?

FIGURE 1.22

Physical Evaluation

- Has this area been cleaned? How often? Method? How long does it take to accumulate?
- Are there gutters? Are they sufficient? Are they clear? How are they cleaned? How frequently? How long does it take to accumulate blockages?
- Is the concrete sealed? What is the product (MSDS, other documentation available)? How often applied?
- To what degree does the walkway cover affect the availability of sunlight on the walkway?

Potential solutions:

- Reduce impermeable surface by replacing concrete with brick. This will allow water to absorb/drain from the area, reduce wet times and standing water.
- Perform a deep cleaning, first using a mild cleaner like muriatic acid, and then performing pressure washing.
- Install a drainage system with recessed drains and grade to achieve water removal.
- Use a multiple mat system (exterior and interior) of beveled mats at each entrance/exit.
- Provide grass coverage for the dirt area adjacent to the walkway to avoid soil runoff onto the walkway.
- Consider fully enclosing the walkway. However, the impact of water accumulation needs to be studied to determine if it creates unacceptable catch areas.
- Grind/roughen the concrete to increase asperities.

1.6 PARKING AREAS

Similar to sidewalks and other level walkway surfaces, cracks, holes, depressions, and other blemishes need to be regularly identified and corrected. The guidelines are the same, in practice ¼ in. (6.4 mm) or more. In addition to the issues related to level walking surfaces mentioned previously, however, parking areas have additional design needs. They are a challenge to maintain primarily because they are subject to heavy, concentrated foot and vehicle traffic during operating hours. Having parking lots sealed reduces the susceptibility of the asphalt to damage, which reduces the potential for falls.

Weather, vehicle accidents, and plows can all cause damage. Left unrepaired, these conditions become progressively more severe, increasing the potential for injury to pedestrians. In addition to assuring asphalt surfaces are in good condition, making sure exterior lighting is maintained can also help reduce the risk (Figure 1.23).

1.6.1 Tire Stops

Parking lots should be designed without tire stops. Tire stops require maintenance, are often damaged by snowplows and other vehicles, and are a common tripping hazard (Figure 1.24).

FIGURE 1.23 An extreme example of conditions that can develop in parking lots.

FIGURE 1.24 Tire stops are subject to substantial punishment and must be inspected and maintained.

In some instances, the presence of tire stops in newly designed parking lots can be considered poor design (Figure 1.25). Where tire stops are present, several precautions should be observed (Figure 1.26):

- The maximum height should be 6½ in. (16.5 cm), with at least 3 ft. (91 cm) between wheel stops.
- To minimize the tripping hazard, tire stops should not protrude beyond the width of tires.
- Tire stops should be well marked with a contrasting color so they do not blend in with the parking lot, a condition that becomes even more hazardous at nightfall.

Physical Evaluation

FIGURE 1.25 This tire stop has been cleverly disguised to blend in with the floor surface, effectively making it invisible to many pedestrians. Although it is not a major concern when a vehicle is parked (as long as the driver can see it while parking), the tire stop presents a serious trip hazard when the spot is unoccupied. (Photograph courtesy of K. Vidal.)

FIGURE 1.26 This tire stop is well marked, strongly contrasting with its surroundings, and is in good repair. Constructed of recycled rubber, it is also more durable and resilient than concrete. (Photograph courtesy of K. Vidal.)

- Tire stops of railroad tie construction should be avoided because they deteriorate quickly, presenting an increased hazard.
- Reinforcing bar and other methods of securement should not extend beyond the tire stop itself.

1.6.2 Speed Bumps

National and professional standards have been established for many components of public transportation roadway systems. Some local municipalities have also developed guidelines. However, no standards have yet been issued from an adoptive national government body or agency for "traffic calming" devices. Two nationally accepted transportation standards, the Manual on Uniform Traffic Control Devices for Streets and Highways (MUTCD), established by the U.S. Federal Highway Administration, and A Policy of Geometric Design of Highways and Streets, produced by the American Association of State Highway and Transportation Officials, are silent as to required design features or placement of traffic calming devices.

After automobiles were invented at the turn of the 20th century, and their use became common, local governments were confronted with the problem of speed control. Clearly, the traffic calming issue of today is nothing new. The concept of using physical barriers began early on with the installation of speed bumps on public streets. Speed "bumps" are typically found in parking lots. They differ greatly from the speed "humps" encountered on some public roadways, which are much wider and more gradual than speed bumps (see Figure 1.27 and Figure 1.28).

A properly designed speed hump should not significantly affect traffic flow and comfort unless the vehicles are traveling faster than 30 mph (48 km/h) or are heavily loaded. Well-designed speed humps generally are 12 ft. (30 cm) to 22 ft. (56 cm) wide, and are generally 3 in. (7.6 cm) or 4 in. (10 cm) in height, whereas older speed bumps were only 3 in. (7.6 cm) to 36 in. (91 cm) wide and 3 in. (7.6 cm) to 6 in. (15 cm) high (Klik and Faghri, 1993). However, their public use was short-lived.

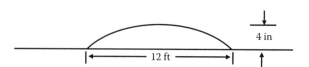

FIGURE 1.27 Speed hump.

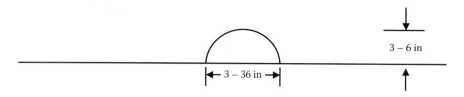

FIGURE 1.28 Speed bump.

Physical Evaluation   23

FIGURE 1.29   Although fairly well marked with yellow stripes, this speed bump also spans a frequent and natural pedestrian walkway area.

Similar to tire stops, speed bumps are usually unnecessary in properly designed parking lots. The layout of the lot should make it impractical to drive at high speeds. Speed bumps can cause damage to vehicles and can be damaged by snowplows. Unrealized by most drivers, heavy automobiles can more smoothly pass over them at 35 mph (56 k) than at slower speeds (Figure 1.29).

Usually constructed from asphalt, severe weather can accelerate deterioration of speed bumps.

Speed bumps can pose a tripping hazard and should be located away from natural and expected pedestrian paths, especially heavily trafficked entrance and exit areas.

Where speed bumps exist, they should be painted white or yellow to contrast with the parking lot. ANSI Z-535.1 Safety Color Coding provides specific guidance on such marking. It is also advisable to provide signage indicating the presence of speed bumps.

The Manual on Uniform Traffic Control Devices for Streets and Highways, 2001 (U.S. Federal Highway Administration) provides specifications regarding marking and signage, but not design or dimensional criteria. This publication can be obtained in PDF format at http://mutcd.fhwa.dot.gov/

ASTM F-1637 (Practice for Safe Walking Surfaces), Sections 7 ("Speed Bumps") and 4.2 ("Walkway Changes in Level"), indicate that speed bumps should be avoided in design. If they are used, they should not be in pedestrian paths. If they are in pedestrian paths, they should be marked according to ANSI Z535.1 with slip-resistant paint; caution signs are also recommended.

Other publications on the topic include:

- Guidelines for the Design and Application of Speed Humps, Institute of Transportation Engineers (ITE) Task Force TENC-5TF-01 Staff, published by the ITE (1993).
- A Policy on Geometric Design of Highways and Streets 2001—This fourth edition "Green Book" contains the latest design practices in universal use as the standard of highway geometric design.

FIGURE 1.30 A speed bump sign is used to give warning of a sharp rise in the profile of the road. This sign may be supplemented with an Advisory Speed plaque.

FIGURE 1.31 A speed hump sign is used to give warning of a vertical deflection in the roadway that is designed to limit the speed of traffic. This sign should be supplemented by an Advisory Speed plaque.

More information is also available at http://www.digitalthreads.com/utarpt/utarpt5.html.

According to the MUTCD, Figure 1.30 and Figure 1.31 illustrate the requirements for signage relating to speed bumps and humps, and Figure 1.32 presents the two common options for marking of speed humps.

1.7 CHANGES IN LEVELS

Unmarked or unidentified floor and walking surface level changes are a source of frequent and severe fall claims. This exposure includes conditions ranging from recessed seating to a subtle height change from one room to another, to a quick step down just outside of a door.

1.7.1 Air Steps

Injuries including fractured or sprained ankles and damaged knees, hips, and spines can occur even if a misstep does not result in a fall. An air step occurs when a person,

Physical Evaluation

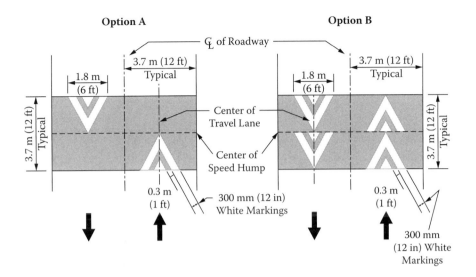

FIGURE 1.32 Shown are MUTCD speed bump marking options.

without uncontrollably losing balance, unexpectedly steps into a hole, depressed part of a walkway, or an unanticipated stair step.

The depth of such steps could be an inch or several inches. If the unexpected drop exceeds about 7 in. (18 cm), the body responds to the drop and takes reflexive action. A fall by Fidel Castro, captured on video and widely seen, is a typical air step that resulted in serious injuries (Pauls, 2007a).

1.7.2 Design

In general, building codes are silent on requirements for short flight designs; however, there is some agreement on designs to avoid an unexpected change in level from a doorway.

The rule of thumb (also spelled out in the International Building Code) is that the landing on both sides of a doorway should be as long as the doorway is wide, and at least 44 in. (112 cm) in the direction of travel (Figure 1.33).

According to ADA Accessibility Guidelines for Buildings and Facilities (Americans with Disabilities Act, ADAAG, 4.5.2):

> Changes in level up to ¼ in. (6 mm) may be vertical and without edge treatment. Changes in level between ¼ in. and ½ in. (6 mm and 13 mm) shall be beveled with a slope no greater than 1:2. Changes in level greater than ½ in. (13 mm) shall be accomplished by means of a ramp that complies with 4.7 or 4.8.

Where practical, short flights should be converted to a qualifying ramp. Otherwise, effective controls can include the use of contrasting colors, special lighting features, and warning signs. The goal is to provide visual cues to make the pedestrian aware of the change. This is particularly important for short flight step changes in which,

FIGURE 1.33 If not for the handrails, there would be no indication to the approaching pedestrian that a change in level was present.

from the perspective of the approaching pedestrian, the change is barely noticeable or not noticeable at all.

1.8 STAIRS

A high school student died after falling down a flight of stairs at her home. The 16-year-old was home alone when she fell. She stayed home from school because she wasn't feeling well. She fell as she was going from the main level of the house to the finished basement. She banged her head on the wall, lost consciousness and landed in a position that caused her to suffocate. "Those stairs are treacherous, and it isn't the first time someone has slipped and fallen," said her father. (Breakey, 2008)

1.8.1 STAIR MISSTEPS—OVERSTEP AND HEEL SCUFFS

Oversteps and heel scuffs involve descending stairs, especially those with undersized treads. People learn to adapt to short stair treads by twisting their feet to one side to maximize foot contact on the tread or by twisting both feet outward. Generally, the ball of the foot must be adequately supported by the tread to safely descend a stair.

There are two types of oversteps:

- If slip resistance is adequate, the foot will continue to rotate until the leg cannot support the weight applied to the leading foot and the individual begins to pitch forward or collapse.
- The foot slips off the nosing rapidly. Whether there is adequate slip resistance on the nosing or not makes little or no difference in this case. Because this happens quickly, loss of body support is difficult to overcome.

Heel scuff missteps occur when overcompensating with foot placement to avoid overstepping a tread and scuffing the heel against the riser. Heel scuffs and overstepping missteps are similar, and can occur due to undersized treads, irregular tread depths, or both. In such cases, the stair must be measured and the way it was visually perceived by the victim and how his/her feet reacted to the geometry of step nosings must be evaluated. Step geometry should be measured as the vertical and horizontal distances between adjacent step nosings (Pauls, 2007a).

Understeps are another type of misstep in which the foot is inadequately placed on a step tread while ascending stairs.

1.8.2 Stair Statistics

Injuries sustained while ascending stairs are generally less severe than those sustained while descending. In ascent, the forward momentum of a fall is arrested by the stair structure itself, while in descent there is potentially a much greater distance to fall (see Table 1.2 for injury statistics).

The number of injuries sustained after a fall while descending stairs grossly outweighs those while in ascent. Templer et al. (1985) reported that 92% of injuries were incurred during stair descent, and Nagata (1995) also reported a similar proportion of injuries experienced in stair descent (78% for males and 92% for females).

Research has shown that pedestrians view only the first three and last three steps, negotiating the remainder of the stairway without looking (Maynard and Brogmus, 2007). This means the design of the top three and bottom three steps is very important. Moreover, 70% of all stair accidents occur on the top three and bottom three steps.

NFPA International (NFPA) 101, the Life Safety Code (National Fire Protection Association, 2006), is a consensus standard—a national voluntary standard that has, in many jurisdictions, become law in some form (Figure 1.34). Compliance (or lack of compliance) with this standard has a strong impact in the courts. Just as

TABLE 1.2
NEISS Survey of Stair Accidents

Incident Type	Incidents Per Year
Flight uses	1,953,000,000,000
Noticeable missteps	264,000,000
Minor accidents	31,000,000
Disabling accidents	2,660,000
Hospital treatments	540,000
Related deaths	3,800

Note: The estimated incident occurrence on stairs (1975) is the result of data compiled by the CPSC from NEISS, from A Survey of Stair Use and Quality conducted by Carson Consultants, Inc., and from videotapes of stair use studied by the NBS. The NBS reports that NEISS compiled information of reported injuries treated in 119 hospital emergency rooms across the United States.

FIGURE 1.34 It is clear that the Great Wall of China, built and maintained from the fifth century B.C. to the 16th century and extending for 4,160 miles (6,695 km), was constructed prior to the establishment of U.S. consensus standards.

importantly, the life safety code specifications are tried-and-true guidelines for safe stairs (Figure 1.35).

Based on a review of prominent literature on stairway safety and a review of findings from a survey of 40 in-depth stair accident cases, it is believed that the overriding issue with stairway accidents is not pedestrian or external variables, but rather the dimensional inconsistency inherent in some stairways (Jackson and Cohen, 1995).

Maintenance and code variations are the main items to evaluate for stairs and landings.

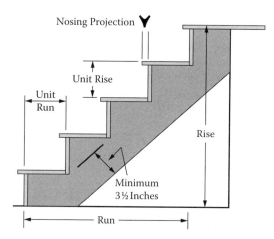

FIGURE 1.35 Stair components and measurements. (Drawing courtesy of www.sizes.com. Used by permission.)

1.8.3 Stair Design

- The rise angle of stairways should be between 30 and 35 degrees of slope.
- Stair riser height should be 7 in. (18 cm) [new] or 8 in. (20 cm) [existing], with no deviation between adjacent risers exceeding 3/16 in. (5 mm) (the International Building Code permits 3/8 in. [9.5 mm])
- Stair riser height should deviate no more than 3/8 in. (9.5 mm) over the entire flight.
- Stair tread depth should be 9 in. (23 cm) or 10 in. (25 cm) [existing], or 11 in. (28 cm) [new], with no deviation exceeding 3/16 in. (5 mm)
- Stair width should be at least 44 in. (112 cm) clear width, 36 in. (91 cm) if serving fewer than 50 occupants.
- Nosing should not protrude more than 1½ in. (38 mm) and should be beveled to reduce trip potential. The surface of treads should be of "nonslip" material.
- Short flights (three risers or less) may require additional design considerations, including increased tread depth for improved visibility as well as supplemental visual cues.

(See Figure 1.36 and Figure 1.37.) Stairs should be uniform in tread width and riser height. The human body is an amazingly precise machine. It can detect subtle changes in elevation and distance. Thus a change in riser height as small as 3/16 in. (5 mm) can disrupt a person's gait or walking rhythm (cadence) and significantly increase the potential for a fall.

In stairway descent, tread depth must be adequate for the ball of the foot to land without extending over the step below. Trips and falls that occur during stairway ascent are often attributed to variation in riser, or vertical surface, height (Maynard

FIGURE 1.36 An example of good design and condition.

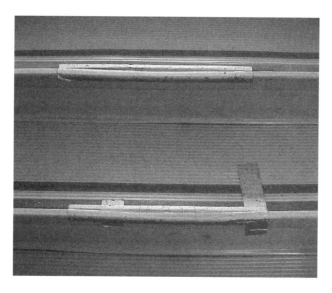

FIGURE 1.37 Stair nosings here are in poor condition with a less than adequate repair attempt.

and Brogmus, 2007). Many in-depth studies on stair safety, including research by J.A. Templer (1975, 1985, 1992) support these stair design criteria:

In ascent:
- Stairs with risers 6.3 in. (16 cm) to 8.9 in. (23 cm) and treads 7.7 in. (20 cm) to 14.2 in. (36 cm) had the fewest missteps.

In descent:
- The larger the tread, the fewer the missteps.
- Treads below about 9 in. (23 cm) performed uniformly poorly.
- Risers measuring 4.6 in. (12 cm) to 7.2 in. (18 cm) had the fewest missteps.
- Missteps increased with steepness. Performance was poorest with small treads.

1.8.3.1 Stair Nosings

According to many researchers (Templer et al., 1985), having the entire tread of uniform slip resistance is optimal to reducing stair falls. Also, when slip-resistant strips are used, they should extend to the edge of the nosings. If the nosings are rounded, the strips should extend over the curvature.

According to Templer, in his 1992 publication *The Staircase*:

> If the tread material has a poor coefficient of friction, then the addition of an abrasive strip to the nosing may be useful in preventing slips at that point only. Its utility will be vitiated if it causes trips because it projects above the tread surface or acts to confuse one visually as to where the edge is. For new stairs, this treatment is no substitute for an adequately abrasive surface for the whole tread, including the nosing. Some

Physical Evaluation

> **Online Resource: Stair Design and Dimensional Criteria**
> Visual Interpretation of the International Building Code 2006 Stair Building Code, Stairway Manufacturers Association, http://www.stairways.org/pdf/2006%20Stair%20IRC%20SCREEN.pdf.

manufacturers do not recognize that it is the coefficient of friction of the nosing that matters most, so many proprietary abrasive tread and nosing systems (and manufactured stair systems) provide tread surfaces with a high coefficient of friction and leave the nosing edge with a potentially hazardous, slick surface.

SIDEBAR: Measuring Walkways and Stairs

To measure the vertical height of a riser or a change in elevation, consider a machinist's square. This device has a head with a 90-degree edge, a level bubble, and a clamping screw in it, which slides on a ruler, or scale. Place the scale end on the base surface to be measured and slide the head down to the higher surface being measured, keeping the bubble in the center of the glass vial (level). When the head is level and in the position of the desired measurement, turn the locking screw to secure the head on the scale, then pick up the machinist square and read the bottom edge of the head against the scale. This instrument can also be used to measure the tread depth (Kaufmann, 2007).

Other component measurement methods are as follows:

- To measure pitch, slope, or angles, an electronic level is most helpful. It can be placed on a walkway or tread surface to measure the angle, and it can be placed on a handrail or tread nosings to measure the angle of the handrail and stairway to determine if they are parallel.
- To measure distance, a tape measure, rolling wheel measurer, or laser beam ruler can be used.
- Measuring radiuses reasonably accurately requires a radius gage.
- For measuring thickness or small distances, use a dial caliper.

1.8.3.2 Stair Visibility

As people descend a stairway, treads are clearly visible but risers are not. This can make it difficult to properly judge the vertical distance the foot must travel to reach the next step. Therefore it is advisable to make the treads more visible using contrasting nosings and adequate lighting.

Install at least 20 foot-candles (200 lux) of illumination to highlight the stairway and the floor approaching it. Make sure the edge of each tread is properly illuminated, that shadows do not impede the view, and that glare does not disrupt visibility (Maynard and Brogmus, 2007).

Another little-discussed issue regarding stair visibility is the wide use of bifocals. The lower portion of these glasses is intended for close-up viewing, such as reading. However, when descending stairs, it is only these lenses through which wearers must attempt to see the more distant stairs below their feet.

Alternative Stairway Safety Features

A review of literature by the National Association of Home Builders (NAHB) Research Center (1992), suggested design enhancements to further improve stair safety were summarized as follows:

Floor Covering

- Provide a stable walking surface
- Use slip-resistant treads
- Ensure a tight, uniform covering

Construction and Design

- Use round nosings
- Ensure consistent stair dimensions
- Ensure structural integrity
- Place doors to avoid striking other people

Surrounding Environment

- Remove glass objects near stairway
- Use energy-absorbing materials

Visual Enhancements

- Make distinct stair edges
- Use contrasting colors between stairs and walls
- Reduce glare
- Reduce shadows
- Shield light sources
- Eliminate visual distractions

Stair Illumination

- Place light switches at top and bottom away from first riser
- Use contrast lighting
- Ensure consistent light levels between stairway and surrounding areas
- Illuminate one- and two-step risers
- Provide low-voltage permanent bulb
- Provide redundancy in light sources

Handrails and Guardrails

- Make continuous
- Make comfortable
- Use dual handrail (where possible)
- Install properly
- Use terminations (where appropriate)
- Provide a child's handrail (where appropriate)
- Ensure proper dimensions, height, materials, wall clearance

1.8.3.3 Stair Landings

Per NFPA 101 (Life Safety Code) 7.2.2.3.2, stairs are required to have landings at door openings. Landings should have a consistent width; the width should be at least that of the stairs to which it is connected.

As specified by most building codes and by NFPA for new construction, stair landings should be at least as long as the width of the stair. Thus, if the stair width is 44 in. (112 cm), the length (in the direction of travel) should be at least 44 in. (112 cm). This provision is made in order to avoid a door that opens directly onto the stair tread.

Landings need to be kept clear of storage or any other materials at all times in order to minimize congestion and ensure that adequate space passage is available (Figure 1.38).

Physical Evaluation 33

FIGURE 1.38 This is a suitable dimensional design for a stair landing, although the horizontal handrail supports are not preferred (see Section 1.9.2).

SIDEBAR: Assessing the Stair Safety Requirements of Model Building Codes

In 2004, a comparison was done of model code requirements with regard to stairways. The 2003-edition, model building codes analyzed were the International Code Council's (ICC) *International Building Code* (IBC) and NFPA International's *Building Construction and Safety Code* (NFPA 5000). Requirements were rated on a scale of A to F for technical quality and the extent of enforcement.

It is important to note that organizations such as NAHB can exert influence over amendments to model code requirements in the local or state adoption process in an effort to make them less onerous and less effective for injury prevention and mitigation. Thus, the content of the national code may not remain intact when adopted by the local jurisdiction.

Marking. NFPA 5000 has better requirements for marking and offers guidance on surface conditions for treads, including nosing marking and prohibition of tripping lips, so it gets a B for the model requirements and F for enforcement. The IBC was graded as C for requirement and enforcement, due mainly to a lack of a uniformity requirement for treads.

Lighting. IBC and NFPA have weak requirements for luminance levels for stairways, 1 foot-candle, though NFPA increased luminance of new stairs to 10 foot-candles in the 2006 edition. Both codes continue a traditional requirement for room lighting to average 10 foot-candles at a height of 30 in. above floor level, but there is no evidence it is enforced.

Stair Step Geometry. Both IBC and NFPA received an A- for requiring the "7–11" stair step.

Handrails. Both IBC and NFPA handrail requirements for height, maximum lateral spacing, terminations, continuity, and graspabilty are generally good, earning them ratings in the B to C range.

Overall Stairway Ratings. The analysis gave the IBC a C+ for model requirements and a C for enforcement. NFPA got a B+ for model requirements, but due to its lack of adoption, an F for enforcement (Pauls, 2007b).

1.9 HANDRAILS

Stair guardrails protect pedestrians from falling off stairs or landings, while handrails provide visual cues, help pedestrians keep their balance, and provide support when using stairs. Lack of, or improperly designed handrails are often a factor in stairway falls. Properly designed handrails can reduce the likelihood of a fall and can limit the distance down the stairs one falls.

Per NFPA 101 7.2.2.4, handrails should be accessible within 30 in. (76 cm) [for new buildings] and 44 in. (112 cm) [for existing buildings] of all portions of stair

FIGURE 1.39 Although dimensionally sound, the condition of the stair nosings is questionable, and the single railing is inadequate.

width. This is to ensure that, even if standing in the most remote part of the stair (farthest away from a handrail), a person should be able to reach out and grasp a handrail (see Figure 1.39 and Figure 1.40).

If a handrail is on only one side of the stair, it is preferable for the handrail to be located on the right side descending the stair. A person should be able to reach it while standing on the other side of the stair. If handrails are on both sides, a person should be able to stand in the center and reach either handrail.

Handrails should be between 34 in. (86 cm) and 38 in. (97 cm) in height. This optimizes the ability of users to grab and exert the necessary force to stop a fall. Maki et al. (1984, 1985) tested the maximum forces exerted in various directions

FIGURE 1.40 Although the railings are well spaced at the base of the stairs, they become progressively less useful higher on the flight. They are also not "graspable." (Photograph courtesy of K. Vidal.)

Physical Evaluation

by individuals on stairs with slopes of 33, 41, and 49 degrees. The optimal height of the handrail midline, as measured above the nosing, was determined to be 38 in. (97 cm). Lower heights require less force exerted against the handrail, so one is less able to stabilize oneself during the start of a fall, and the fall is more likely to occur.

1.9.1 Graspability

Unfortunately, designers often see handrails as aesthetic elements of building construction instead of essential safety features. Thus, handrails are frequently too wide or otherwise poorly designed for building occupants to grasp (see Figure 1.41 and Figure 1.42). Handrails should allow continuous holding without being impeded by supports or other obstacles.

Handrails with a circular cross section should have a diameter between 1¼ to 2 in. (32 to 51 mm). This design is preferred because it allows most users to employ a

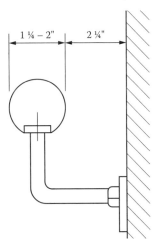

FIGURE 1.41 This is a diagram showing the preferred dimensional criteria for handrails, including distance from the wall.

FIGURE 1.42 A graspable handrail extending beyond the stair constitutes a sound design.

FIGURE 1.43 A squared cross section of handrail is difficult to grasp.

power grip (rather than a pinch grip), in which fingers can curl around the rail. For other shapes, the perimeter should be between 4 to 6¼ in. (102 to 159 mm) with the largest cross section no larger than 2¼ in. (57 mm) and with rounded edges (NFPA 101 7.2.2.4.4.6). This specification considers the varied hand size of men and women, as well as the reduced hand strength of older individuals, and is intended to assure that the handrail can be fully grasped in normal and emergency situations. The profile of the handrail should match hand grip (Figure 1.43). Handrails for use primarily by children should be on the smaller side of these ranges. Functional tests have shown that handrails of larger dimensions that do not match hand grip perform poorly.

An extensive review of studies performed regarding the efficacy of handrails indicated that optimal shape and size of the handrail is circular with a 32–50 mm (1.26–2.0 in.) diameter or oval with a thickness 18–37 mm (0.71–1.46 in.) horizontally and 32–50 mm (1.26–2.0 in.) vertically. In addition, the height recommended "with confidence" for adults was 935–1000 mm (36.8–39.4 in.) above step nosings (Feeney-Webber, 1994).

New handrails are to be installed with a clearance of at least 2¼ in. (57 mm) between the handrail and the wall. Although permitted by some codes, handrails as close as 1½ in. (38 mm) from the wall are less desirable (NFPA 101 7.2.2.4.4.5).

Handrails should extend at least 12 in. (30 cm) beyond the bottom of the stairway when descending so the user can maintain a hold on it while taking the last steps to the floor. They should also extend at least 12 in. (30 cm) beyond the top of the stairway to provide a visual cue to the presence of stair and permit them to grasp the rail before beginning their descent (see Figure 1.44).

1.9.2 SUPPORTS

Handrail supports should be vertical, not horizontal. In essence, horizontal rails constitute a ladder to young children, who have been known to climb them and fall over the other side (Figure 1.45).

Physical Evaluation 37

FIGURE 1.44 Here, the handrails stop where the stairs cease. This requires the user to step forward without the benefit of handrail support. These should extend at least 12 in. past the end of the stair on the top and the bottom.

FIGURE 1.45 An unintended application, horizontal supports can be used for climbing.

Handrail supports should not permit the passage of a 4-in. (102 cm) sphere—this precludes children from getting their heads stuck between the rails.

CASE STUDY: Handrail Design and Placement

Situation 1. The design of many exterior handrails is inadequate. They do not meet the requirements of NFPA 101 Life Safety Code for graspability, which is needed not only for any component in a means of egress but has been long established as safe design. There are two handrail designs in question. One is a long oval which is too wide to grasp. The other is a large squared-off design with sharp angles. These were noted in most side entrances with exterior stairs.

Discussion. The handrail is the last opportunity for an individual to retard or prevent a fall. As such, handrails must be readily graspable. Unfortunately, designers often see handrails as aesthetic elements of building construction rather than essential safety features. Thus, handrails are frequently too wide or otherwise poorly designed for building occupants to grasp. Such appears to be the case with many of these handrails.

Recommendation. Schedule and budget for replacement of existing handrails with circular handrails having a diameter of 1¼–1½ in. (32–38 cm). This takes into account the anthropometry of hand size in men and women, as well as the reduced hand strength of older individuals, and is intended to assure that the handrail can be gripped properly by most people.

FIGURE 1.46

Physical Evaluation

FIGURE 1.47

NFPA 101 7.2.2.4.4.6 also permits noncircular handrails which must have a perimeter dimension of between 4 and 6¼ in. (102 and 159 mm) with the largest cross sectional dimension no larger than 2¼ in. (57 cm) as long as graspable edges are rounded to provide a radius of at least $1^1/_8$ in. (3 cm). For additional clarification on option two, see the NFPA 101 Appendix for 7.2.2.4.4.6. This replacement should be completed along with relocated/supplemental handrails as discussed below.

Situation 2. Handrails provided at most exterior stairs are spaced too far apart to be accessible to pedestrians using stairs, according to NFPA 101 and safe practice and design. This condition was noted at most wide exterior stairs in which two or more handrails are present.

Discussion. Per NFPA 101 7.2.2.4.1.2, handrails should be accessible within 30 in. (76 cm) of all portions of stair width. This is to assure that, even if standing in the most remote part of the stair (farthest away from a handrail), a person should be able to reach out and grasp a handrail.

Recommendation. Schedule and budget for replacement of handrails spaced to be accessible in accordance with the above criteria. This will require one or two additional handrails at many locations.

1.10 GUARDS (OR GUARDRAILS)

Per NFPA 101 7.2.2.4.6, in order to reduce the potential for falls, guards should be provided where stairs are at a height 30 in. (76 cm) or more above the floor level. Located above handrails, guards should be at least 42 in. (107 cm) in height (Figure 1.48).

Anthropometrics appears to be relevant to this specification. A person leaning on a guardrail or inadvertently walking into it might be less likely to go over it if the top of the rail is close to his center of gravity. The Whole Body Centroid Height of the 97.5 percentile male (published in a 1974 reference) is 41.2 in. (105 cm).

FIGURE 1.48 Handrails are shown here with guardrails above.

1.11 RAMPS

Many ramps are intended to provide access to facilities for persons with disabilities, particularly those in wheelchairs. As such, specific requirements must be followed regarding dimensions, slope, and handrails for these components.

Optimally, facilities should be designed to avoid (or minimize) such changes in level. By situating handicap parking spaces on opposite sides of an entrance from the primary pedestrian path, mobility-impaired individuals can often be accommodated while minimizing the exposure of pedestrians to such ramps.

Ramp slopes are calculated by rise (vertical distance) over run (the length of the ramp) (Figure 1.49).

The slope should be no greater than one (vertical) × eight (horizontal), or 7.1°. Ramps used for individuals with disabilities should have a slope no greater than 1 × 12 (4.8°), with slopes between 1:16 (3.6°) and 1:20 (2.9°) being optimal (NFPA, 2000; ADA, 1991). Figure 1.50 and Table 1.3 describe ramp slope and ramp slope conversions, respectively.

1.11.1 Ramp Design

The following additional guidelines apply:

- In general, a clear width of at least 36 in. (91 cm) should be present (ANSI A117.1), and no projections should extend into that space. This provides sufficient width to permit the passage of a wheelchair. NFPA 101 requires new ramps to be 44 in. (112 cm) minimum clear width, but permits some existing ramps to be as narrow as 30 in. (91 cm). Clear width and other ramp dimensions can also vary according to NFPA 101, depending on the occupancy.

Physical Evaluation

FIGURE 1.49 An example of a good ramp design, retrofitted to an existing building.

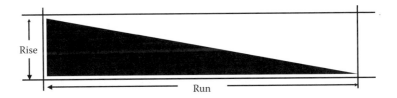

FIGURE 1.50 Slope is rise (height) over run (distance).

TABLE 1.3
Ramp Slope Conversions

Slope Ratio	Degrees	Notes
1:4	14	
1:6	9.5	
1:8	7.1	Maximum pedestrian ramp slope
1:10	5.7	
1:12	4.8	Maximum handicap new construction
1:15	3.8	
1:20	2.9	Walkway

FIGURE 1.51 This is a poorly designed ramp. The slope is uneven and a tripping hazard is present at the base of the ramp.

- For 180° turns, ramp width must be at least 42 in. (107 cm) and at least 48 in. (122 cm) at the turn per ANSI A117.1 (see Figure 1.51).
- Unless a landing is present (Figure 1.52), the ramp must be a continuous, uniform slope.
- Handrails should be provided for ramps with a rise over 6 in. (15 cm).

Figure 1.51 and Figure 1.53 illustrate poorly designed ramps.

FIGURE 1.52 This is a good example of a ramp with a single turn.

Physical Evaluation

FIGURE 1.53 This is another poorly designed ramp—no handrails, and an uneven and excessive slope.

1.11.2 RAMP LANDINGS

- Ramp landings must be present at the top and bottom, and at doors opening onto the ramp.
- The landing slope should be no steeper than 1:48 and should be at least the same width as the ramp. The landing should be at least 60 in. long (152 cm) [in the direction of travel]. For NPFA 101, existing ramp landings need not meet this requirement.

Figure 1.52 is an example of a ramp with a single turn.

1.12 FLOOR MATS/ENTRANCES AND EXITS

Entrance and exit areas are critical in controlling slip, trips, and falls, because the concentrations of traffic and surface transitions are usually high. The exposure is significant because they constitute bottlenecks. Congestion is greater, contributing to the potential for pushing and tripping. Also, a high degree of surface wear occurs (which makes the walkway surfaces smoother and less tractive), as well as a higher concentration of water, dirt, and other contaminants from the footwear of those traversing this finite area.

CASE STUDY: Slippery Supermarket Entrance in Wet Weather

A recently opened supermarket was experiencing a high frequency of slip/fall accidents, mostly on wet terrazzo floor tiles at the entrance area and in the first few aisles of the supermarket nearest the entrance. It was determined that water from rain was entering through the foyer. Entrance mats were too small to absorb the amount of water deposited by footwear of customers.

Short-term solutions:
- An increase in the frequency of cleaning in the foyer at times of wet weather.
- A system of staff vigilance for signs of water on the supermarket floor. When water was identified, store cleaning would follow.
- Rather than mopping (which left the floor surface wet), staff used a wet vac, which left the floor dry.
- Existing recessed matting was complemented by additional mats during wet conditions.

In the longer term, the supermarket built a canopy over the entrance to further inhibit water entry.

Cost of interventions:
- Staff training on the frequency of cleaning, approximately half day per employee.
- Purchase of wet vac (less than £500).
- Purchase of supplementary entrance mats (about £20 each).
- Construction of the canopy was absorbed during store refurbishment, because it was in the original plan anyway.

After 18 months of these changes being in place, there had not been a serious slipping accident.

(http://www.hse.gov.uk/slips/experience/store-entrance.htm)

1.12.1 GENERAL PRECAUTIONS

To reduce the potential for falls in these areas, consider the following:

- Identify surfaces with excessive smoothness and other defects. Changes in surface transitions can be of more concern than changes in levels, because it is more subtle. This affects the "expectation" of the individual and often leads to stumbles and falls.
- Ensure door closure mechanisms are not forceful enough to knock someone over.
- Saddle (doorsill) height is a factor. To minimize an unexpected condition, it should be flush with the floor. The maximum safe height is ¾ in. (19 mm). Saddles should slope at the edges, to avoid tripping hazards, and should be grooved or otherwise made slip resistant. Saddles should be secured tightly to the floor.

Physical Evaluation

- Consider the impact of transitions between different types of floor surfaces, particularly transitions to extremes (e.g., from excessively low to very high traction or the reverse), which may present the pedestrian with an unexpected condition, thus increasing the likelihood of a fall.

1.12.2 Minimizing Contaminants

Entrances and exits should be designed to minimize slip and fall potential caused by ice and snow tracked in on visitors' footwear. Ideally, a grate system with a catch basin should be used in high-traffic situations to allow moisture removal from footwear. Grates should be placed perpendicular to the direction of travel (Figure 1.54).

If this is not possible, mats should be used. The importance of proper mat selection and maintenance is crucial. If floor mats are not engineered for wear or properly cleaned on a consistent schedule, their ability to control the entry of contaminants is significantly diminished because:

- As much as 80% of all dust, dirt, and other contaminants found in buildings are tracked in from the outside through entrances.
- One square yard of commercial carpet can accumulate one pound (453 g) of dirt a week and up to two pounds a week in wet weather.
- A vacuum cleaning removes only 10% of dirt from floor mats.
- One square yard of entrance matting removes, traps, and holds one pound of dirt per week.
- Over a 20-day period, 1,000 people will deposit 24 pounds (11 kg) of dirt.
- Within the first 6 ft. (1.8 m), 42% of a floor's finish will be removed after only 1,500 people have entered.

FIGURE 1.54 This is an example of a well-installed recessed foyer grate system.

Properly cleaned floor mats can catch approximately 70% of the dust tracked into a facility. Floor mats protect high-traffic areas and carpets from wear and tear. Because they absorb water, mud, and internally generated soil, floor mats also minimize conditions that can contribute to slips and falls. Floor mats can be used effectively in many areas, including front and rear entrance and exit doors, checkout counter areas, restrooms, kitchens, and around vending machines, ice machines/refrigerators, and water fountains.

1.12.3 Mat Design

Floor mats that are designed for removal of dust, dirt, and moisture from the footwear bottom are distinctly different from other types of floor coverings. Scatter rugs, carpet remnants, or mats with borderless backing, vinyl backing, or no backing do not efficiently remove contaminants that accumulate on footwear bottoms, nor are they designed to minimize tripping or sliding underfoot.

The edging of floor mats should be beveled in order to provide a smooth transition from the floor to the mat. Mats should be replaced before they become dog-eared or otherwise damaged or overly worn. In most cases, the best solution is to use heavy-duty entrance mats constructed with a type of wire mesh that allows moisture and contaminants to drop away from the footwear.

Low-density rubber mats are more susceptible to bunching, rippling, and other undesirable movement. Well-constructed high-density rubber mats are less prone to such conditions, making them worth the additional expense.

Fabric and rubber mats are appropriate at entrance doors when light water loads are expected. Cocoa fiber mat in a plush or woven finish can effectively absorb liquids. Manufacturing and office buildings often use linked, solid, or perforated rubber mats for their durability. Extreme temperatures may require vinyl or neoprene mats with a carpet or ribbed thread finish for traction.

1.12.4 Mat Design and Arrangement

Entry mats should not be approached in a piecemeal manner. Rather, such protection should be thought of as a system. An industry-recommended approach involves a three-component approach, involving a 15-foot mat run (Commercial Entrance Building Care):

1. **Scraper Mats** are designed to be used outdoors (under cover), but are commonly used in entrances. Their abrasive surface creates a scraping or raking effect which is designed to remove heavy debris. Nonabsorbing, they allow moisture to run off or evaporate, and the debris settles to the bottom of the mat. The first 5 ft. (1.5 m) of scraping action from a scraper mat can reduce the volume of dirt entering the building by up to 40%.
2. **Wiper/Scrapers** are designed to moderately dry and wipe the footwear bottom and scrape away remaining debris. An irregular mat surface forces dirt particles to fall down into the crevices, allowing a continuing cleaning

FIGURE 1.55 Strategically placed wiper matting is arranged to cover natural and expected pedestrian paths.

process. The 5 ft. (1.5 m) of scraping/wiping action from a wiper/scraper mat can further reduce the volume of dirt entering the building by up to 30%.

3. **Wipers** dry the footwear bottom. Because the heavy dirt and debris has been removed by the scraping and wiper/scraping mats, the wiper mat can make contact with the footwear to thoroughly dry the footwear bottom.

This approach allows the bulk of contaminants to be removed by abrasive mats, with absorptive mats removing supplemental moisture (Figure 1.55).

Some mats are designed with cleats, or small rubber protrusions on the underside. These are intended to inhibit movement when used over carpet. Such designs are not appropriate on hard, smooth surfaces because the cleats can compress from pressure and create ripples in the mat. This is especially the case when heavy objects, such as carts, are moved across the mat.

Avoid stacking mats in use. This option might be considered when a mat has become saturated and, instead of removing and replacing it, another is placed on top. This increases the potential for tripping due to the substantially higher or uneven edge against the floor which can result in rippling.

Mats are more likely to slide when dust and dirt accumulate under the mat due to inadequate cleaning of the floor underneath. Smooth-backed mats are the most susceptible to sliding under these conditions. The primary solutions are to ensure that mats are of the appropriate design for the flooring, and that cleaning is sufficient to preclude an accumulation of dust and dirt under the mat.

1.12.5 Mat Storage

Curling of in-service mats presents a tripping hazard. To reduce the potential for mat curling, mats should be rolled properly. Storing mats flat is optimal, but often impractical due to space limitations. Another effective method is to alternate rolling: each time the mat is rolled, alternate between front out and back out.

1.12.6 MAT SIZE

The Carpet and Rug Institute (CRI) Commercial Carpet Maintenance Manual *generally* defines an entrance (or soil wipe-off) area as the 90 ft^2 (8.36 m^2) or 6 ft. (1.8 m) × 15 ft. (4.6 m) at building exterior entrances, where most tracked-in soil is deposited. CRI research shows that 80% of the soil brought into any building is trapped within the first 15 ft (4.6 m) on a carpeted surface. A 10-ft^2 (0.93-m^2) soil track-off region can also be found where the carpet and hard surfaces meet at main internal doorways and a 40-ft^2 (12 m^2) area in 6 ft. (1.8 m)-wide main corridors.

Unfortunately, many commercial and institutional property owners and managers buy or rent mats that are too small to be effective.

CASE STUDY 1: Entrance Floor Mats

The single 6-ft. × 9-ft. (1.8 m × 2.7 m) mat is too small. The mat slides easily due to dirt/dust accumulation underneath. The area immediately inside the door has no mat. Due to the security device at the entryway, a single runoff mat cannot be laid flat. Two options are as follows:

- Have at least two mats custom made for the area immediately inside the door, and change out as needed.
- Cut and secure carpet squares in this area, clean in place, and replace as needed.

FIGURE 1.56

Physical Evaluation

FIGURE 1.57

A standard carpet mat can be used for the area immediately past the security device.

CASE STUDY 2: Entrance Floor Mats

Situation. There are several improvement opportunities to reduce slips/falls at entryways. There have been at least two such accidents over the past 12 months. The exposure is high in these areas, since they constitute bottlenecks.

FIGURE 1.58

FIGURE 1.59

FIGURE 1.60

While some entrances have sound multiple mat systems, others are lacking. In some instances, the mats are too small or numerous, permitting insufficient walk-off space to remove moisture and contaminants from footwear bottoms. In other cases, placement or housekeeping is lacking.

Recommendation.
- To prevent sliding, assure that cleaning under the mat is sufficient to preclude an accumulation of dust and dirt underneath.

Physical Evaluation   51

- Provide at least one extra mat for each mat in service, so it can be readily changed out when it becomes contaminated, misshapen, or damaged.
- According to CRI best practice, provide walk-off mats long enough to take two full steps before stepping onto other floor surfaces.
- Establish a cleaning and change-out schedule for mats commensurate with the extent of use and prevailing weather conditions.
- Add mat conditions to the list of scheduled inspection items.

1.12.7 Protection of Hard Flooring Surfaces

Hard-surface floor coverings are unable to trap dirt and moisture, and allow such contaminants to spread. The Resilient Floor Covering Institute (RFCI) states that accumulated soil diminishes the appearance of resilient flooring. RFCI's maintenance guide reports:

> Preventing abrasive dirt from being tracked onto the surface of flooring is critical to prolong the life of resilient flooring. The use of walk-off mats at entrances is an effective way to prevent scratches and stains caused by tracked-in material from outside.

CRI's Floor Covering Maintenance for School Facilities states that daily maintenance of hard-surface floor covering presents a greater challenge than carpeted surfaces. Hard-surface flooring must receive constant care because of its inability to hide soil; the finish is also easily damaged by dry, gritty particulate soil and dirt. Recommended daily maintenance includes dry and wet mopping as well as spot mopping, and high-traffic areas may require dust mopping several times a day. Strategically placed floor mats, designed to remove dirt and moisture deposits from footwear bottoms, reduce the need for cleanup labor.

1.12.8 Mat Cleaning Guidelines

Mats should receive scheduled cleanings of appropriate frequency based on the conditions to which they are subject. CRI's manual for school facilities states:

> Walk-off mats should be cleaned frequently. Once a walk-off mat becomes filled with soil, the soil will then transfer to the soles of shoes and spread throughout the facility.

Entrance mats—Mats should be shaken outside to remove excess dirt and debris. If necessary, use a hose (avoid high pressure or high temperatures) to wash them. Allow carpeted mats to dry before bringing them inside.
Carpet mats—Carpet mats can be cleaned the same way as carpeting. Vacuum daily, and extract or shampoo when dirt builds up.
Molded rubber and PVC antifatigue mats (designed for wet areas)—Use a high-pressure hose (not to exceed 1800 psi) and hot water (max 160 °F) to remove oils. For best results, use a mild soap or detergent with a pH between 4.0 and

> **Online Resource**
>
> For more information about floor mat design, maintenance, and professional mat services, visit the Textile Rental Services Association (TRSA) at http://www.trsa.org.

9.0 to clean the mats. Do not use steam, degreasers, or caustic chemicals. Although it is common to place them in commercial dishwashers, avoid machine washing or mechanical scrubbing.

PVC sponges and urethane mats (designed for dry areas)—Sweep regularly or dry mop the surface. These mats can be wet mopped with mild soap or detergent. For best results, use a detergent with a pH between 4.0 and 9.0.

Runner mats—Sweep the surface with a broom or vacuum. Some runners can also be wet mopped with mild soap.

Electrostatic discharge (ESD) mats—ESD mats are designed to control the generation of static electricity. It is important to regularly sweep or dry mop the surface of ESD mats. Also, wet mop or wipe off with mild soap or a static control cleaning solution that will not leave a residue, so the mat will continue to work well.

1.13 BATHROOMS

For durability, appearance, and ease of cleaning, glazed ceramic or similar tile is the flooring of choice in many bathrooms. Untreated, this type of flooring commonly exhibits low slip resistance in wet conditions. In some cases, mats are placed at sinks where water is most likely to reach the floor; however, rarely are the facilities to dry hands similarly adjacent to the sink area. Unfortunately, paper towel dispensers and blowers are usually near the entrance or elsewhere, requiring travel with wet hands to that location, trailing water along the way. A simple change in arrangement can significantly reduce the exposure to falls (Figure 1.61).

The bottom of bathtubs and showers are frequent sites of falls. These areas should not be smooth. Instead, they should be textured with appliqués, treatments, or embedded patterns. Optimally, rubber bath mats should be routinely used.

1.14 ELEVATORS

- Elevators need to be adjusted to be even and level with the floor surface on each floor at which they stop.
- The space between the elevator car and the floor should be minimal (see Figure 1.62).
- Changes in floor surfaces in the elevator/lobby area should be minimal and, if present, clearly marked or differentiated.

Physical Evaluation

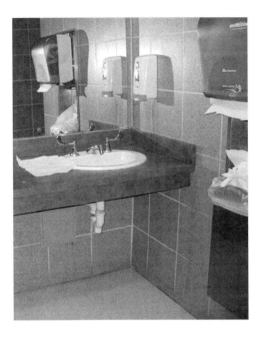

FIGURE 1.61 In this case, the paper towels are situated close to the sink, minimizing the potential for dripping water spreading across the area. It would be helpful to have a floor mat in this area as well.

FIGURE 1.62 There is a fairly wide space between the elevator car and the floor. Multiple floor types can increase the potential for falls.

1.15 ESCALATORS

The U.S. Consumer Product Safety Commission (CPSC) estimates that 11,000 escalator-related injuries occurred in 2007. While most of these injuries are from falls, 10% occur when hands, feet, or shoes become trapped. The most common entrapment is to the foot, with soft-sided shoes most likely to get stuck and result in injury. CPSC has reports of 77 such entrapment incidents since January 2006, with about half resulting in injury, all but two of which involved popular soft-sided flexible clogs and slides.

- Warning signs should be provided to discourage or prohibit the use of strollers, instruct users to use extra caution if wearing high-heeled shoes, and to advise users to hold onto the handrail (see Figure 1.64).
- Emergency shutdown procedures should be in place.
- Where no other physical barrier is present, child guards should be provided at entrance points to each escalator. These are designed to prevent children from holding onto and attempting to ride up on the outside of the escalator and becoming injured (see Figure 1.63 and Figure 1.65).
- ANSI A117.1 specifies that escalator handrails should be between 33 in. (84 cm) and 42 in. (107 cm).

FIGURE 1.63 Here is a well-arranged escalator. The barriers at the rear act as child guards.

Physical Evaluation

FIGURE 1.64 Here are well-presented posted safety warnings.

FIGURE 1.65 Properly placed child guards are a requirement for escalators.

FIGURE 1.66 Carpeting in poor stages of repair.

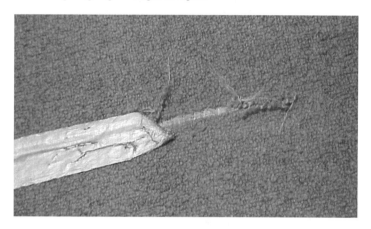

FIGURE 1.67 Carpeting in poor stages of repair.

1.16 CARPET

In terms of slip resistance, carpet is one of the best walking surfaces that can be employed. Carpets have the added benefit of eliminating the problem of reflected glare associated with hard surface floors, reducing visual disorientation and subsequent falls (see Figure 1.66 and Figure 1.67).

Some precautions should be kept in mind when using carpet:

- Carpets should be checked for wear and looseness and repaired as necessary. Optimally, such conditions should not be permitted to develop.
- Just as with any other type of floor covering, carpets need to kept clean and free of debris, including granular material (e.g., sand, dirt) and wet contamination (e.g., water, oil).
- Many public facilities use dark carpeting on stairs, perhaps as a means of keeping the carpet from appearing soiled. A dark carpet may reflect only about 5% of available light, which can be as low as the minimum required one foot-candle. In such cases, nosings are hard to see regardless of their color.

Physical Evaluation

- Stairs nosings should have contrasting colors, and perhaps incorporate alternating carpet colors per John Templer's work in *The Staircase*.
- Avoid "busy" carpet patterns. Instead, patterns should be subtle with low-contrast colors. High-contrast patterns tend to diminish an individual's ability to judge perspective, masking the approach to short-run stairs and other changes in the level of walking surfaces.

For the purposes of accessibility, the ADA Accessibility Guidelines for Buildings and Facilities (ADAAG, 4.5.3) states:

> If carpet or carpet tile is used on a ground or floor surface, then it shall be securely attached; have a firm cushion, pad, or backing, or no cushion or pad; and have a level loop, textured loop, level cut pile, or level cut/uncut pile texture. The maximum pile thickness shall be ½ in. (13 mm). Exposed edges of carpet shall be fastened to floor surfaces and have trim along the entire length of the exposed edge.

1.17 ACCESSIBILITY

Special access for individuals with disabilities should be evaluated for slip, trip, and fall hazards. Instead of just performing ADA compliance assessments, identify the potential for loss due to the inability of the facility to provide for adequate safe access of the disabled. For instance, the lack of entrance ramps and passenger loading areas may result in a person using a wheelchair to attempt to go over curbing to get to a sidewalk or entrance; this may result in a fall. Lack of room accommodations in a hotel, such as showers equipped for the handicapped, may force use of a conventional shower, which may contribute to a fall for a handicapped person.

Extensive design specifications are available on accessibility from references on the ADA (http://www.usdoj.gov/crt/ada/adahom1.htm) and ANSI A117.1 (http://www.ansi.org).

1.18 SIGNAGE

Signage should be provided where appropriate, considering changes in the type of floor surface, levels, trip hazards, and other conditions that the pedestrian may not expect to encounter.

Unfortunately, property owners often go to great expense to provide signs that blend into the environment. This negates the effectiveness of signs because they must be visible in order to be observed.

In selecting effective signage, signs must be understandable, legible, visible, and in compliance with legal standards. Consider the likelihood of multilingual and illiterate individuals. In these instances pictorial or multilingual signs may be most appropriate.

The Occupational Safety and Health Administration (OSHA) (29 CFR 1910.144) provides guidance on the proper design for safety signage:

- Danger signs indicate immediate dangers and whether special precautions are necessary. Colors should be red, black, and white.
- Caution signs warn against potential hazards or caution against unsafe practices. Colors should be yellow (background) and black with yellow letters (panel). Letters used against the yellow background must be black.
- Safety instruction signs should be used where a need exists for general instructions related to safety. Colors should be white (background) and green with white letters (panel). Letters against the white background must be black.

OSHA 29 CFR 1910.145 states that yellow shall be used for designating caution and for marking hazards, including stumbling, falling, and tripping.

Special consideration should also be given to those areas where guests need special warnings. For example, these areas might include prohibiting the use of strollers and the wearing of high heels. Other examples include signs reminding building occupants to use handrails, warning of icy conditions or a slippery ramp, pointing to a step up or down, and general guidance in congested or otherwise more hazardous areas to watch one's step. Signs of this type and many others are widely available (see Figure 1.68).

FIGURE 1.68 This is one of the many commonly used designs for warning cones to identify wet floor conditions.

Physical Evaluation

1.19 REFERENCE STANDARDS

1.19.1 NFPA International (Formerly National Fire Protection Association)

For information about NFPA (http://www.nfpa.org), see Chapter 5, U.S. Standards and Guidelines.

1.19.1.1 NFPA 101 Life Safety Code (2006)

Although it is designed for building fire safety, this code also includes many well-founded principles of stair, ramp, and other walkway surface component geometry. It is also a nationally recognized consensus standard that can carry much weight in the courtroom. It provides requirements based on the occupancy (e.g., health care, places of assembly, mercantile, residential) and whether the facility is new or existing construction.

NPFA 101 is updated every three years. Although the requirements of existing construction rarely change, those for new construction may change significantly.

1.19.2 ASTM International (Formerly American Society of Testing and Materials)

For information about ASTM (http://www.astm.org), see Chapter 5, U.S. Standards and Guidelines.

1.19.2.1 ASTM F1637 Practice for Safe Walking Surfaces

This is a brief but useful document providing good design guidelines for all types of walkway surface components. It is updated periodically by ASTM under the jurisdiction of Technical Committee F13 Pedestrian/Walkway Safety and Footwear.

1.19.3 American National Standards Institute (ANSI)

For information about ANSI (http://www.ansi.org), see Chapter 5, U.S. Standards and Guidelines.

1.19.3.1 ANSI A1264.1

This standard provides minimum safety requirements in industrial and workplace situations for protecting persons in areas/places where danger exists of persons or object falling through floor, roof, or wall openings, or from platforms, runways, ramps and fixed stairs, or roof edges in normal, temporary, and emergency conditions. Areas covered are as follows:

- *Protection of floor openings and floor holes*—Includes stairway floor and ladderway floor openings; hatchway and chute floor openings; skylights; pit, trap door, and manhole floor openings; and temporary floor openings.
- *Protection of wall openings and wall holes*—Includes wall openings; chute wall openings; window wall openings; and temporary wall openings.
- *Protection of open-sided floors, platforms, and runways (ramps)*—Includes runways; hazardous locations; railing systems; stair railing systems;

toeboards; handrails; clearances; floor opening covers; skylight screens; and barrier and screens for wall openings.
- *Requirement for fixed stairs*—Includes fixed stairs for access; load criteria; clearance; slope; tread width and riser height; hosing; slip resistance; uniformity of risers and treads; long flights of stairs; stair landings; door and gate openings; and vertical clearances.
- *Requirements for use of railing systems, rails and handrails*—Includes provision and design; handrails; and stair railing system/handrail required use.

1.19.3.2   ANSI A1264.2

A1264.2-2006 is the Standard for the Provision of Slip Resistance on Walking/Working Surfaces. For more information on this standard, see Chapter 5, U.S. Guidelines and Standards.

1.19.3.3   ICC/ANSI A117.1

The International Code Council (see Section 1.19.5) acts as secretariat for this standard, known as ADA-ANSI A117.1-1980, Accessible and Usable Buildings and Facilities. It provides specifications on making facilities accessible by persons with disabilities. There are two versions of the same standard with clear design guidelines for such components as ramps, landings, and curb cutouts. The ANSI 2003 edition is the most current (as of this writing, the 2008 edition is under development), but the ADA edition is the most enforceable.

1.19.4   OCCUPATIONAL SAFETY AND HEALTH ADMINISTRATION (OSHA)

For more information about OSHA, go to http://www.osha.gov.

1.19.4.1   1910, Subpart D—Workplace Walking and Working Surfaces

Subpart D (http://www.osha.gov/pls/oshaweb/owastand.display_standard_group?p_toc_level=1&p_part_number=1910) provides dimensional criteria for walkway components, including ramps, stairs, railings, and ladders:

- 1910.23 Guarding floor and wall openings and holes
- 1910.24 Fixed industrial stairs
- 1910.25 Portable wood ladders
- 1910.26 Portable metal ladders
- 1910.27 Fixed ladders
- 1910.28 Safety requirements for scaffolding
- 1910.29 Manually propelled mobile ladder stands and scaffolds (towers)
- 1910.30 Other working surfaces

1.19.5   MODEL BUILDING CODES

The primary building code organizations in the United States are:

Physical Evaluation

- International Code Council (ICC), which promulgates the International Building Code (IBC)—The IBC was developed in a joint effort of the three primary U.S. building code organizations. Established in 1994 as a nonprofit organization, the objective of the International Code Council is to develop a single set of comprehensive and coordinated national model construction codes (http://www.ICCsafe.org/). The organizations and codes that are no longer maintained, but to which many existing buildings have been constructed are as follows:
    - Building Officials and Code Administrators, which promulgated the BOCA Code
    - Southern Building Code Congress International, which promulgated the Southern Building Code (SBC)
    - International Conference of Building Officials, which promulgated the Uniform Building Code (UBC)
- National Fire Protection Association (NFPA 5000) (http://www.nfpa.org)

Primary specifications relating to slip resistance and dimensional criteria for pedestrian safety are contained in Section 10—Means of Egress. In general, the dimensional guidelines for ramps, stairs, handrails, guards, and other building features related to pedestrian safety are consistent among the various building codes, and with the NFPA Life Safety Code—Means of Egress.

However, unlike NFPA 101, the model building codes address only new construction. In addition, states and even local jurisdictions may adopt different codes and different versions as well. The specific version of the model building code may have undergone modifications unique to the municipality. See Exhibit 1.2 for a summary of state-mandated building codes. These are perhaps the most important standards from a litigation standpoint, so it is essential that each facility comply with the appropriate building code.

1.19.6 ASME International (Formerly American Society of Mechanical Engineers)

Founded in 1880, ASME (http://www.asme.org) is a nonprofit educational and technical organization servicing a worldwide membership of 125,000. It promulgates and maintains numerous standards involving the design, inspection, maintenance, and testing of escalators, elevators, and moving walks.

EXHIBIT 1.1

SLIP AND FALL SELF-ASSESSMENT CHECKLIST

Facility Name: **Completed By:**   **Date:**

	Good	Fair	Poor	N/A
Interior				
Facility Design and Arrangement				
• Stairs (nonslip, well maintained, uniform tread and riser)	☐	☐	☐	☐
• Handrails (34–42" high, graspable, always within reach, sturdy)	☐	☐	☐	☐
• Elevators (level, minimal space between car/floor)	☐	☐	☐	☐
• Entrances/exits (nonslip, level, inclement weather measures)	☐	☐	☐	☐
• Entrance mats (recessed, appropriate size, good condition)	☐	☐	☐	☐
• Lighting (adequate, maintained)	☐	☐	☐	☐
• Floor levels/surface transitions (marked, unobstructed)	☐	☐	☐	☐
Spill Program/Housekeeping				
• Response time standard reasonable (specify):	☐	☐	☐	☐
• Appropriate barriers/signage	☐	☐	☐	☐
• Department responsible appropriate (e.g., housekeeping, security)	☐	☐	☐	☐
• Interim measures prior to spill cleanup in effect	☐	☐	☐	☐
Incident Reporting/Investigation				
• Response time standard reasonable (specify):	☐	☐	☐	☐
• Department responsible appropriate/ trained in first aid/CPR	☐	☐	☐	☐
• First priority—attend to injured person	☐	☐	☐	☐
• Obtain facts about the accident	☐	☐	☐	☐
• Obtain witness names, addresses, telephone numbers	☐	☐	☐	☐
• Take photographs of accident scene	☐	☐	☐	☐
• Complete written report of findings, including recommendations	☐	☐	☐	☐
• Take corrective action	☐	☐	☐	☐

EXHIBIT 1.1 (*Continued*)

SLIP AND FALL SELF-ASSESSMENT CHECKLIST

	Good	Fair	Poor	N/A
Construction, Renovation, Special Event Planning				
• Department/personnel responsible for overseeing activities (specify):	☐	☐	☐	☐
• Methods for securing areas from public access	☐	☐	☐	☐
• Adequacy of nonpublic areas (e.g., hazards to contractors, employees)	☐	☐	☐	☐
Preventive/Other Programs				
• Identify areas of facility with high frequency of slip/falls	☐	☐	☐	☐
• Slip meter readings/documentation (specify):	☐	☐	☐	☐
• Purchasing controls in place/adequate (e.g., floor materials/finishes, footwear)	☐	☐	☐	☐
• Subcontracted housekeeping/maintenance (hold harmless, certificates of insurance)	☐	☐	☐	☐
• Other measures (specify):	☐	☐	☐	☐
Exterior				
Design and Arrangement				
• Grading/drainage adequate (ice and water)	☐	☐	☐	☐
• Sidewalks/steps (level, nonslip, changes marked, handrails, condition)	☐	☐	☐	☐
• Parking lots (level, condition, arrangement)	☐	☐	☐	☐
• Lighting adequate (foot-candle measurements documented)	☐	☐	☐	☐
• Shortcuts minimized (landscape discourage access beyond sidewalks)	☐	☐	☐	☐
Inspection and Maintenance				
• Self-inspection and corrective action	☐	☐	☐	☐
• Maintenance (scheduled, prioritized action on work orders)	☐	☐	☐	☐
• Snow/ice removal procedures	☐	☐	☐	☐

Write Additional Comments On Other Side

EXHIBIT 1.2

STATE-MANDATED BUILDING CODES

State	State Code Name	Code Basis/ Edition
Alabama	None currently, but many jurisdictions have adopted building codes.	
Alaska	State funded residential construction only	2003 International Building Code
Arizona	None currently, but individual jurisdictions have adopted building codes.	
Arkansas	2000 International Building Code w/Arkansas amendments. Adopts mandatory minimum codes but local jurisdictions may amend them to be more stringent.	Same
California	2001 California Building Code	1997 Uniform Building Code
Colorado	No state code	
Connecticut	2005 State Building Code (effective 2006). 2003 International Building Code with amendments.	2003 International Building Code
Delaware	No state code	
Florida	2003 International Building Code with Florida supplement	Same
Georgia	2006 International Building Code, w/ amendments. Mandatory, however, it is local option whether to enforce.	Same
Hawaii	In 2007 a state code council was established requiring implementing a state building code, to include the latest edition of the International Building Code.	2006 International Building Code
Idaho	Buildings owned or occupied by the state and public schools are required to be in compliance with the 2006 International Building Code and appendices thereto pertaining to building accessibility. City and county building codes are adopted locally by ordinance.	Same
Illinois	No state code	
Indiana	2000 International Building Code w/Indiana amendments. Local jurisdictions may adopt more stringent provisions with approval from the state. Local enforcement is not required.	Same
Iowa	2006 International Building Code. Applies only to buildings that are state-owned or buildings partially or wholly funded by state funds. Local jurisdictions may adopt the state code or others.	Same
Kansas	No state code	

EXHIBIT 1.2 (*Continued*)

STATE-MANDATED BUILDING CODES

State	State Code Name	Code Basis/ Edition
Kentucky	2006 International Building Code w/ amendments. If there is no local inspection department to enforce the commercial codes, the State Office of Housing, Building and Construction will perform the inspections.	Same
Louisiana	2006 International Building Code	Same
Maine	Maine Model Building Code. Voluntary, but a locality may not adopt a building code other than this one.	2003 International Building Code
Maryland	International Building Code 2003. Municipalities may vary from these codes.	Same
Massachusetts	Massachusetts State Building Code (residential only)	1993 BOCA National Building Code
Michigan	2003 International Building Code with amendments	Same
Minnesota	Minnesota State Building Code (residential only)	2000 International Residential Code
Mississippi	Adopts by reference the latest editions of the International Building Code. Local jurisdictions may adopt the codes adopted by Mississippi.	2006 International Building Code
Missouri	No state code	
Montana	2006 International Building Code (effective January 26, 2007)	Same
Nebraska	Nebraska State Residential Code (residential only)	2000 International Residential Code
Nevada	No state code	
New Hampshire	2000 International Building Code with amendments	Same
New Jersey	2006 International Building Code with amendments	Same
New Mexico	New Mexico Building Code (residential only)	2003 International Residential Code
New York	2000 International Building Codes with 2001 amendments (effective 2003). The State may be empowered to adopt higher or more restrictive standards upon the recommendation of local governments.	Same
North Carolina	2003 International Building Code with amendments	Same
North Dakota	North Dakota State Building Code	2003 International Building Code

Continued

EXHIBIT 1.2 (*Continued*)

STATE-MANDATED BUILDING CODES

State	State Code Name	Code Basis/ Edition
Ohio	No state code	
Oklahoma	Oklahoma Building Code applies only to schools, hospitals, nursing homes, child care facilities, state buildings, mental health facilities, correctional facilities, county/city jails, and horse race tracks.	2003 International Building Code
Oregon	2007 Oregon Structural Specialty Code (2006 IBC with amendments)	2006 International Building Code
Pennsylvania	International Building Code 2006 (base code for commercial construction)	Same
Rhode Island	2003 International Building Code with amendments	Same
South Carolina	2003 International Building Code with S.C. modifications	Same
South Dakota	As of 2004, any local government wishing to adopt ordinances prescribing standards for new construction must adopt ordinances in compliance with either the 1997 UBC or the 2003 IBC.	1997 Uniform Building Code or 2003 International Building Code
Tennessee	No state code	
Texas	No state code. Senate bill 1458 requires compliance with the 2003 IBC, but allows amendments.	2003 International Building Code
Utah	2006 International Building Code effective January 1, 2007	Same
Vermont	The Life Safety Code (NFPA 101) applies to all buildings and premises regulated under the Code (Effective 2005). The International Building Code (IBC) applies to new construction and structural requirements.	International Building Code
Virginia	2003 International Building Code	Same
Washington	2006 International Building Code with amendments	Same
West Virginia	West Virginia State Residential Code (residential only)	2000 International Residential Code
Wisconsin	Uniform Dwelling Code (residential only)	None
Wyoming	2003 International Building Code	Same

Source: The Institute for Business and Safety Building Codes (http://www.disastersafety.org/building_codes/map.asp).

2 Management Controls

Organizational Responsibilities

Component	Potential Responsibility
Behavioral Safety/Traffic Flow	Safety, Design/Construction
Building Inspection and Maintenance	Maintenance, Safety, Operations
Slip and Fall Self-Inspection	Maintenance, Safety, Operations
Spill and Wet Program	Housekeeping, Safety, Operations
Snow/Ice Removal Guidelines	Maintenance, Safety, Operations
Snow Removal	Maintenance, Housekeeping
Lighting Design	Maintenance, Purchasing, Safety
Lighting Maintenance	Maintenance
Contractual Risk Transfer	Legal, Safety, Operations
Construction/Special Event Planning	Design/Construction, Housekeeping
Loss Analysis	Safety, Operations, Security

Don't fall before you're pushed.

English Proverb

2.1 INTRODUCTION

A well-designed facility can still be subject to frequent fall accidents if appropriate management controls are not in place. The purpose of management controls is to maintain the facility in a condition as free from hazards as reasonably possible. The extent of management controls considered adequate depends on a variety of factors, including:

- The size of the facility
- The amount of foot traffic
- The familiarity of building occupants with their surroundings
- The types of hazards
- The scope of programs implemented by other similar facilities

Programs considered adequate for a small manufacturing facility would be vastly different from what might be suitable for a large hotel. The manufacturing facility likely has low foot traffic, mostly from employees familiar with the facility, while the hotel likely has very high foot traffic, mostly from visitors who are unfamiliar with their surroundings.

2.2 ENGINEERING PRECEPT

When designing your facility and developing management control programs, it is helpful to remember this engineering precept: *"design for the worst-case scenario."*

Slips and falls are the most common type of loss not only for many workers, but also for visitors and customers. It must be assumed that anyone in any physical or mental condition could be on the premises. Once someone is on site, management assumes some degree of responsibility for the safety of the surroundings. Consider that someone who is disabled or impaired, including the nonambulatory, blind, or deaf, may be present. Consider that someone under the influence of alcohol or drugs, prescription, over-the-counter, or illegal, may be present. Consider that one or more people are distracted or stressed with other things on their mind.

Any of these risk factors represent an increased risk of injury, particularly for slips and falls. Therefore, an effective method is to take into account all of these variables when designing, implementing, and testing your program components.

2.2.1 NO "SILVER BULLET"

Many managers want to believe there is a single, quick and simple solution for reducing or eliminating slips and falls, and this belief is often reinforced by vendors of products that promise to do so. But a good footwear program alone, or an effective floor care procedure in itself cannot eliminate all slip and fall accidents. Controlling slip and fall risks requires a multifaceted methodology.

2.3 INTEGRATE CONTROLS

Optimally, you should not aim for a distinct slip/fall prevention program. The elements that make up slip/fall prevention are best integrated into existing programs and functions. Rather than having a separate slip/fall initiative to cover components including footwear, cleaning regimen, and reporting/investigation, integrate these components with a corresponding, existing program. Purchasing should be responsible for buying footwear that is specified by the safety department. Floor cleaning specifications should remain with maintenance, using products specified by purchasing. Accident reporting and investigation should be coordinated by security, which should follow procedures that include supplemental information addressing how to respond to a slip/fall incident.

Electing to develop yet another separate safety program introduces more work for everyone. Required controls for slip, trip, and fall prevention must be developed and reviewed by each function just to determine which items may involve work for them. A gap analysis of the scope and procedures for each discipline (e.g., housekeeping, maintenance, purchasing) must be completed, and missing controls integrated with the appropriate function.

Management and employees are also responsible for implementing a variety of other stand-alone programs such as electrical safety, blood borne pathogens, machine guarding, emergency preparedness, material handling, and a host of other safety plans. These responsibilities result in increased workloads that may impact the integrity of the programs and hinder successful outcomes.

An integrated approach to implementing risk management procedures makes it more manageable for those responsible for specific areas of work to focus on reducing risk to a particular exposures; integrating these functions also significantly increases the likelihood of success.

2.3.1 Program Component Design

Even with an integrated approach, it is still necessary to plan carefully and develop a strategy to ensure that key elements of slip and fall prevention are addressed. Existing functions need to be mapped out to determine where responsibilities lay, critical loss areas must be identified and included, and existing policies and procedures need to be updated to include the required slip and fall exposures and controls. Finally, orientation, training, and follow-up/enforcement are required.

A successful effort to managing slip and fall risk requires an integrated approach to designing and implementing the components. Affected departments need to be involved in the revision of their policies and procedures. This not only ensures that the result is a practical approach, but also helps improve buy-in and observance once implemented, because there is an increased sense of ownership by the departments involved.

Examples of departments that should be involved in the design and implementation of slip and fall prevention program components are detailed at the top of each chapter as applicable.

2.3.2 Hierarchy of Controls

When determining how to best control an exposure, the most effective control measure is to eliminate the risk. If it is not reasonably feasible to do so, identify effective measures to reduce the risk. Safety professionals advise against depending solely on the use of administrative controls (such as training) or personal protective equipment (such as footwear) to reduce the risk unless it is clear that more effective controls are not practical. In some cases, these controls need to be used in combination for maximum effectiveness.

2.3.2.1 Elimination

Elimination of hazards refers to total removal of the hazard and, therefore, complete reduction of potential for accidents. If the hazard is truly eliminated, other management controls, such as workplace monitoring, training, safety auditing, and record keeping, should no longer be required.

2.3.2.2 Engineering Controls

Engineering controls are physical means taken to limit the hazard. These include structural changes to the work environment or processes, creating a barrier between the worker and the hazard. For slips and falls, engineering controls include items such as proper selection and design of flooring, steps, ramps, and handrails.

2.3.2.3 Administrative Controls

These reduce or eliminate exposure to a hazard by following specified procedures or instructions. Documentation must emphasize all steps to be taken and the controls to be used in carrying out the activity safely. Training, maintenance, and inspection are examples of administrative controls.

2.3.2.4 Personal Protective Equipment (PPE)

Safety gear should be used as a last resort, after all other control measures have been considered, as a short-term contingency, or as a supplemental protective measure. The success of PPE, such as footwear, depends on the proper selection and maintenance, and that it is worn when required.

See Exhibit 2.3, Sample Walkway Surface Design Audit.

2.4 BEHAVIORAL SAFETY AND PEDESTRIAN TRAFFIC FLOW

High-traffic-flow areas are especially susceptible to trips/falls for two reasons: (1) there is more congestion, and (2) there is more exposure (e.g., more people present that could fall). These areas should be observed closely, and traffic flows should be determined for various times of the day. Management procedures should be evaluated to ensure that adequate controls are in place to address the various issues that may arise.

Analyze losses according to time of day, day of the week, relation to activities or events taking place, and similar criteria to determine if congestion related to traffic flow is a factor in slip/fall incidents.

Assess the adequacy of measures taken to manage large crowds. Consider putting controls in place to guide large crowds safely through walkway areas. Use fencing for outside conditions or other forms of guide rails inside and outside. Understand that barriers and visual cues, such as lighting and signage, may not be sufficient to produce the desired behavior. In general, it is human nature to take the shortest and easiest path to a destination.

2.4.1 Natural Paths/Observation

After identifying the time periods representing the highest exposure (based on volume and/or incidents), spend some time watching how people move. Also observe other evidence of heavy pedestrian traffic such as worn pathways cutting across open areas where the sidewalk changes directions, wear and tear on floor surfaces resulting in depressions, and areas of worn carpeting (see Figure 2.1).

2.4.2 Adapt or Adopt

Where safe and practical, it is easiest and most effective to simply adopt the path created by human behavior. Thus, as presented in Figure 2.2, consider paving that worn path and making it suitable for pedestrian use.

In the design and construction phase of a facility, sidewalks and other pedestrian routes should be arranged to use the most efficient path. This will reduce or eliminate the potential for pedestrians to shortcut into unintended and possibly dangerous areas.

Management Controls 71

FIGURE 2.1 Actual pedestrian traffic flow is clearly indicated by this well-worn shortcut from parking lot to the sidewalk.

FIGURE 2.2 A more direct path that avoids walking around the corner has been created by pedestrians.

It is optimal to determine where and how people tend to walk and design these routes based on that behavior. Once known, alternatives can be developed and again tested by observation.

2.5 BUILDING INSPECTION AND MAINTENANCE PROGRAMS

One of the most frequent causes of falls relates to pedestrians encountering an unexpectedly slippery (usually wet) surface. This is often the result of inadequate maintenance of the building or equipment (e.g., leaking of pipes, hoses, tanks, roofs) or by delayed detection of a condition (e.g., spills, overspray, infiltration of moisture at entryways). Thus, it could be argued that, in conjunction with proper facility design, superior maintenance and inspection programs are the most effective means of preventing conditions that result in slips and falls.

Management's inspection and maintenance programs must meet several criteria to be considered adequate. These programs must:

- Be of a scope commensurate to the size and type of the facility—A large facility with vague, incomplete, or non-hazard-specific programs will likely be unsuccessful in preventing or quickly correcting hazardous conditions.

- Be scheduled at a reasonable frequency—The longer hazardous conditions are permitted to remain unabated, the greater the potential for an incident.
- Be consistently and uniformly applied—What is put in writing must be followed and completed. If, for example, your inspection program calls for three inspections a day of a given area, monitor to assure that three—not two—inspections are being performed.
- Include all employees, not just those directly involved in performing inspection and maintenance activities. The more eyes looking at the facility, the more quickly and thoroughly substandard conditions will be detected and corrected.
- Be well documented—If inspection is performed three times a day, there must also be documentation to confirm it.

2.6 SLIP/FALL HAZARD SELF-INSPECTION PROGRAMS

A jury returned a slip and fall verdict in the amount of a half million dollars to Tod Flemke in a case against a pharmacy. Mr. Flemke sustained injuries to his lower back and underwent surgery to repair herniated discs. Flemke tripped and fell over a plug in a store. The store did not place caution or warning signs in the aisle to alert customers to the fact that they were cleaning, despite it being an unwritten policy to place caution signs in front of outlets.

For most facilities, there should be a formal mechanism in place to inspect for slip and fall–related exposures. This control may be integrated with a facility inspection or review that covers slip and fall hazards as well as other issues (e.g., security, housekeeping, life safety). Consistently documented inspections can help establish the presence or absence of those hazards.

Mapping of slip, trip, fall incidents and/or hazards is a simple and relatively easy way to identify problem areas (see Figure 2.3). It is designed to pinpoint where slips and trips have been occurring so that effective solutions to conditions identified can be developed. This can also serve to help establish a reasonable schedule for self-inspections.

- Sketch or photocopy a diagram of the area.
- Mark slips, trips, and falls reported in the prior 12 months (or other period) with crosses. Larger numbers of accidents (by using longer periods) are more likely to yield reliable results.
- Ask workers about "near misses" and add them to the diagram.
- Probe workers to determine what they believe are causing people to slip, trip, or fall in their area.
- Identify the "hotspots" that are likely to appear on the diagram. This will allow you to take corrective action and monitor those areas more closely.

Other ways of establishing reasonable self-inspection schedules include:

- Obtain and adopt industry-specific standards developed and promoted by industry associations.

Management Controls

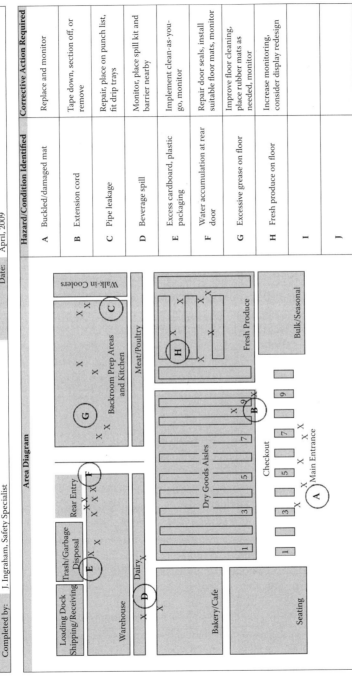

FIGURE 2.3 Slip, trip, and fall hazard mapping can be valuable in identifying areas of concern.

- Consult with other similar facilities nearby to determine their frequency and scope of self-inspection.

These two options can provide a significant additional benefit. In the event of legal action, a plaintiff's counsel may compare the quality of your program with that of your industry and other facilities like yours.

For more information on inspection and maintenance, see Chapter 6, Flooring and Floor Maintenance. Also see Exhibit 2.4, Sample Walkway Surfaces Inspection.

2.7 SPILL AND WET PROGRAM

Former NBA All Star Robert Reid explains that in 14 years of playing professional ball he never slipped and fell while running up and down the court. But when Reid came off an elevator in a Houston hotel, he ran into a wet spot and fell. "My right foot hit that wet spot and next thing you know I'm flying up in the air and I came down and I blacked out." Reid, who needed two surgeries to repair his knee, sued the hotel and was awarded $285,000.

According to a study of supermarket documentation of walkway inspections by the Gleason Group:

- While most use written sweep/aisle walk logs to monitor the condition of floors, 40% consider them to be less than accurate.
- For spills and subsequent cleanup, the amount of detail recorded varies by store.
- The frequency with which logs are verified by managers varies considerably.
- Three out of four respondents who maintain logs use them when handling slip/fall liability claims.

Seven out of ten stores surveyed use sweep/aisle walk logs. These are handwritten logs that employees prepare when they have walked or swept store aisles. Of the stores using written logs, 40% consider their records to be somewhat or not very accurate. The amount of detail recorded varies by store—43% note the type of hazards that exist (e.g., spills), while 54% do not keep such detail or do not know.

According to Gleason's study, less than half verify the logs daily (46%), and other managers only verify the logs weekly (40%), monthly (11%), or less than monthly (3%). Most stores perform regular walks to monitor aisles, but the frequency varies. In fact, 61% walk aisles at least every hour; 11% every two hours; 11% every three hours.

Three out of four respondents indicated that logs are used adjusting claims, and 88% say their insurance administrator has access to log records. Most keep log reports for more than three years (46%), but a significant number only keep the logs from one to three years (29%), or less than a year (23%) (Research Shows, 2001).

Because slips on wet surfaces are a leading cause of severe injuries, operations with this substantial exposure should have policies and procedures to deal with spills, wet surfaces from inclement weather, and any other source of water or wet walkway surfaces (Figure 2.4).

Management Controls   75

FIGURE 2.4 Using umbrella stands or racks at entryways can help prevent water from entering the building and reduce inspection and housekeeping efforts. (Courtesy of Gleason Technology.)

A spill and wet program is essentially a focused housekeeping program designed to provide prompt detection, response, and corrective action for wet conditions.

Sufficient equipment and materials should be available to handle any wet problem that is likely to be encountered. Signs and barriers should be employed. For spills, there should be interim measures in effect to temporarily protect the area until cleanup is complete. This will usually consist of yellow cones, triangles, or other placards. These barriers should mark off the area boundaries in order to notify pedestrians, coming from any direction, of the hazard. Barriers should not be removed until the area is not only clean, but is also completely dry (see Figure 2.5 and Figure 2.6).

For facilities with a frequency of spill and wet exposures, standards for response time and cleanup should be established, measured, and documented. Once put in writing, it is essential that those procedures be observed.

2.7.1 Electronic Inspection System

Used in many retail supermarkets in the U.S. and Canada, GleasonESP® is a floor safety inspection system designed to systematize and accurately document floor safety walk-around inspection tours by employees and to report results remotely.

Buttons are strategically placed throughout the facility. A hand-held data reader collects information each time it is touched to a button. These silver buttons contain a data chip that holds one piece of information, such as a location in the facility or a description of a hazard (see Figure 2.7).

FIGURE 2.5 Signs like this are not acceptable alternatives to proper design, inspection, and maintenance.

FIGURE 2.6 A basic component of an effective spill and wet program is to preclude pedestrian access by fully sectioning off wet floor areas by using readily identifiable barriers such as wet floor signs/cones. Note that this sign does not mean "produce slips and falls"; the produce department of the supermarket simply did not want their sign taken by another department.

Management Controls 77

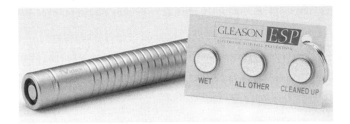

FIGURE 2.7 Hand-held reader and buttons describe conditions identified. (Courtesy of Gleason Technology.)

An employee conducts a walk at a predetermined schedule. The walk begins when the employee touches the data reader to a button matched with his/her name. The employee follows a predetermined route around the facility and touches the data reader to location buttons along the way. The route is strategically designed to take employees past high-risk areas, such as the produce department (when being used in a grocery store). If a hazard is found, the employee touches one of the hazard buttons on a touch card attached to the data reader that describes the hazard (see Figure 2.8).

When the walk is completed, the data reader is placed in a cradle, where a log of these activities is transmitted electronically to an online database. The data is processed to provide chronological, management, and exception reports, for auditing compliance with assigned floor inspection tours and for statistical analysis of hazards.

FIGURE 2.8 Placement of buttons strategically throughout the facility, reported through a hand-held reader. (Courtesy of Gleason Technology.)

There are strong indications of the efficacy of such a system:

- According to an independent study by Shelter Island Risk Services, this system reduced the probability of incurring a slip and fall claim over 30%. Results were consistent, regardless of facility location (urban or rural), size, or season.
- Georgia-based Harvey's Supermarkets, a 45-store chain, reportedly reduced its slip/fall incidents and claim costs by 72% within a year.
- High-volume Giant Eagle store in Pennsylvania reported an 87% reduction in the number of slip/fall claims.
- In three case studies completed by Gleason tracking 54 supermarkets over a two-year period, there was a significant reduction in claims.

2.8 RECOMMENDED PRACTICES FOR SNOW REMOVAL

Two Framingham police officers were injured when they slipped on some icy stairs at an apartment. When Officer Kim Kelleher began to slip, she tried to grasp the handrail, but that, too, was slippery. Officer Rob Kelleher tried to stop her, and he slipped. Kim Kelleher fell on her back, wrenching it and bruising a kidney. Rob slid down the stairs on his back, striking his head on all seven stairs as he went down.

Ice and snow management should be considered emergency work, and the property should be cleared day or night. Removal of snow from walkways and parking lots is essential to the safety of employees and visitors. Snow removal efforts should be performed as expeditiously as conditions and resources allow.

The planning and preparation process can be made difficult by the harsh conditions that occur during and after each storm. Variables in moisture content, temperature, wind, depth of snow, and rate of snowfall all influence the speed and effectiveness of removal efforts. For example, if there is ice buildup under the snow, it may be prudent to leave the snow on the ice until warmer weather arrives. If the wind is blowing strongly, the removal effort is a judgment call. Light snow can be more readily removed in greater depths with a broom than can wet, heavy snow.

2.8.1 OBJECTIVES

Understanding that employees and visitors expect a broom-clean and dry pavement 24 hours a day, 365 days a year, the goal should be to achieve this condition as frequently and closely as reasonably possible. When grounds maintenance receives a complaint, it should be investigated and efforts made to improve the situation in a reasonable time frame. Primary objectives of snow and ice removal are:

- Safety for motorists and pedestrians
- Safety for the grounds maintenance personnel performing the work
- Protection of landscaping and property

Management Controls

If parking areas and walkways cannot be cleared properly prior to the start of the work day, the opening of the facility should be delayed to allow the removal to be completed.

2.8.2 Personnel and Responsibilities

Contingency plans and inclement weather policies must be updated and communicated to all personnel prior to the start of the season. The plan should include procedures and assigned responsibilities for each contingency.

2.8.2.1 Facility Manager

The facility manager (or designee) should make decisions about snow removal efforts after careful consideration of all variables. The facility manager may declare an ice/snow removal emergency. This may require the closure of parking facilities and the removal of parked vehicles that interfere with ice and snow removal operations.

2.8.2.2 Custodians

Custodians in each building should be responsible for removing snow at building entrances for a distance of about 6 feet from the buildings. The custodial staff should also be responsible for removal of snow from steps at main entrances to buildings. Steps should be kept as ice-free as possible.

2.8.2.3 Grounds Maintenance Staff

Grounds maintenance staff should be responsible for deicing and snow removal of primary and secondary sidewalks, Americans with Disabilities Act (ADA) ramps and curb cuts, weather-exposed stairwells, and parking lots. The grounds maintenance staff should assure that, once the melt has started, curb inlets, catch basins, and trench drains are unobstructed to speed the melt runoff and reduce the potential for refreezing.

2.8.2.4 Contracted Snow Removal

Caution should be exercised when contracting snow removal. Inquiries should be made regarding the workload of the contractor. Obtain reasonable assurance that the contractor has not secured so many jobs that it is unable to fulfill its responsibilities in a timely manner, or to the specifications of the contract. Verify that the contractor maintains adequate limits of general liability insurance by obtaining a Certificate of Insurance. Optimally, contracts should use the best practices specified in Section 2.10, Contractual Risk Transfer.

If engaged, snow removal contractors should be responsible by contract to remove snow from parking lots, parking lot entrances, and loading docks. The loading and hauling of snow may be required to clear adequate parking spaces.

2.8.3 Priorities for Removal

1. Fire lanes must be open for emergency equipment. Fire hydrants must be free of snow and accessible at all times.

2. Main entrances, ADA ramps and curb cuts, weather-exposed stairs, and primary sidewalks and parking lots should be cleaned before the building opens.
3. Parking lots, secondary entrances, and other low-usage areas should be cleaned by noon.

A hot list of problem wet/icy areas should be kept, and added attention and priority should be given to these sections. For problem areas that cannot be sufficiently cleared to be considered safe, close the area or provide warning of the problem. For adverse situations, reasonable care and a special effort are expected.

When snow is removed during any time other than the normal working hours (e.g., weekend snow removal), the grounds maintenance supervisor should call all personnel to inform them of the best time to report to work. Contractors should remove snow from sidewalks on weekends on an as-needed basis. The grounds crew should be responsible for weekend snow removal if contractors are not engaged or not available.

Where practical (especially at facilities in which there are employee parking lots, and shiftwork is being performed), temporarily close off portions of the parking lot so they can be cleared thoroughly and efficiently.

2.8.4 Guidelines for Removal

The facilities manager should call for removal operations when there is more than one inch of snow accumulation, or sleet and iced-over conditions.

- Use brooms to clean sidewalks and walkways for snowfalls of less than 4 in. (10 cm).
- Use mechanical means (such as a skid steer loader) on sidewalks and walkways if snowfall is 4 in. (10 cm) or greater.
- Sanding roadways should be performed at the discretion of the grounds maintenance supervisor in consultation with the facility manager and the police department. Note that sanding should be used when ice cannot be promptly removed. Sand can impair drainage, so these systems need to be properly maintained and kept free flowing.
- Consider the use of "self-service" treatment by strategically placing a dispenser for salt, sand, or ice-melt chemical in the parking lot and at key walkway locations.
- Snowfalls greater than 18 in. (46 cm) may require the assistance of a contractor to remove snow from parking lots (Figure 2.9). The facility manager and grounds maintenance supervisor should make decisions about contracted snow removal.

After the snow stops falling, and after the major walks and roadways are cleared, crews should concentrate on clearing snow from remaining areas and sanding as needed. Additional precautions include:

- Monitor walkways and parking lots for refreezing of melted snow and ice.

Management Controls

FIGURE 2.9 Heavy equipment is required for efficient removal of large amounts of snow.

- Control the buildup of icicles and frozen snow accumulations on roof overhangs, rain gutters, and overhead fixtures.
- Ensure that concrete floor surfaces of parking garages are sealed to barrier road salt and other materials that can spall (e.g., chip or flake) or otherwise degrade concrete.
- Maintain walkway and parking lot surfaces in good condition to prevent ice and snow-filled holes from creating concealed tripping hazards.
- Keep outside recreation areas closed until free of ice and snow.
- Close long, steep-slope walking surfaces from use during inclement weather.

2.8.5 Snow Storage

Generally, snow can be piled in a section of the parking lots that have been cleared.

- Snow should be piled at locations and in heights where it does not impair visibility.
- Avoid blocking storm drainage with snow piles.
- Keep ADA curb cuts and parking spots clear.
- Do not trap water around buildings with snow piles.
- Do not pile snow at a higher elevation or adjacent to pedestrian walkways, such that the melting snow would create runoff and present additional hazards upon refreezing.

2.8.6 Application of Anti-Icing, Deicing, and Sand

The use of ice-melt chemicals has, in some cases, replaced the traditional salt or sand. Although some of these chemicals are effective and appropriate, others can create an unexpected slipping hazard. Anhydrous chemicals are hydrophilic and

TABLE 2.1
Melt Times for Salt (NaCl) at Different Pavement Temperatures

Pavement Temp. °F	One Pound of Salt (NaCl) Melts	Melt Times
30° (272 K)	46.3 lbs (21 kg) of ice	5 min
25° (269 K)	14.4 lbs (6.5 kg) of ice	10 min
20° (266 K)	8.6 lbs (3.9 kg) of ice	20 min
15° (264 K)	6.3 lbs (2.9 kg) of ice	1 hour
10° (261 K)	4.9 lbs (2.2 kg) of ice	Dry salt will blow away

TABLE 2.2
Melting Characteristics

Chemical	Lowest Practical Melting Temp.
$CaCl_2$ (calcium chloride)	−20°F (−266 K)
KAc (potassium acetate)	−15°F (−264 K)
$MgCl_2$ (magnesium chloride)	−10°F (−261 K)
NaCl (sodium chloride)	15°F (264 K)
Blends	Check product specifications
Winter sand/abrasives	Only provides traction

Note: The deicing material should be picked based on lowest practical melting temperature, not eutectic temperature which is often listed on the bag.

should be avoided. Under certain conditions, as the ice is melted, these chemicals can combine with the water to become a slippery surface.

An ice control product will work until product dilution causes the freeze point of the brine to equal the pavement temperature. At this point, the material will stop melting and may refreeze if pavement temperatures are dropping. How long an application will last depends on five factors: pavement temperature, application rate, precipitation, beginning concentration, and chemical type. These factors explain why one application rate will not fit all storm events (see Table 2.1 and Table 2.2).

Apply environmental and concrete-friendly ice-melting materials near trees and shrubs and on steps, stairs, and landings.

2.8.6.1 Temperature and Ice Melting

Most weather stations measure temperature 30 feet (9 m) above ground. Because air temperature can differ substantially from pavement temperatures, be sure to use pavement temperature to determine the correct application rate. Changes in

FIGURE 2.10 Anti-icing is a liquid applied to roadways and walkway surfaces before snow and ice accumulations develop.

pavement temperature first affect bridge decks; pavement temperatures will also be lower in shady areas. Note that NaCl (road salt) does not work on cold days, when the temperature is less than 15 °F (264 K).

2.8.6.2 Before the Storm

Anti-icing should be considered as part of a prevention strategy for dealing with expected snow/ice accumulations. Applying chemical freezing point–depressant materials before a storm can prevent snow and ice from bonding to the pavement (Figure 2.10). Anti-icing requires about 25% of the material of deicing at 10% of the overall cost, making it a cost-effective option.

The anti-icing application rate must be correct. Too low a rate will not be effective, and too high a rate can result in a slippery roadway. Anti-icing application rate guidelines can be obtained from Minnesota Snow and Ice Control—Field Handbook for Snowplow Operators (see URL at end of section). Note that guidelines are a starting point; reduce or increase rates incrementally based on experience (see Table 2.3).

TABLE 2.3
Variables Affecting Application Rate

Increase Rates If:	Decrease Rates If:
Compaction occurs and cannot be removed mechanically	Light snow or light freezing rain
There is a lot of snow left behind	Pavement temperature is rising

2.8.6.2.1 Anti-Icing Guidelines
- Early application is particularly important for frost or light freezing drizzle.
- Liquids are most efficient and may be applied days in advance.
- Pretreated salts work at lower applications closer to the expected event.
- Schedule applications on bridge decks and critical areas if conditions may produce frost or black ice.
- Consider spot applications on inclines, curves, and intersections if predicted conditions warrant.
- Do not anti-ice under blowing conditions or in areas prone to drifting.
- Reapplication may not be necessary if there are still residual materials/chemicals. The residual effect can remain for up to five days if precipitation or traffic wear does not dilute initial application.
- The surface can refreeze when precipitation or moisture in the air dilutes the chemical.
- Do not apply $MgCl_2$ or $CaCl_2$ to a warm road (above 28°F [271 K]) pavement temperature). It can become slippery and cause accidents.
- Do not apply before expected rain.
- For the first application or after a prolonged dry spell, apply liquids at half the rate (not half the concentration). On dry roads, liquids tend to mix with oil from vehicles and result in slippery conditions.
- Do not exceed application recommendations. Excess amounts may result in a slippery roadway.

2.8.6.2.2 Pretreating and Prewetting Salt and Sand
Dry material blows off the road, so pretreat or prewet dry material. Liquids also increase the effectiveness of salt by jump-starting the melting process. Depending on the liquid, it can also lower salt's working temperature. Because pretreating and prewetting cause material to stick to the road, 20% to 30% less material is used—saving money and reducing environmental impacts.

2.8.6.3 During the Storm

2.8.6.3.1 Deicing
Deicing is a reactive operation in which a deicer is applied to the top of an accumulation of snow, ice, or frost that is already bonded to the pavement surface. Deicing generally costs more than anti-icing in materials, time, equipment, and environmental damage.

Removing ice that has already bonded to the pavement can be difficult, and removing it mechanically can damage equipment and roads. Generally, enough ice must be melted chemically to break the bond between the ice and the pavement, which requires larger quantities of chemical than anti-icing.

- Use an appropriate amount of salt. Most oversalting can be prevented by using calibrated, speed-synchronized spreaders and good judgment in selecting application rates and truck speed.

Management Controls

- It is necessary to apply just enough salt to loosen the bond between the road and the ice so it can be plowed off.
- Apply deicers in the center of the road or high side of the curve.

2.8.6.3.2 50/50 Salt/Sand Mix

Using a 50/50 salt/sand mix makes you either half right or half wrong. Using a salt/sand mix leads to over-application of both materials. Salt reduces the effectiveness of sand, and sand reduces the effectiveness of salt. However, a salt/sand mix may be helpful under narrow circumstances such as a long freezing rain event where the salt is washed away quickly. A 25% to 50% sand/salt mix has been documented as effective in increasing friction by sticking the sand to the surface.

2.8.6.3.3 Abrasives

Use winter sand and other abrasives when temperatures are too cold for deicing chemicals to be effective. But be aware that sand only provides temporary traction, and only when it is on top. Sand also clogs sewers, ditches, and streams.

- Use abrasives in slow-moving traffic areas such as intersections and curves.
- If your intention is to melt, use salt only.
- Salt is ineffective in cold weather, so use sand or an alternative chemical.
- Sand is not cheap when you consider handling, cleanup, and disposal costs.
- Sweep up sand frequently; after each event if practical.

2.8.6.3.3.1 Efficacy of Sand on Icy Surfaces   A study (Johnson, 2005) was performed to determine the concentration of sand needed to increase the slip resistance of a flat icy surface to 0.5. The surfaces studied simulate smooth icy surfaces pedestrians encounter in cold weather. The results were:

- The two brands of sand used in the study achieved the same level of slip resistance.
- The slip resistance of ice with no sand was measured in warm air (70°F/294 K) and cool air (36°F/275 K) over 28 trials. An average slip resistance of 0.19 was recorded in each temperature condition; the results were not statistically different.
- Sand increased the slip resistance of ice (see Table 2.4). Air temperatures during these tests ranged from warm (70°F/294 K) to cold (25°F/269 K). Analyses did not find significant differences in slip resistance for a given concentration of sand within this range of air temperatures. Sand applied at a concentration of 2.2 lb/100 ft^2 (10.8 kg/100 m^2) was enough to produce a slip resistance of 0.50.

Cold sand is less likely to increase slip resistance than sand applied at room temperature or warmer. Slip resistance was measured using granules cooled to 45°F (280 K) and compared to granules that were 75°F (297 K) when applied. When put

TABLE 2.4
Amount of Sand and Slip Resistance

Sand (lb /100 ft²)	Sand (kg /100 m²)	Mean Slip Resistance (Range)
0.3–1.0	2.0–4.9	0.3 (0.27–0.37)
1.0–2.0	5.0–9.9	0.4 (0.34–0.48)
2.2	10.8	0.5 (0.48–0.55)

onto ice in a concentration of 0.75 lb/100 ft² (3.7 kg/100 m²), and keeping all other conditions identical, slip resistance for the cold sand was 0.25, significantly less than the slip resistance of 0.40 for the warm sand. Apparently, relatively warm sand melts into the ice and, when the ice refreezes, sticks more firmly than cold sand, thereby producing a greater slip resistance.

Extensive pretesting found that slip resistance did not vary with the surface temperature of ice; cold ice (e.g., 0°F/255 K) did not exhibit a different slip resistance than when the surface temperature stabilized during phase transition (32°F /273 K).

2.8.6.4 After the Storm

After the storm, when snow and ice control operations have ended, evaluate what was done, how well it worked, and what could be changed to improve operations.

- Accurately record material use at the end of each shift.
- Conduct a post-storm meeting to evaluate your operations.
- Look for opportunities to try new and improved practices.
- Clean and check all equipment.
- Report hazards such as low-hanging branches, raised utilities, snow accumulation on bridges, or other potential problems.
- At the end of the season, clean and maintain the truck, tanks, brine-making systems, and pumps according to manufacturer specifications.

Online Resources

For more extensive discussion on this topic, see:
- "Using Salt and Sand for Winter Road Maintenance," Wisconsin Transportation Bulletin No. 6, Revised August 2005, http://epdfiles.engr.wisc.edu/pdf_web_files/tic/bulletins/Bltn_006_SaltNSand.pdf
- "Minnesota Snow and Ice Control—Field Handbook for Snowplow Operators," Manual Number 2005-1, Minnesota Local Road Research Board, August 2005, http://www.ttap.mtu.edu/publications/2005/snowicecontrolhandbook.pdf

2.9 LIGHTING

A study (Davies et al., 2001) indicated that poor lighting significantly increases the risk of an underfoot accident. Gender was found to be an indicator, with women experiencing a substantially higher risk of these accidents than men. One of the conclusions was that wearing bifocal/variable-focal eyeglasses increased the risk of "missed edge of" (step) accidents. Researchers suggested that two sets of eyewear (one for walking, another for reading and close work) might be effective in reducing this risk.

Most people lose visual acuity with age. By age 40, visual acuity is reduced by 10% and is down 26% by age 60. Between ages 70 and 79, only 25% of people demonstrate normal 20/20 vision, and only 14% of those over age 80 have normal vision. This loss is more pronounced under low-light conditions. Color and depth perception also decrease with age. These conditions lessen the ability to recognize and adapt to nonstandard stairs and ramps. Avoiding glare is particularly important for older individuals, because the aging eye is less able to detect contrasts and does not adjust as rapidly to changes from light to dark. The ability to focus on things that are close as well as perceiving color contrasts is more difficult for older individuals, so more light is needed.

Poor lighting plays an increasing role in litigation for slip, trip, and fall accidents. Experts refer to the standards published by the Illuminating Engineering Society of North America (IESNA, 2000), which provides guidelines in foot-candles for most occupancies and activities (http://www.iesna.org). The IESNA notes that the guidelines do not contemplate the age of occupants, acknowledging that requirements of older individuals are markedly different for two reasons:

(1) There is a thickening of the yellow crystalline lens, which decreases the amount of light reaching the retina, increases scatter within the eye, and reduces the range of distances that can be properly focused, and (2) there is a reduction of pupil size, decreasing the amount of light reaching the retina.

2.9.1 LIGHT SOURCES

Selecting the right type of light source is an important consideration. Studies show that the eye's response to color depends upon the amount of light available. The color sensitivity of the eye changes at different light levels. Under low light levels, the eye's sensitivity to yellow and red light is greatly reduced, while the response to blue light is greatly increased.

2.9.1.1 Mercury Vapor

Mercury vapor lamps are known for their long life (24,000+ h) and good efficiency (31 to 63 LPW) as compared with incandescent lamps. Because of their long life, mercury vapor lamps are widely used in street lighting; approximately 75% of all street lighting is mercury vapor.

2.9.1.2 Metal Halide

Metal halide lamps are similar in design and operation to mercury vapor lamps. The efficiency of metal halide lamps (80 to 115 LPW) is approximately 50% higher than mercury vapor lamps, but metal halide has a much shorter lamp life (6000 h).

Metal halide lamps have strong light output in the blue, green, and yellow regions. The result is a high lumen output at all light levels. The blue light output of metal halide is in the high sensitivity region of the eye for low light levels. This means that the effective lumens actually increase for a metal halide lamp as the amount of light reduces and the eye shifts to a high blue or green sensitivity. Metal halide lamps are also less sensitive to temperature than, for example, fluorescents. Fluorescents are designed to perform optimally at around 70°F (294 K), and will show measurable decline in lumens in higher or lower temperatures. Metal halide lamps, on the other hand, operate efficiently down to −40°F (233 K) and are relatively unaffected by wide variations in ambient temperature.

2.9.1.3 High-Pressure Sodium (HPS)

HPS lights have rapidly gained acceptance in the exterior lighting of parking areas, roadways, and building exteriors because of their high efficiency. They operate on the same principles as mercury vapor and metal halide lamps. This type of lamp has a high efficiency (80 to 140 LPW), relatively good color rendition, long lamp life (24,000 h), and an excellent lumen depreciation factor that averages about 90% throughout its rated life.

Most of the output from HPS light sources is in the yellow range, giving very high lumen output under higher light levels but poor output under lower light levels. Consequently, HPS lamps have high lumen ratings as perceived by the eye. On the other hand, under low light conditions, the effective lumens are greatly reduced because sodium produces little blue and green light.

2.9.2 Lighting Levels—Safety Only

Following are general guidelines on lighting levels, considered by the IESNA as the "absolute minimum for safety alone." Substantially higher levels of lighting are appropriate for many reasons, such as that needed for work activities and security. These bare minimum requirements will vary based upon the occupancy/activities and site conditions.

Lighting is normally measured in terms of "foot-candles" (fc), which is the amount of light given off by a candle at a distance of one foot. It can also be measured in "lux" (lx), in which one fc is equivalent to 10.76 lx (see Figure 2.11).

Interior lighting standards are relative to the hazard. The IESNA specifies the following "absolute minimums for safety alone":

- Two fc (22 lx) for areas where there is a low level of activity;
- Five fc (54 lx) for areas where there is a high level of activity.

For exterior lighting standards, light meter readings must be done at night. In general, exterior areas of high pedestrian usage should be at least 0.9 fc (10 lx), and 0.2

Management Controls

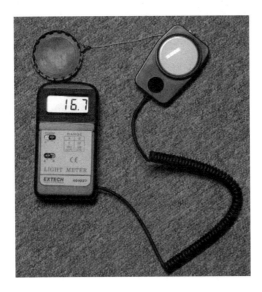

FIGURE 2.11 Digital light meters can provide light level readings in foot-candle (fc) or lux (lx) scales.

fc (2 lx) should be provided in areas of less use. Enclosed garages should be provided with at least 5 fc (54 lx).

High-intensity discharge (HID) lamps are recommended, which include mercury vapor, metal halide, and HPS.

2.9.3 Lighting Levels—Categories

In 1979, the IESNA developed recommended lighting levels based on a judgment of best practice for typical applications. Originally comprising nine categories, this tiered classification system was later revised to seven categories.

- Categories A, B, and C are for the purposes of "orientation and simple visual tasks," and specify the lowest levels of lighting.
- Categories D, E, and F are intended for "common visual tasks."
- Category G was created to address "visually demanding" tasks and specify the highest levels of illumination.

Categories A through C are of most interest with respect to fall prevention, as presented in Table 2.5. Examples of activities/areas within these categories are listed in Table 2.6.

2.9.4 Lighting Transitions

Transitioning between well-lit and low-light areas is another important consideration. Time is required for the human eye to adjust to different lighting levels, even

TABLE 2.5
IESNA Lighting Categories A, B, and C

Category	Activity	Foot-Candles
For purposes of orientation and simple visual tasks		
A	Public spaces	3 (32 lx)
B	Basic orientation for short visits	5 (54 lx)
C	Working spaces for visual tasks	10 (108 lx)

TABLE 2.6
IESNA Examples of Lighting Categories

Category	Activity and Occupancy
A	Dance halls
	Hospital corridors
	Houses of worship Congregational areas
	Museum lobbies
	Toilets and washrooms
B	Hospital patient rooms
	Library bookstacks (inactive)
	Merchandising dressing areas
	Service space stairways and corridors
	Shopping mall main concourse areas
C	Educational corridors
	Hotel lobbies
	Merchandising general display areas
	Office filing areas
	Supermarkets

for individuals with excellent vision. Transitions between bright and dark areas should be as gradual as possible. Visual cues and barriers designed to slow pedestrian traffic flow can also be useful by giving pedestrians more time to adjust to new light levels.

Another aspect of lighting transition is the effective control of shadows and glare. Shadows can mask hazardous conditions, making them unexpected to the pedestrian. Glare, similar to transition extremes, can temporarily impair sight, increasing the potential for falling.

2.9.5 Maintenance

An effective lighting maintenance program is an important part of providing good visibility. Gradual accumulation of dust and dirt on fixture lenses and lamps reduces light output. Regularly scheduled cleaning and re-lamping, as well as prompt

Management Controls

FIGURE 2.12 Of the 16 light sockets, only one has a bulb, significantly reducing lighting levels at night. Also, protective covers are not used.

replacement of defective lenses and ballasts, can increase efficiency. Lighting maintenance should include the following elements:

- The extent and frequency of maintenance should be consistent with the manufacturer's specifications. Cleaning luminaires (casing and reflector) and replacing bulbs is essential to achieving the maximum efficiency and brightness (Figure 2.12).
- Where possible, weekly inspections of lighting should be performed, especially where maintenance logs reveal a trend of frequent outages. An alternative is to provide monitoring by a computerized environmental control system that indicates when maintenance is required.
- It is also important that lighting be overdesigned for the area. The provision of supplemental or additional lights or bulbs will help ensure that sufficient lighting is available even when one or two bulbs are out (Figure 2.13).
- Use of timers rather than photoelectric sensors to control lighting requires additional procedures, i.e., adjustments for seasonal time and light changes.

2.10 CONTRACTUAL RISK TRANSFER

Risk transfer or risk sharing is an effective tool. Some facilities subcontract all or part of their floor housekeeping/care or snow removal activities. Key elements of these contracts include a "hold harmless" clause with favorable language and specified limits of liability. Claim activity should be monitored to ensure claims for which the contractor is responsible are not being erroneously posted to your policy.

A limited risk transfer can be achieved by subcontracting normal maintenance and verifying liability insurance of adequate limits. When renovating properties or expanding operations, adherence to specifications for slip resistance in purchasing contracts and certification of materials used should be required.

Other risk transfer issues may also be related to slips, trips, and falls, for example, when the tenant, rather than the property owner, is held liable for hazardous conditions

FIGURE 2.13 Lighting in parking lots should be overdesigned to allow for time to change burned out bulbs and provide redundant light sources to avoid dependence on a single lighting unit.

in parking lots or other walkway surfaces. Responsibilities regarding maintenance, inspection, housekeeping, insurance, and related responsibilities should be clearly assigned in contractual agreements.

2.10.1 GENERAL ADMINISTRATIVE MEASURES

Regardless of the size of the organization, a structured approach to controlling contractual risk transfers is essential. Risk transfer agreements are not always easy to identify or evaluate. It is recommended that all contracts be reviewed by legal counsel. In addition, when changes are made to standard contracts, an initial in-depth review and follow-up audits and monitoring activities may be required.

2.10.2 SPECIFIC CONTROL MEASURES

Another benefit of having all contracts reviewed is the opportunity to submit recommendations to management on how to best handle the exposure posed by provisions contained in the contract.

Reduce exposures assumed—Whenever possible, narrow the scope of the transfer provision to what you know can be accommodated by insurance coverage. Be cautious about the number of contractual transfers in effect simultaneously, because an excessive number may complicate business relationships and increase costs.

Judicious retention of exposures—Agreements may present exposures that the organization is willing to assume without calling upon another party's insurance. If retained, the exposures should be limited to those known to be low in expected frequency and severity, and with a low catastrophe potential.

Insurance transfer—To evaluate whether exposures will be covered by insurance, first be certain coverage is available. The other party's applicable insurance coverage should be examined and verified.

Dealing with "unmanageable" provisions—Loss reduction, risk retention, and insurance may not be adequate to control all transferred exposures. Instead of rejecting the contract, consider requesting change or removal of objectionable provisions. Success largely depends upon the relative bargaining strength of the parties and the importance of the contractual relationship to each of them.

Clarity of contract language—Contract language should always be written as clearly as possible; if the language is unclear, it must be revised.

2.10.3 FUNDAMENTAL GUIDELINES

Make sure the indemnitor is financially able to stand behind its commitment. In considering court decisions, it is almost essential that the commitment be backed by adequate limits of insurance. Consult a licensed insurance professional for advice as to how much is adequate in a given situation.

Require a certificate of insurance for contractual liability coverage before contract operations begin. The certificate should clearly state that the issuer (or insurer) provide at least 30 days automatic notice of cancellation, nonrenewal, or material change in coverage directly to you, the certificate holder. Merely obtaining a copy of the insurance policy will not provide notice of cancellation or coverage change. It is advisable to establish a system for updating certificates prior to expiration, when a contract is renewed or open-ended, or when the contract exceeds the period covered by the certificate.

Where possible, be named as an additional insured and obtain a waiver of subrogation from insurers. Becoming an additional insured may obligate the organization to pay premiums or require compliance with certain conditions following a loss, but the advantages can far outweigh the disadvantages.

2.11 CONSTRUCTION, RENOVATION, AND SPECIAL EVENT PLANNING

It is important to adequately plan for construction and special event activities in order to help reduce the potential for slips and falls. This includes employing adequate staff to promptly identify and correct hazardous conditions and to control access to nonpublic and hazardous areas. In addition, adequate physical barriers should be in place to secure hazardous areas from public access.

Guidance on appropriate controls for this type of exposure can be found in Occupational Safety and Health Administration (OSHA) 1926, Subpart M on Fall Protection. Although much of this section has to do with fall arrest systems, there is also information on perimeter protection for vertical openings and other prevention measures that should be used in construction environments. Another resource is OSHA 1926 Subpart X on Stairways and Ladders.

2.12 LOSS ANALYSIS

With a sufficiently large sample of loss experience, accident potential can be readily identified without the need for floor testing. For example, consider a large number of similar facilities such as a chain of restaurants, which have any of several types of floors: terrazzo, smooth quarry tile, and abrasive quarry tile. To determine the degree of hazard by floor type, first identify the type of floor in each facility and track accidents by floor type and exposure.

2.12.1 TRACKING EXPOSURE

Determine the exposure basis for each floor type. In most facilities, business activity indicators can be used to quantify the exposure. Restaurant operations may use transaction counts or man-hours to determine the number of people potentially exposed to fall hazards. Exposure is determined by multiplying the unit exposure times the number of units. Thus, if the exposure unit is store receipts, it is fairly easy to track the number of falls per million dollars of sales or per million man-hours.

The accident rate for each type of floor can readily reveal the extent of the fall hazard for each operation. If 9.6 falls occur per million dollars of sales on terrazzo, 7.3 falls per million dollars on smooth quarry tile, and 4.7 per million dollars on abrasive quarry tile, action plans for accident reduction can be developed accordingly.

2.12.2 GATHERING THE DATA

Unless the right information concerning the circumstances of each fall has been collected, it is unlikely that an analysis can be performed to support a rationale for meaningful corrective action. To capture relevant data, accidents must first be well investigated.

Fall patterns normally arise from the type of hazards, the facility layout and design, and the activities of personnel. A fall-specific investigation report format may be needed to gather essential information for each event (see the Chapter 12, Accident Investigation and Mitigation).

To determine what information is important, first select the most severe fall cases. Using the loss run from an insurance carrier, the amount paid plus reserve (which is the additional amount expected to be paid) is a good indicator of severity, so look at the highest-cost cases. It may be necessary to reinvestigate cases to develop an accurate profile. After reviewing enough losses to get a good picture of the patterns, a focused report format can be developed that prompts staff to collect pertinent information.

Management Controls

Establish a coding system to gather relevant information in a database for meaningful analysis. Information may include time of day, activity being performed by the victim, area of the facility, and condition of the floor. Be aware of multiple causation and the difference between a cause and an injury when selecting codes.

Once the proper data is gathered and entered into the database, fall trends can be readily identified. The analysis will suggest obvious solutions and help prioritize issues so that management can achieve significant improvement quickly.

2.12.3 Developing Solutions

After identifying the critical few problem areas, the next step is to develop solutions. For example, a hotel chain realized that most falls occurred in kitchen and dining room areas. Falls were reduced 50% by using epoxy concrete coatings on floors in the kitchen areas, and carpeting in dining room areas.

Through analysis and testing, a large museum determined that pedestrian traffic on soft stone flooring was creating dust. Falls were occurring as pedestrians moved from these areas into other galleries that had waxed flooring. Sealing the stone flooring prevented the accumulation of dust on shoe soles and significantly reduced the number of falls.

EXHIBIT 2.1

Facility and Parking Lots: Hourly Inspection Log

Facility Name:	Area Surveyed:

Instructions:

This log is a tool to help control slip and fall hazards.

The person conducting the inspection should initial the box after all the hazards identified during the inspection are eliminated and/or corrected.

Parking lots and sidewalks should be surveyed at least daily to identify unsafe conditions that can lead to falls.

Special emphasis should be placed on curbs, potholes, changes in walking levels, lighting, and traffic control.

The time completed and the employees' initials should be in the boxes marked "wash" and "wax."

The department or facility manager should review the log daily and initial the box provided.

Reminder:

"Wet Floor" signs should be used when conditions warrant. Signs should be at least 36" high and positioned to visibly indicate the location of the spill.

Time	Dates (show month/day below)									
6 am										
7 am										
8 am										
9 am										
10 am										
11 am										
noon										
1 pm										
2 pm										
3 pm										
4 pm										
5 pm										
6 pm										
7 pm										
8 pm										
9 pm										
10 pm										
11 pm										
12 pm										
1 am										
2 am										
3 am										
4 am										
5 am										
Wash										
Wax										
Park										
Mgr										

Management Controls

EXHIBIT 2.2
Sample Snow Removal Report

Facility: _____ Location: _____

Date: _____ Day of Week: _____ Temperature: _____ F–C _____

Weather: ☐ Snow ☐ Sleet ☐ Hail ☐ Rain ☐ Cloudy ☐ Clear
Amount: _____ *ft inches* _____ Employee: _____

Area	Direction				Treatment				Time		
	N	S	E	W	Plow	Blow	Salt	Sand	Time	am	pm
Steps, Front:	☐	☐	☐	☐	☐	☐	☐	☐	_____	☐	☐
Steps, Rear:	☐	☐	☐	☐	☐	☐	☐	☐	_____	☐	☐
Steps, L Side:	☐	☐	☐	☐	☐	☐	☐	☐	_____	☐	☐
Steps, R Side:	☐	☐	☐	☐	☐	☐	☐	☐	_____	☐	☐
Ramps:	☐	☐	☐	☐	☐	☐	☐	☐	_____	☐	☐
Access to Lots:	☐	☐	☐	☐	☐	☐	☐	☐	_____	☐	☐
Walkways, Front:	☐	☐	☐	☐	☐	☐	☐	☐	_____	☐	☐
Walkways, Rear:	☐	☐	☐	☐	☐	☐	☐	☐	_____	☐	☐
Walkways, L Side:	☐	☐	☐	☐	☐	☐	☐	☐	_____	☐	☐
Walkways, R Side:	☐	☐	☐	☐	☐	☐	☐	☐	_____	☐	☐

Employee: _____ Title: _____

Comments

_____
_____
_____
_____
_____
_____
_____

© ESIS, Inc. All rights reserved.

EXHIBIT 2.3

Sample Walkway Surface Design Audit

Building Section:		Completed By:	Date/Time:

INTERIOR AREAS

Level Walkway Surfaces	Good	Fair	Poor	N/A
Access covers in hallways are flush with walkway surface.	☐	☐	☐	☐
The height of doorsills does not exceed 3/4, with sloping saddles.	☐	☐	☐	☐
Transitions between different floor surfaces are readily observed.	☐	☐	☐	☐
There are no blind turns.	☐	☐	☐	☐

Stairs	Good	Fair	Poor	N/A
Stairs have 10" or 11" treads and 9" or 10" risers.	☐	☐	☐	☐
Stairs do not deviate between risers more than 3/16."	☐	☐	☐	☐
Stair treads are slip resistant.	☐	☐	☐	☐
Nosings are clearly visible to persons descending the stairway.	☐	☐	☐	☐
Stairways with two or more 2 steps are provided with handrail.	☐	☐	☐	☐
There are no single step changes in level.	☐	☐	☐	☐
Lighting is adequate.	☐	☐	☐	☐

Handrails	Good	Fair	Poor	N/A
Handrails are provided for ramps and where there are two or more stairs.	☐	☐	☐	☐
Handrails are circular and graspable.	☐	☐	☐	☐
Handrails are accessible within 30" of all portions of stair width.	☐	☐	☐	☐
Handrail height is between 34" and 42."	☐	☐	☐	☐

Other Conditions	Good	Fair	Poor	N/A
Areas with shared pedestrian and vehicular traffic are properly marked.	☐	☐	☐	☐
There is sufficient room for storage of materials off the floor.	☐	☐	☐	☐
Sufficient drainage is provided where wet processes are used.	☐	☐	☐	☐
Stools are equipped with appropriate casters.	☐	☐	☐	☐
Design precludes computer and electrical cables from crossing pedestrian paths.	☐	☐	☐	☐

Management Controls 99

EXHIBIT 2.3 (*Continued*)
Sample Walkway Surface Design Audit

COMMENTS ON INTERIOR AREAS

Level Walkway Surfaces

Stairs

Handrails

Other Conditions

EXTERIOR AREAS

Stairs/Handrails	Good	Fair	Poor	N/A
Landings are provided with overhangs.	☐	☐	☐	☐
Stairs have 10" or 11" treads and 9" or 10" risers.	☐	☐	☐	☐
Stairs do not deviate between risers more than 3/16."	☐	☐	☐	☐
Stair treads are slip resistant.	☐	☐	☐	☐
Nosings are clearly visible to persons descending the stairway.	☐	☐	☐	☐
Stairways with two or more steps are provided with handrails.	☐	☐	☐	☐
There are no single step changes in level.	☐	☐	☐	☐
Handrails are provided for ramps and where there are two or more stairs.	☐	☐	☐	☐
Handrails are circular and graspable.	☐	☐	☐	☐
Handrails are accessible within 30" of all portions of stair width.	☐	☐	☐	☐
Handrail height is between 34" and 42."	☐	☐	☐	☐
Ramp slopes should be between 1:16 and 1:20.	☐	☐	☐	☐
Level Walkways/Curbing	**Good**	**Fair**	**Poor**	**N/A**
Curbing is no higher than 6."	☐	☐	☐	☐
Curbing height is consistent except for curb ramps.	☐	☐	☐	☐
Walkway surfaces are level with no protrusions above ¼."	☐	☐	☐	☐
Access covers located away from pedestrian paths.	☐	☐	☐	☐
Parking Lots	**Good**	**Fair**	**Poor**	**N/A**
Drainage is located at the low point of areas close to the building.	☐	☐	☐	☐

Continued

EXHIBIT 2.3 (*Continued*)

Sample Walkway Surface Design Audit

	Good	Fair	Poor	N/A
Drainage grates have openings no wider than ½."	☐	☐	☐	☐
Parking lot is graded down and away from the building.	☐	☐	☐	☐
Lighting is adequate.	☐	☐	☐	☐

COMMENTS ON EXTERIOR AREAS

Stairs/Handrails

_____

Level Walkways/Curbing

_____

Parking Lots

_____

_____

EXHIBIT 2.4

Sample Walkway Surfaces Inspection

Building Section:	Completed By:	Date/Time:

INTERIOR AREAS

Entrance Areas	Good	Fair	Poor	N/A
Outside mats are properly placed/oriented, coarse, and in good repair.	☐	☐	☐	☐
Outside mats are clean on top and on the underside.	☐	☐	☐	☐
Foyer mats are coarse, in good repair, and clean.	☐	☐	☐	☐
Foyer mats are clean on top and on the underside.	☐	☐	☐	☐
Walk off mats are properly placed/oriented, in good repair, and clean.	☐	☐	☐	☐
Walk off mats are the proper size (large and long enough) for the area served.	☐	☐	☐	☐
Walk off mats are clean on top and on the underside.	☐	☐	☐	☐
There are no signs of excess water/contaminants beyond walk off mats.	☐	☐	☐	☐
All mats lie flat.	☐	☐	☐	☐
Level Walkway Surfaces	**Good**	**Fair**	**Poor**	**N/A**
Floor levels/surface transitions are marked, unobstructed, and in good condition.	☐	☐	☐	☐
Walkways, including hallways, are free of storage.	☐	☐	☐	☐
There are interim measures prior to spill cleanup In effect.	☐	☐	☐	☐
Floor access/drain covers and screws are flush and secure.	☐	☐	☐	☐
Carpet is in good repair with no wear, looseness, ripping/tearing, or other damage.	☐	☐	☐	☐
Epoxy flooring is unwaxed.	☐	☐	☐	☐
Cafeteria/Kitchen	**Good**	**Fair**	**Poor**	**N/A**
Serving line mat is in good condition, properly placed.	☐	☐	☐	☐
Kitchen flooring has no significant grease accumulation.	☐	☐	☐	☐
Dining room floor free of contaminants.	☐	☐	☐	☐
Stairs/Handrails	**Good**	**Fair**	**Poor**	**N/A**
Treads and nosings sturdy, secure, and in good condition.	☐	☐	☐	☐
Handrails are sturdy.	☐	☐	☐	☐
Stairwell lighting is adequate.	☐	☐	☐	☐
Footwear/Other Conditions	**Good**	**Fair**	**Poor**	**N/A**
Appropriate footwear (per guidelines) is in use.	☐	☐	☐	☐

Continued

EXHIBIT 2.4 (*Continued*)

Sample Walkway Surfaces Inspection

	Good	Fair	Poor	N/A
Computer, electrical, and other cables do not cross pedestrian paths.	☐	☐	☐	☐
There are no sources of leaks or moisture reaching the floor.	☐	☐	☐	☐

COMMENTS ON INTERIOR AREAS

Entrance areas

___

Level walkway surfaces

___

Cafeteria/kitchen

___

Stairs/handrails

___

Footwear/other conditions

___

EXTERIOR AREAS

Stairs/Handrails	Good	Fair	Poor	N/A
Stairs are in good condition, with no unrepaired damage.	☐	☐	☐	☐
Handrails are sturdy and secure.	☐	☐	☐	☐
Lighting is adequate.	☐	☐	☐	☐
Level Walkways/Curbing	Good	Fair	Poor	N/A
Surface is level with no protrusions above ¼."	☐	☐	☐	☐
Surface is in good condition with no major cracks, breaks, or other damage.	☐	☐	☐	☐
Access/drain covers are flush and secure.	☐	☐	☐	☐
Curbs are level with the corresponding sidewalk.	☐	☐	☐	☐
Parking Lots	Good	Fair	Poor	N/A
Surface is level with no major cracks, potholes/depressions, or other damage.	☐	☐	☐	☐
There is no ice, snow, or water accumulation at access ramps or major walkways.	☐	☐	☐	☐
Lighting is adequate, all units are operational.	☐	☐	☐	☐

COMMENTS ON EXTERIOR AREAS

Stairs/handrails
_____

Level walkways/curbing
_____

Parking lots
_____
_____

EXHIBIT 2.5
Sample Slip Resistance Test Results Report

Company:		Tribometer Operator:	
Location/Address:		Address/Telephone Number:	
Date/Time of Tests:		Tribometer/Serial #:	
Test Method:		Test Foot, Preparation:	

Location/Orientation	Surface Slope	Floor Material & Texture	Floor Finish	Floor Condition	Wet/Dry	Reading 1	Reading 2	Reading 3	Reading 4	Average

Comments

© ESIS, Inc. All rights reserved.

3 Principles of Slip Resistance

You might as well fall flat on your face as lean over too far backward.

James Thurber

3.1 INTRODUCTION

It could be argued that falls due to slips are the most frequent type of fall accident. In order to effectively apply assessment and measurement techniques for slip resistance, it is essential to have a working knowledge of the underlying principles of friction, factors that affect traction on walkway surfaces, and the origins of measurement instruments and related thresholds of safety. Knowledge of these principles permits informed decision making regarding appropriate methods of measuring traction, critical factors that contribute to or hamper pedestrian traction, and meaningful interpretation of slip resistance test results.

3.2 PRINCIPLES OF FRICTION

Leonardo da Vinci stated two of the three laws of measuring forces contributing to friction. The first is that the frictional force is proportional to the load (i.e., the *static coefficient of friction* [SCOF] is equal to the horizontal force needed to just start the object in motion, divided by the vertical force [weight] of the object, or SCOF = H/V [horizontal over vertical]). The second principle is that the coefficient of friction is independent of the area of contact.

The third principle, first expressed by Coulomb, is that the friction is independent of the sliding velocity. Friction, then, depends only on the applied load. The coefficient of friction, which is the ratio of the load to the force required for movement, should be constant under all conditions. In practice, the first two laws are generally true with only about a 10% variation, but it has been recognized that friction is not independent of the sliding velocity (speed).

In 1835, A. Morin proposed that, because the force resisting the start of sliding was obviously not the same as the force required to maintain sliding, two different coefficients of friction were operating. The first is the SCOF, which is the relative force (H/V) required to start motion in a body at rest. The second coefficient is the *dynamic coefficient of friction* (DCOF), which is the relative force required to maintain motion in a sliding body. Generally, SCOF produces a higher number than

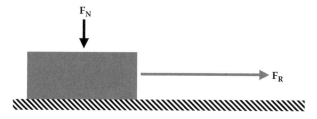

FIGURE 3.1 The classic physics definition of SCOF.

DCOF, because it takes more force to initiate a sliding motion than to keep a sliding object in motion.

Given the three laws of friction, measuring SCOF can be considered a simple procedure: place a weighed block on a surface, attach a strain gauge, and measure the "pull" (in pounds) required to move the block, divided by the weight of the block in pounds (see Figure 3.1).

In the 1940s, it became clear that the results of this simple method of measuring SCOF did not correlate to the perception of human ambulation. A rethinking of the interrelation of friction and velocity occurred in 1939 when F. P. Bowden and L. L. Leben, physical chemists at the University of Cambridge, identified the "stick-slip" phenomenon. It was established that friction first increases with velocity and then declines. Ernest Rabinowicz puts forth a clear explanation of this phenomenon in his 1956 article in *Scientific American*, titled "Stick and Slip":

> ... if the slide of one surface over another slows down, friction increases. However, at extremely slow speeds the situation is reversed: as friction increases the sliding velocity also increases. The most plausible explanation seems to lie in the phenomenon called creep. All materials slowly change shape ("creep") even under moderate forces. An increase in force will increase the rate of creep.
>
> Thus in the case of surfaces sliding very slowly over each other, an increase in frictional force may produce a perceptible acceleration of the slide in the form of creep of one surface past the other. The limit of speed attained by the creep mechanism varies with the material, because soft materials creep faster than hard ones.
>
> These considerations present us with the paradoxical conclusion that there is really no such thing as a static coefficient of friction for most materials. Any frictional force applied to them will produce some creep, i.e., motion. (Rabinowicz, 1956, p. 115)

3.3 SLIP RESISTANCE DEFINED

When discussing the *coefficient of friction* (COF) and/or slip resistance, it is essential to understand that a walkway surface does not have its own COF. Friction is a function of the interaction between two surfaces: the walkway surface (or surrogate) and the footwear bottom (or surrogate). When using these terms, the composition, type, and condition of both surfaces are relevant. Because slip resistance is an interaction between two variables (e.g., walkway and footwear bottom), it is essential that when evaluating one (e.g., the walkway), the other must be a constant or control material (e.g., footwear bottom surrogate material).

Principles of Slip Resistance

The term "slip resistance" is generally favored by organizations that write standards for human ambulation over the term "coefficient of friction." COF measurements can be made for many reasons (e.g., ball-bearing properties, manufacturing of sheet plastic), while slip resistance measurements are generally understood to be COF measurements used exclusively to assess pedestrian walkway and footwear safety.

So, whereas the classic physics type of equipment does measure SCOF as expressed in the three laws (SCOF = H/V), this system does not incorporate the host of other variables associated with human ambulation. People do not walk by sliding their feet. Additional biomechanical factors of walking have been incorporated into some slip meters that have been developed more recently (see Chapter 4, U.S. Tribometers). The more closely an instrument is able to emulate human ambulation, the more closely the instrument results are likely to correlate to the human perception of slipperiness.

3.4 SLIP RESISTANCE FACTORS

Four primary physical factors affect the amount of traction between the shoe and the floor surface:

1. Material and finish of the floor (controllable)
2. Footwear-bottom material and condition (generally uncontrollable, except in the workplace)
3. Environmental contaminants (controllable in some situations)
4. Gait dynamics, i.e., how an individual walks (uncontrollable and infinitely variable)

Whereas floor design and condition (see 1 and 3 above) are controllable, the types of footwear worn by individuals as well as how these individuals walk usually are not. The exception to this is employee safety, especially when the type of footwear is specified for a given occupation or task.

3.5 THE MECHANICS OF WALKING

In simple terms, walking is a method of locomotion in which the body weight (or center of gravity) is carried alternatively by the right and left foot. Contact with the walkway occurs in four distinct phases:

1. Heel strike—the initial touchdown of the footwear on the surface in which only the back edge of the heel is in contact with the walkway surface
2. Stance—the point at which the foot has become flat on the walkway surface
3. Take-off—when only the front portion (or ball) of the foot is in contact with the walkway surface and it pushes off to move forward
4. Swing—when the rear foot is moving forward off the ground

It is understandable and well known that the majority of slip and fall occurrences are the result of the heel-strike phase of walking. At that point, body weight is being transferred from the trailing leg to the leading leg, which is contacting the walkway

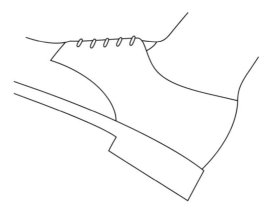

FIGURE 3.2 An illustration of the heel-strike phase of human ambulation.

surface only by the edge of the heel. The swing leg becomes impeded, resulting in the body's center of mass continuing forward beyond its base of support, leading to a loss of equilibrium.

Under low-traction conditions, the heel edge begins to slide before an opportunity arises for the walker to make full contact with the heel or to compensate for the slide (see Figure 3.2).

A slip leaves only about a quarter second to save oneself from falling. One-tenth to two-tenths of a second after the foot contacts a slippery spot, knee and hip joints react to try to bring the skidding foot back. An attempt is made to push the hip forward to regain balance. If unsuccessful, the other foot is pulled forward.

A study at the University of Illinois sent 52 adults of various ages for a walk over artificial ice to see what reactions make a difference between falling and just wobbling. Two key observations were made:

- Those able to slow down the slipping foot were more likely to recover.
- How far the nonslipping foot was to the left or right of center was a critical factor. Study analysis suggested that if that foot could be moved as little as 4 cm (about 1.5 in.) further out, the likelihood of recovery from a slip increases by half (Ritter, 2007).

Research suggests that people can be trained to react differently to events as sudden as a slip. Studies suggest people who practice recovery moves can improve their reactions to slipping.

3.6 CAUSATION BETWEEN INCIDENT AND INJURY

From a biomechanical perspective, causation between an incident and an injury is determined by addressing two questions:

- Did the event create a means by which injury is known to occur?
- If so, was it sufficient to cause injury?

Principles of Slip Resistance

Establishing a causal relationship between injuries and a fall requires an analysis of the incident, an understanding of the individual's unique tolerance level, and a biomechanical analysis of the injury mechanisms and forces. This job is often left to a physician who may not be knowledgeable in these areas. A biomechanical engineer trained in the application of mechanical engineering and physical sciences to medicine and the human body is best suited to determine a correlation of injuries to the kinematics of the fall (Gushue et al., 2007).

3.7 SLIP RESISTANCE SCALE

Under normal conditions, the slip resistance scale ranges from a minimum of zero to a maximum of one, measured in tenths of a point. The closer the rating is to zero, the less slip-resistant the surface. For example, a rating of 0.1 is low traction, while a rating of 0.9 is considered high traction. Levels above 1.0 can be measured when mechanical interlocking takes place.

Studies have indicated that ice measures at about 0.20, while the minimum amount of traction needed for walking is 0.25. The often-mentioned 0.50 guideline includes a factor for safety needed, because conditions other than the walkway surface (such as footwear) could increase the hazard. Whereas the United States generally favors 0.50 (based on a SCOF approach), a threshold of 0.40 (based on a DCOF approach) is most often cited overseas. These guidelines are discussed in Chapter 4, U.S. Tribometers, and Chapter 5, U.S. Standards and Guidelines.

It appears that a recommendation from Sidney James of Underwriters Laboratory to the Casualty Council of Underwriters Laboratories in 1945 was the first mention of the 0.5 threshold. This was probably based on a combination of laboratory test results using the James Machine and field experience over several years. This guideline made its way into the floor wax and floor polish industry, although no known accident statistics to support this threshold have ever been made public.

In 1953, the Federal Trade Commission (FTC) published a set of 20 Proposed Trade Rules for the Floor Wax and Floor Polish Industry, which included the following:

Rule 5–Improper Use of the Terms "Slip Resistant" (See Definitions), "Slip Retardant," Anti-slip," etc.

Note: Subject to the development and acceptance of improved testing methods, either or both of the following tests with resultant coefficients of friction may be employed for the purpose of compliance with this rule:

1. A [dynamic] coefficient of friction of not less than 0.40 ... Sigler test ...
2. A [static] coefficient of friction of not less than 0.5 ... by the test for slip resistance ... by Underwriters Laboratories Inc. [James Machine] ..."

This and other proposed rules were never issued as final rules.

ASTM International (ASTM) Committee D21 on Polishes (formerly Wax Polishes and Related Materials) was formed in 1950 and immediately began work on the slip resistance issue, considering the James and Sigler instruments. It was not until 1964, however, that tentative method D2047–64T was issued, calling for the use of the James Machine only. In 1969, ASTM D-2047 was adopted, which specified the use of the James Machine for polishes; however, it contained no threshold. In 1970, the Chemical Specialties Manufacturers Association (CSMA), now the Consumer Specialty Products Association (CSPA), adopted ASTM D-2047 with the provision that 0.5 be the threshold. In 1974, ASTM D-2047 incorporated the threshold of 0.5. For more details, see Chapter 5, U.S. Standards and Guidelines.

Currently, no standard or law requires that floors must have a certain level of slip resistance. Historical precedent, standards, guidelines, and case law all point to 0.5 as a reasonable threshold, however, and it is the most generally recognized and accepted value in the Unites States. Research into slip resistance has suggested that no single threshold can be reasonably applied to all activities and conditions. For example, the traction demand for someone running around a corner would be higher than someone walking at an average pace in a straight line. This concept has been used in some overseas standards, but research is ongoing in the Unites States.

3.8 SURFACE ROUGHNESS

The surface roughness of a floor is the property that makes it slip resistant in the presence of contaminants such as water or oil. Several measures are employed to determine surface roughness (Table 3.1). Two key factors for the purposes of slip resistance are the sharpness of the asperities (i.e., peaks and valleys) and the depth of the asperities.

Long before instruments for measuring roughness became available, contractors increased the roughness of sidewalks, highways, and exterior stairs by broom finishing, rock-salt finishing, machine grinding, and other means. The purpose was

TABLE 3.1
Some Surface Roughness Metrics

Parameter	Name	Description
R_a	Roughness average	Absolute value of the surface height averaged over the surface.
R_z	Determined roughness	Average of n individual roughness depths over specified length ($n = 5$ is a common value).
R_z (ISO), R_{tm}	Roughness height	
R_q	Root mean square (RMS) roughness	
R_v, R_m	Maximum profile valley depth	
R_p	Maximum profile peak height	
R_{pm}	Average maximum profile peak height	
R_t, R_y	Maximum height of the profile	
R_{max}	Maximum roughness depth	
R_c	Mean height of profile irregularities	
P_c	Peak count (peak density)	

Principles of Slip Resistance

to provide increased traction when these surfaces were wet without changing the composition of the base material.

Surface metrology is the science of measuring small-scale features on surfaces. Surface primary form, waviness, and roughness are the parameters most commonly measured.

3.8.1 Height or Sharpness

Some researchers believe that calculating only surface heights using roughness meters directly correlates with measured slip resistance.

Others believe that measuring with roughness meters as a way of assessing slip resistance is limited because of their inability to take into account anything but the distance between high to low points on the surface. Slip resistance is as much a function of the sharpness of asperities as it is of the mean peak-to-valley distance. By experimentation, it can be demonstrated that a flat surface with pits in it (which would register a certain "roughness" with a surface comparator) performs similarly to the same surface without the pits.

Likewise, surface peaks that are rounded off from surface wear may still show a high peak-to-valley distance, but traction performance is reduced. It is common to measure a difference in traction on ramps based on the orientation (direction) of the testing. Normally, floor traffic will wear the peaks off on the high side (as would be expected in downhill traffic) so that higher traction numbers are calculated when measuring uphill than downhill.

Sharpness is the major contributor to slip resistance. The sharper the peaks, the more slip-resistant the surface. On a microscopic scale, all surfaces have asperities. However, they must be tall and sharp enough to extend up through a contaminant to engage the shoe bottom. Therefore, it is important to select flooring materials that are sufficiently rough with durable micropeaks that will protrude through a contaminant and engage or "dig into" the shoe bottom over the expected life of the material.

3.8.2 Liquid Dispersion

The ability of a floor to disperse liquid quickly is greatly affected by the surface texture of the floor. Many slips occur because of the build-up of what is known as a hydrodynamic squeeze film. This is similar to what happens to a car tire when it hydroplanes on a wet road. The peak-to-valley depth also has some impact. The higher the asperities are, the greater the ability of the surface to channel water away from the footwear–walkway interface, which reduces the potential for hydroplaning.

3.8.3 Measuring Roughness

A profilometer is often used to measure a small-scale profile of the surface. These traditionally used a stylus and worked much like a phonograph. Newer, noncontact versions often employ optical interferometry, confocal microscopy, optical triangulation (triangulation sensor), and digital holography.

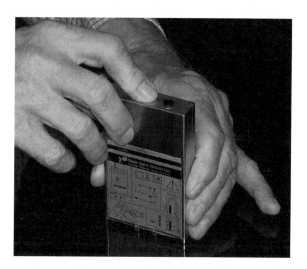

FIGURE 3.3 A Taylor Hobson roughness meter in use.

Like a profilometer, a skid type surface roughness tester uses a diamond stylus that moves across the specimen, and a piezoelectric pickup records vertical movement. Peaks and valleys are recorded and converted into a known value of a given parameter. Parameters differ in how they interpret peaks and valleys.

A precision workshop using Taylor Hobson Surtronic roughness meters (Figure 3.3) was conducted in 2000 by the U.K. Slip Resistance Group (see Chapter 7, Overseas Standards). Nine instruments were tested on five surface textures. One such instrument obtained results ranging from 0.5 to 1.5 on the same surface; another yielded a still greater range on the same surface, suggesting that roughness meters may not be as precise and sensitive as needed for the purposes of the fine measurement of roughness related to slip resistance.

3.8.4 ROUGHNESS THRESHOLDS

G. W. Harris and S. R. Shaw first published the relationship of human traction to the combination of roughness peak-to-valley height (Rtm) and dynamic friction in the *Journal of Occupational Accidents* in 1988. They suggested 10 microns as a minimum for Rtm. An accompanying paper by Proctor and Coleman in the same issue quoted the same 10-micron minimum.

In 1993, Proctor, in *Safety Science*, reviewed research in the field and confirmed the finding of Harris and Shaw that an Rtm of 10 microns or more is necessary to ensure the safety of pedestrians on wet walkways. In 1995, a paper by Rowland et al. at Polymer Testing '95 reiterated the 10-micron Rtm minimum for wet walkways.

3.8.4.1 HSE Roughness Thresholds

According to the U.K. Health and Safety Executive (HSE), "*High surface roughness is obtained from larger and sharper grains making up the floor surface.*" HSE

TABLE 3.2
HSE Minimum Levels of Rtm Roughness Required to Allow Satisfactory COF Values

Contaminant Viscosity (cPs)	Workplace Analogue	Minimum Rtm Floor
<1	Clean Water	20 um
1–5	Milk	45 um
5–30	Stock	60 um
30–50	Olive Oil	70 um
>50	Margarine	>70 um

emphasizes the importance of roughness peak height (Rpm) over peak-to-valley height (Rtm). HSE research has identified the levels of Rtm roughness required to penetrate squeeze films of differing viscosities sufficiently to allow "satisfactory" COF values (per U.K. Slip Resistance Group guidelines). These suggested minimum roughness values are shown in Table 3.2.

HSE believes the height of roughness peaks (Rp) on a floor surface is responsible for most frictional properties (e.g., slip resistance) under wet conditions. They identified Rpm, which is the mean of several maximum peak height measurements, as a suitable measure of peak roughness. However, measurement of Rpm is significantly more difficult to obtain than Rtm. The Rtm parameter is therefore measured far more often in the field.

Additionally, HSE found a strong correlation between Rpm roughness measurements and human-based floor surface COF testing. However, the correlation varies depending on the hardness of the floor material. Harder floors generally require greater peak roughness to achieve adequate slip resistance. Laboratory measurements suggest that "soft" flooring materials should have Rpm values of no less than eight μm, and "hard" flooring materials should have Rpm values of no less than 25 μm for enhanced slip resistance under wet conditions.

The most commonly used and accepted measure of surface roughness (Rtm) is the mean of several peak-to-valley height (Rtm) measurements. The HSE suggests that a floor surface Rtm roughness level of at least 20 μm is required to enhance the slip-resistance of hard floor materials (e.g., ceramics, concrete) under wet conditions.

The U.K. Slip Resistance Group has established a suite of measurements including floor surface roughness in *Guidelines Recommended by the United Kingdom Slip Resistance Group* (U.K. Slip Resistance Group).

A 2004 study evaluated surface microscopic geometric features that could increase friction on liquid-contaminated surfaces. Three types of surface features were identified as preferred features for higher friction:

- The average of the maximum height above the mean line in each cut-off length (R[pm])
- The arithmetical average of surface slope (Delta[a])
- The kernel roughness depth (R[k])

Although surface roughness is important in determining slipperiness, the study concluded that there is still insufficient information to establish a safety criterion based on roughness (Chang, 2004).

SIDEBAR: Roughness Study—Supported but Not Required

Research published in 2007 analyzed the effects of surface roughness of floors under various conditions in an attempt to identify effective levels of surface roughness. Friction values dropped to a dangerously low level on oil-covered surfaces unless the floors were particularly rough. A medium rough floor with 33 μm in roughness showed the highest slip resistance with all the shoes and surface conditions tested. Smooth floors (less than 10 μm in *Ra* roughness) also yielded effective slip resistance under all conditions tested except the oily surface.

However, it was also found that slip resistance was not linearly correlated with surface roughness in the case of clean and dry surface and some contaminated surface conditions such as wet and soapy. Those findings imply that rough floors are not necessarily required to enhance the slip resistance in even the most dangerous situation such as the oily surface (Kim and Nagata, 2007).

3.8.5 Roughness Measurement Standards

ISO Technical Committee 213 dimensional and geometrical product specifications and verification is responsible for several roughness measurement standards, including:

- *ISO 4287-1 Geometrical Product Specifications (GPS)—Surface texture: Profile method—Terms, definitions and surface texture parameters* discusses the terminology of surface roughness, including the surface and its parameters.

Other ISO standards applicable to surface texture measurement include:

- *ISO 4288: 1996 Geometric Product Specification (GPS)—Surface texture: Profile method—Rules and procedures for the assessment of surface texture*
- *ISO 5436-1: 2000 Geometric Product Specification (GPS)—Surface texture: Profile method–calibration—Part 1: Measurement standards*
- *ISO 5436-2: 2000 Geometric Product Specification (GPS)—Surface texture: Profile method–Calibration—Part 2: Soft gauges*
- *ISO 11562: 1996 Geometric Product Specification (GPS)—Surface texture: Profile method—Metrological characteristics of phase corrected filters*
- *ISO 12085: 1996 Geometric Product Specification (GPS)—Surface texture: Profile method—Motif parameters*
- *ISO 12179: 2000 Geometric Product Specification (GPS)—Surface texture: Profile method calibration of contact (stylus) instruments*
- *ISO 13565-1: 1996 Geometric Product Specification (GPS)—Surface texture: Profile method—Surfaces having stratified functional properties—Part 1. Filtering and general conditions*
- *ISO 13565-2: 1996 Geometric Product Specification (GPS)—Surface texture: Profile method—Surfaces having stratified functional properties—Part 2: Height characterization using the linear material ratio curve*

Principles of Slip Resistance

- *ISO 13565-3: 1996 Geometric Product Specification (GPS)—Surface texture: Profile method—Surfaces having stratified functional properties—Part 3: Height characterization using the material probability curve*

Germany surface roughness measurement standards include:

- DIN 4760: 1982 Form deviations: Concepts: Classification system
- DIN 4761: 1978 Surface character; Geometric characteristics of Surface Texture Terms, Definitions, Symbols
- DIN 4762: 1989 Surface roughness; terminology
- DIN 4763: 1981 Progressive ratio of number values of surface roughness parameters
- DIN 4768: 1990 Determination of roughness parameters Ra, Rz, Rmax by means of stylus instruments; terms, measuring conditions

3.9 WET SURFACES

A majority of slip/fall incidents occur as a result of contact with a spot on the floor surface that is unexpectedly slippery, usually due to moisture. It is important to determine how slip-resistant the surface is under dry and wet conditions because of pedestrian "expectation."

3.9.1 Hydroplaning

The problem of hydroplaning exists with wet surfaces, where the foot is momentarily supported by the film of liquid on the surface and not the surface below. When hydroplaning occurs, the slip resistance of the surface becomes irrelevant. This phenomenon occurs most often when the liquid has a low viscosity (e.g., oil) and when the application force is minimal (e.g., the instant the heel touches down). Aside from good housekeeping, the best way to minimize this risk is to increase the roughness of the floor surface, which mitigates conditions that contribute to hydroplaning.

3.9.2 Sticktion

Sticktion is the name that safety professionals give to a temporary bond created between the test foot of a slip meter and the walkway surface when contact is made. Often referred to in the literature as "stick-slip," this phenomenon was well discussed by Rabinowicz as early as 1956. Sticktion creates unrealistically high slip resistance readings on wet surfaces, sometimes producing results even higher than the same surface when tested in dry conditions. The proximate cause of sticktion is believed to be "residence time," or the delay between the time the test foot of the slip meter contacts the floor (i.e., vertical force due to gravity) and the application of the horizontal force. This delay results in microscopic bonding. In his research in 1929, R. B. Hunter pointed out that, on wet surfaces, the *"apparently abnormally high values of coefficients of friction under these conditions may not be due to an actual increase in friction. They may be explained on the assumption that perfect contact is made, and that a seal is formed between the wet or oily surfaces giving the effect of a partial vacuum under the shoe."*

In 1956, Rabinowicz explained:

> . . . In any adhesive process the bond becomes stronger the longer it is left undisturbed. This is why the static coefficient of friction increases with time of contact. In the case of sliding surfaces, the period of contact between points on the two surfaces is, of course, longer when the surfaces slide slowly than when they move rapidly. Consequently, if the slide of one surfaced over another slows down, friction increases. This is the situation that favors stick slip. However, laboratory tests have developed the unexpected finding that at extremely slow speeds the situation is reversed: as friction increases the sliding velocity also increases. (p. 112–113)

Studies have shown there is no correlation between sticktion and the length of residence time. Residence times as short as 16 milliseconds before the application of horizontal force yield the same degree of sticktion generated by longer times. From this, it can be concluded that any test method that begins with vertical force prior to the application of horizontal force is subject to sticktion and unreliably high results on wet surfaces (High, 2007).

Examples of tribometers that can produce sticktion are the horizontal pull slip meter and the Tortus-type instruments, in which a weight is placed on the surface and is pulled (or dragged) across the floor (see Figure 3.4). However, slip meters that apply horizontal and vertical forces simultaneously generate no residence time. As a result, they can avoid sticktion.

3.10 HUMAN PERCEPTION OF SLIPPERINESS

Two studies published in 2007 evaluated the ability of people to visually assess the slip resistance of a given walkway surface accurately.

The first study investigated the use of visual cues in forming judgments of slipperiness as well as the consistency of those judgments over time. In this study, 31 participants viewed 38 different floor surfaces under controlled viewing conditions. The floor surfaces varied on cues to slipperiness and the validity of those cues relative to COF. Participants rated floor surfaces on slipperiness, reflectiveness, light/dark, traction, texture, and likelihood of slipping. Rated reflectiveness correlated most strongly with ratings of perceived slipperiness. Slipperiness ratings correlated most strongly with measured COF. Additionally, judgments of slipperiness were consistent over time and different response measures. The results suggest there is potential for designing floor surfaces against slips and falls by incorporating visual cues into the environment to alert pedestrians (Lesch et al., 2007).

In the second study, researchers assessed the visual perception of floor surface slipperiness in a real-world environment and using a large sample size for increased reliability. The results suggest that humans are able to estimate floor surface slipperiness with reasonable accuracy. However, they also indicate that, when the surface is

Principles of Slip Resistance

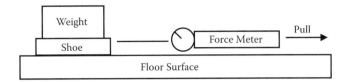

FIGURE 3.4 Horizontal pull tribometer.

wet and glossy, they make significant underestimates about slip resistance (Kuzel et al. 2007).

3.11 CLASSES OF TRIBOMETERS

The three classes of slip resistance testers are the horizontal pull (or dragsled), the pendulum, and the articulated strut.

3.11.1 Horizontal Pull (Dragsled)

The basic principle of the horizontal pull, or dragsled, meter is the pulling of footwear or surrogate material across a walkway surface under a fixed load (or weight) at a constant speed. Some devices are manually pulled, and some are motorized. Dragsled-type meters measure friction in the classic physics sense, SCOF = H/V. Included in this class are motorized dragsleds, which present challenges in assuring calibration and reliable readings. Figure 3.4 depicts the horizontal pull tribometer.

3.11.2 Pendulum

The basic principle of the pendulum class of slip resistance tester involves the calculation of energy loss as an indirect measurement of slip resistance. The pendulum is raised to a fixed height above the surface and is swung across it. As the test foot crosses the walkway, a spring presses the foot material against the surface. The rubbing of the foot on the surface results in a loss of energy due to friction determined by the reduced length of the swing. Figure 3.5 depicts a pendulum tribometer.

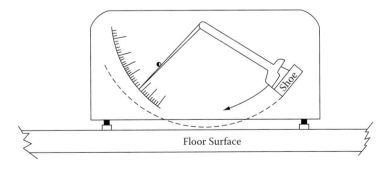

FIGURE 3.5 Pendulum tribometer.

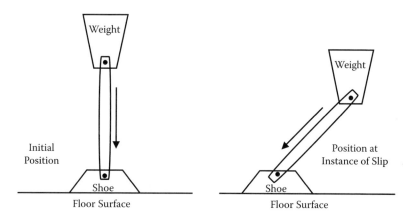

FIGURE 3.6 Articulated strut tribometer.

3.11.3 ARTICULATED STRUT

Some articulated strut tribometers are considered by many experts to be the most reliable and accurate variety of slip resistance testers. These devices apply a fixed speed and angle, representing the closest approximation of human ambulation of the three classes (see Figure 3.6).

3.12 HUNTER MACHINE

The Hunter Machine (see Figure 3.7) is the grandfather of slip resistance test instruments. R. B. Hunter developed it as a result of his work under project A-22 of the American Standards Association (now the American National Standards Institute [ANSI]) during his research in the field from 1924 to 1929. In his 1929 article, "A Method of Measuring Frictional Coefficients of Walk-Way Materials," in the *Bureau of Standards Journal of Research*, Hunter describes the instrument:

> It operates on an oblique thrust principle corresponding to the thrust on the shoe in walking and consists of a right-angled frame carrying a slotted 75-pound weight between two vertical bars of the frame which serve as guides to the weight. A 10-inch thrust arm is pivoted at one end near the center of gravity of the weight and at the other end through the center of area of a 3 by 3 inch shoe. The weight . . . is supported in the raised position by the friction of the shoe on the surface under test. By means of a screw and lug the shoe may be drawn forward by small increments, increasing the horizontal component of the force until the shoe slips on the surface, letting the weight drop . . . The lug carries an index which shows the horizontal distance of the shoe from its position when the thrust arm is vertical. (p. 333)

Subsequently, in 1940, the laboratories of Liberty Mutual Insurance Company developed a portable version of the Hunter Machine; however, this instrument, similar to the original, was never widely used or accepted.

Although several subsequent inventions were laboratory-only devices that had an adhesion and stiction problem, the Hunter Machine was a portable field device

Principles of Slip Resistance

FIGURE 3.7 Hunter Machine.

that overcame this problem by avoiding residence time. Considered a dynamic COF tester, the Hunter Machine dragged a test pad across a walkway specimen until a slip occurred. The literature does not indicate whether it was ever commercially produced. Photographs of the instrument are also hard to find, although a more portable variation of the Hunter Machine was apparently made.

4 U.S. Tribometers

Fall seven times, stand up eight.

Japanese proverb

4.1 INTRODUCTION

Tribometers sold and used in the United States include the build-it-yourself horizontal dynamometer pull meter method, the 1960s-era horizontal pull slipmeter (HPS), the laboratory-only James Machine, and the proprietary portable inclinable articulated strut tester (PIAST) and Variable Incidence Tribometer (VIT) devices. Some methods are only approved for specific uses. With few exceptions (polishes and bathtubs), U.S. slipmeter standards specify no threshold of safety. They are test methods, or steps to follow to arrive at presumably valid results using the specified test device.

Rarely do any two slipmeters agree, even on dry surfaces, and many have proven unreliable for wet testing. No known correlation exists between most of the devices. This is in part because most of these instruments have their own set of biases and operator variability issues, and because friction is a property of the system used to measure it.

4.2 JAMES MACHINE

Sidney James of Underwriters Laboratories developed one of the earliest slipmeters, the James Machine, in the 1940s. The James Machine is a laboratory apparatus for dry testing only (Figure 4.1). As an articulated strut class of tribometer, the James Machine applies a known constant vertical force to a test pad (e.g., a leather pad when testing flooring materials), and then applies an increasing lateral force until a slip occurs. A more detailed description follows.

4.2.1 OPERATION

Similar to its predecessor the Hunter Machine, the test foot is a 3 in. × 3 in. (76 mm × 76 mm) piece of standard leather secured to a flat steel plate. The plate is hinged to an articulated strut 10 in. (25 cm) long. A 75- to 80-lb (34- to 36-kg) weight is situated to apply a vertical and downward force to the strut. The hand wheel is released, and the load is transmitted to the test foot through the strut. The test foot remains in stationary contact with the test surface. The angle of inclination of the strut to the vertical is increased gradually, until it reaches the value at which the test foot slips on

FIGURE 4.1 The manually propelled version of the James Machine.

the test surface. The test foot carries a pointer that indicates the coefficient of friction (COF) at the point of slip on a scale attached to the machine's frame.

The James Machine functions in a way similar to a dragsled with one notable difference: The James Machine applies the friction force in an operator-insensitive manner which reduces the extent of operator influence on results.

4.2.2 SUBSEQUENT VERSIONS

The original versions of the James Machine had a hand-driven transport table. These machines were produced by a number of manufacturers, including TMI, Gardner, and Timken. These are the machines illustrated in the drawings in ASTM D2047 and F489.

By the early 1950s, some of the owners of these machines modified them for motorized test table transport, and motorized machines were made commercially available from the manufacturers just mentioned. The interlaboratory study for the original precision and bias statement for ASTM D2047 was developed with this set of machines.

In the early 1990s, R. Jablonski manufactured the first physical modification of the James Machine. The modifications included shock absorbers to cushion the falling weight, and improved bearings and bearing surfaces. Committee D21.06 (Slip Resistance) and the Maintenance Division of the Chemical Specialties Manufacturers

Association (CSMA), now the Consumer Specialty Products Association (CSPA), studied these modifications for three years in order to be certain that the modified design would produce data statistically consistent with that of other James Machines. Reportedly it did, with a higher degree of repeatability and reproducibility (i.e., narrower error limits), so it was accepted as being a "James Machine" for use in D2047; however, only nine production machines were made before Professor Jablonski withdrew his machine from the market.

In 1997, the Michelman Corporation developed additional modifications: computer interface, electronic self-diagnosis, and digital readout, along with the standard CSPA paper charts.

4.2.3 Operational Issues

The challenges facing users of the James Machine include:

- The instrument has several inherent biases, which prompted users to make modifications in order to achieve good repeatability on a single instrument and good correlation between several similar machines.
- The device needs continuous maintenance and adjustment, in part due to the required release of a heavy weight. ASTM D6205 (see Section 4.2.4.1.1) lists ten reasons why the instrument can yield faulty results, including:
  - Irregular transport of the test table—Especially in the case of manually propelled versions, the manual cranking must be done smoothly and uniformly for accurate readings.
  - Improper rate of transport of test table—The rate of travel is a key factor in obtaining valid results. In the case of manually propelled versions, this is difficult to judge subjectively.
  - Wear or binding of bearings, pivots, and other components.
  - Flat and levelness of test table—No specified method is used for evaluating flatness.
  - Excessive movement in the strut rack gear.
- Warped or out-of-line back plate, chart board, strut arm, or strut rack gear, often caused by improper maintenance or storage, or by the result of the impact of the heavy weight applied.

As far back as 1966, Underwriters Laboratory recognized these variables and cautioned the adoption and widespread use of the James Machine:

- Due to the heavy weight applied, nonsmooth surfaces (i.e., textured, rough) cannot be evaluated because the roughness can result in damage to the test foot material.
- As a laboratory-based machine, it can only be used on flooring material samples, not in-service floors (see Section 4.2.4.1.1 on "D2047").
- Because the device is subject to stiction (see Chapter 3, Principles of Slip Resistance) and it specifies the use of leather (leather's properties change

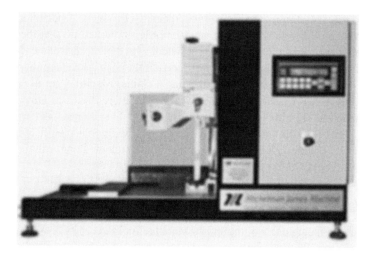

FIGURE 4.2 The Michelman computerized version of the James Machine.

when wet, delivering overly optimistic readings), this device should only be used to test dry surfaces.
- Meaningful precision and bias testing to validate repeatability and reproducibility have yet to be conducted as of this writing. Although both James Machine Standards (F489 and D2047) tacitly include four different versions of the machine, no known works have been published about the correlation of results to each other.

In addition to the manually operated original, a hydraulically operated version, known as the Jablonsky low-friction model, is available (currently in production by Quadra Corporation and sells for approximately $15,000) as well as the Michelman computerized version (see Figure 4.2), which currently sells for approximately $23,000. On the product Web page for the Michelman James Machine (http://www.michem.com/Equipment/Testing-Equipment/Floor-Care-Testing-Equipment/), the wording suggests that the focus of the advertising effort is not safety:

> Protect yourself from liability by testing products' static coefficient of friction (COF) on your Michelman-James Machine.

In addition, machine shop drawings of the apparatus are available from the CSPA, permitting any organization to build the original manually propelled version.

While at Bucknell University in Lewisburg, Pennsylvania, Dr. Robert Brungraber made the following observation regarding other unresolved concerns with this test apparatus:

> ... the tangent of the angle of inclination of the articulated strut is equal to the static coefficient of friction only if both the shoe and the articulated strut are weightless ... It is my opinion that either James never analyzed his device or else he assumed that his use of an 85-pound superimposed weight made the weights of his shoe and strut

negligible. I have never analyzed the James Machine but the ones that I have seen use rather heavy steel struts and shoes, that may very well not be negligible. James also appears to have ignored any friction losses in his device which I also believe to be unwarranted. (Brungraber et al., 1992, p. 23)

4.2.4 STANDARDS

There have been two ASTM standards for operation of the James Machine (for more on ASTM, see Chapter 5, U.S. Standards and Guidelines). F489-96e1—Standard Test Method for Using a James Machine (under the jurisdiction of ASTM F-13), withdrawn in 2004, was used widely in the United States for product testing of footwear slip resistance (see http://www.astm.org/cgi-bin/SoftCart.exe/DATABASE.CART/WITHDRAWN/D5859.htm?E+mystore). The other is D2047, which is under the jurisdiction of ASTM D-21. At the time of this writing, another ASTM committee, F-06 Resilient Flooring, is considering development of a standard for the testing of resilient flooring using the James Machine.

4.2.4.1 Committee D21 Polishes

Formed in 1950, D21 is responsible for 46 standards for characteristics and performance of raw materials, physical and chemical testing, and performance and specifications relating to polishes. The committee scope states that D21 is concerned primarily with floor, furniture, automotive, metal, and plastic polishes. Subcommittee D21.06 Slip Resistance publishes three standards—D2047, D4103, and D6205—which relate to testing using the James Machine.

4.2.4.1.1 D2047

D2047-04 (2004) is the Standard Test Method for Static Coefficient of Friction of Polish-Coated Floor Surfaces as Measured by the James Machine (http://www.astm.org/cgi-bin/SoftCart.exe/DATABASE.CART/REDLINE_PAGES/D2047.htm?E+mystore), which is under the jurisdiction of technical committee D21 Polishes.

D2047 was the first ASTM standard referring to the James Machine. One of the first orders of business of D21 when it was formed in 1950 was the topic of slip resistance. In 1964, tentative method ASTM D2047-64T was issued without a threshold specification; it became a standard in 1969. It was not until 1974 that the threshold of 0.5 was added.

D2047 and F462 are the only ASTM test methods that currently include a threshold upon which to base a quantitative assessment of slip resistance. Related to this standard are:

- D4103-90 (2002), Standard Practice for Preparation of Substrate Surfaces for Coefficient of Friction Testing, which specifies the process of preparing Official Vinyl Composition Tile (OVCT) and wood panels for use in tests to measure COF
- D6205-06 (2006), Standard Practice for Calibration of the James Static Coefficient of Friction Machine, which provides guidance on determining if the test instrument is mechanically calibrated and properly aligned

Correlation of the James Machine to actual usage is limited to several optimal conditions. These assumptions essentially involve unencumbered walking at an average pace of three mph (five km/h) on well-maintained level surfaces, free of gross debris or contamination of any type (e.g., liquid, grit). Despite its shortcomings, this instrument continues to be used to validate the merchantability of new flooring materials and treatments under D2047.

An extensive discussion of the pitfalls of using this apparatus is discussed above. Although the proponents of this test method cite research done to establish a correlation between a James Machine reading of 0.5 and the human perception of a slip-resistant surface, it is important to remember that the James Machine can only test flooring materials in the laboratory, not actual floors *in situ*. Once flooring material measured by the James Machine is placed in service its properties are immediately altered by such factors as wear, contaminants, and cleaning. Any such correlations between the material in pristine condition and those subject to a variety of other unaccounted variables are at best questionable. In addition, work dating back as far as 1972 suggests that several manufacturers of polish found that their least slip-resistant products obtained the highest James Machine COF readings. As discussed in ASTM D21.06 technical committee meetings, representatives of the polish industry have indicated that extensive, long-term experimental data underpins the 0.5 criteria for the James Machine, but that, due to potential liability, the industry is unable to produce it. The author knows of no published accident statistics or peer-reviewed research that supports this threshold.

The research report of the precision workshop conducted to validate this test method can be obtained from ASTM for a nominal fee.

4.2.4.1.2 D3758

D3758-95 (2003), the *Standard Practice for Evaluation of Spray-Buff Products on Test Floors* (http://www.astm.org/cgi-bin/SoftCart.exe/DATABASE.CART/REDLINE_PAGES/D3758.htm?E+mystore) under the jurisdiction of technical committee D21 Polishes, was originally published in 1979. This document covers the comparison of spray-buff products (for use to maintain base floor-polish films) on test floors against a reference material. It specifies the use of a test method from the Chemical Specialties Manufacturers Association (CSMA) Bulletin Number 245–70 Comparative Determination Slip Resistance of Floor Polishes, now CSPA 0202 (see Chapter 5, U.S. Standards and Guidelines).

This standard is under the jurisdiction of D21.04, the Performance Tests subcommittee.

4.3 HORIZONTAL PULL SLIPMETER (HPS)

The HPS, approved for dry testing only, was developed by Charles Irvine in 1965, when he was working for Liberty Mutual Insurance Company. The basic principle of the HPS, a dragsled class of slipmeter, is the pulling of footwear or surrogate material against a walkway surface under a fixed load at a constant velocity. The HPS consists of a 6-lb (2.7-kg) weight onto which is attached a force gauge with a slip index meter. This component is attached to a nylon string and pulled by a capstan-

U.S. Tribometers

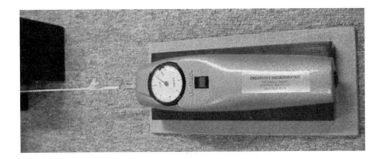

FIGURE 4.3 The Whitely Industries version of the ASTM horizontal pull slipmeter.

headed motor. As the weight/meter is pulled across the walkway surface, a direct reading from the meter can be obtained (Figure 4.3).

The output of the HPS is referred to as the slip index, which is ten times the SCOF. For example, a SCOF of 0.4 is shown as a slip index of 4.0.

Aside from the problem of sticktion, which makes this device unreliable on wet surfaces, other concerns include:

- The use of a spring combined with the analog indicator makes it difficult to obtain a definitive reading.
- The lack of structure between the motor and the meter/weight (a nylon string) results in potential operator variances in the application of lateral forces.
- It is difficult to obtain valid readings on interlocking (i.e., tiled) surfaces, because the sensors on the HPS are so small they can become caught or influenced by these surface irregularities. Thus, the HPS is best suited to linoleum and other uniform resilient floor surfaces (Figure 4.4).

The corresponding U.S. standard for the HPS is ASTM *F609-05 Standard Test Method for Using a Horizontal Pull Slipmeter (HPS)* (http://www.astm.org/cgi-bin/SoftCart.exe/DATABASE.CART/REDLINE_PAGES/F609.htm?E+mystore), under the jurisdiction of F13.10. In 2005, the standard was reapproved with a new precision statement based on recently completed ruggedness testing and an interlaboratory study, which is available from ASTM for a nominal fee. The most notable change was the removal in the scope to permit its use for wet testing. The Significance and Use section of the standard states:

> Slip index, as determined by the HPS, most likely will not give useful information for evaluating liquid contaminated surfaces, and therefore, will not provide an effective assessment of a potential slipping hazard on a walkway surface under these conditions. (ASTM F609-05, 2005)

Although certain devices are based on similar dragsled technology, the only ASTM-compliant versions of the HPS (manufactured by Whitely Industries and Creativity Inc.) were no longer commercially available until recently. Two manufacturers, Trusty-Step Products and CSC Force Measurement, have begun producing

FIGURE 4.4 The underside of the HPS, showing three ½-in. circular test pads.

the HPS. The Trusty-Step instrument (TSI-9010 Slipmeter, http://www.trusty-step.com/9010.php) currently sells for $995. The CSC instrument is being marketed as the HPS III (http://www.cscforce.com/main.htm).

4.4 HORIZONTAL DYNAMOMETER PULL-METER (C21 CERAMIC WHITEWARES AND RELATED PRODUCTS)

Established in 1948, C21 is responsible for 58 standards, primarily related to raw materials, physical and chemical testing, and performance and specifications relating to products of ceramic materials. Subcommittee C21.06 Ceramic Tile has jurisdiction over 17 standards involving such properties as bond strength, warpage, color, abrasion resistance, shock resistance, and electrical resistance.

C1028 is the *Standard Test Method for Determining the Static COF of Ceramic Tile and Other Like Surfaces by the Horizontal Dynamometer Pull-Meter Method* under the jurisdiction of technical committee C21.06, Ceramic Tile (see Figure 4.5).

Although often confused with the F609 HPS device because it operates in a similar way, this is a different instrument. ASTM first published the document in 1984. A do-it-yourself instrument, the C1028 method consists of instructions on how to construct and operate the device, calling for an analog dynamometer, a Neolite test pad, an aluminum base, and a 50-lb (23-kg) weight.

4.4.1 Issues

Because it is not a manufactured device, almost every C1028 unit made is different from another, increasing the potential for variability of results. And although it

U.S. Tribometers

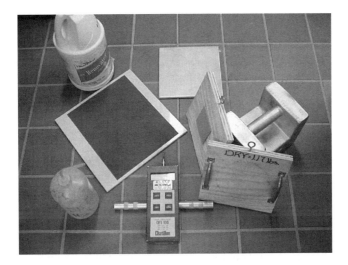

FIGURE 4.5 The materials needed to construct a horizontal pull dynamometer pull meter.

is approved for wet testing, it has been long known that, similar to other dragsled technologies, the C1028 method produces excessively high traction numbers on wet surfaces. As a manually operated instrument, it is more subject to operator influence than mechanically operated apparatus.

4.4.1.1 Scope

The ceramic tile industry and others use this test method to promote their products as slip resistant. However, to avoid the technical and procedural failings of doing this, they claim it is only a merchantability test. The committee has repeatedly refused to align the scope with their claims. In finding a proposal to remove language involving pedestrian slip resistance unpersuasive, the committee opined that ". . . the use of this test method in the field is required to verify that no post production modification of the material has occurred. Additionally, determination of slip resistance is not covered under the scope of this test method" (ASTM C21.06).

In neither the Scope nor the Significance and Use sections does it state that this test method is solely for the purposes of the quality control of flooring products. Contrary to the stated committee position, the Significance and Use section of the standard does not specify this test for "post production modification of material" evaluation or static coefficient of friction testing in the field; however, it does mention its use for slip resistance testing in the field.

4.4.1.2 Significance and Use

According to ASTM regulations, the significance and use statement of a test method must include information on limitations and inappropriate uses of the instrument. For example, because the C1028 device is a manually pulled apparatus, there is clearly a high potential for undue operator influence. Another example is the inappropriateness of the instrument for use on inclines due to its dependence on gravity. The committee found this proposal unconvincing, "... because the test method already addresses these issues and other factors in Section 4.2."

However, the referenced section does not address the issues raised. In fact, Section 4.2 mentions nothing whatsoever about the inherent variability due to manual operation. In essence, by referencing an irrelevant passage, the committee response failed to address the issue while tacitly agreeing to its validity.

4.4.1.3 Apparatus Issues

Two options for the configuration of the weight are provided. However, no assessment has ever been performed to validate that these configurations or numerous other variables impact results. The committee found this proposal unconvincing, "... because weight is a measurement of the force of gravity which does not need ruggedness testing."

The committee response assumes that the only variable at work is gravity. If that were the case, there would be no need to describe the conditions and components of a test method. The influence of the configuration and placement of the weight cannot be known unless it is evaluated. The design of the apparatus should be as specific as possible to minimize variability in results. Design details that are not specified include:

- Location of eyelet (e.g., dead center on a side of aluminum plate)
- Positioning of the Neolite pad on the bottom of the base (e.g., dead center) and method of attachment (e.g., epoxy, double-sided tape)
- Positioning of the weight on the aluminum plate (e.g., dead center, with narrow end facing toward front)
- Method of securing sandpaper when preparing Neolite and when sandpaper should be replaced (i.e., when noticeably worn)

Again the committee found this proposal unconvincing, "... because he provides no data that any of these suggestions will affect the variability of the test method and in the experience of the committee, the test method is sufficiently clear as written." Of course, there is no data, because the committee has never evaluated the impact of these variables.

Considering that the apparatus involved has no manufacturer from which quality control can be exercised, and the fact that as a manually operated instrument it is particularly subject to operator error, ruggedness testing is essential to determine the influence of each of the variables on test results.

Tests in 2003 at the University of Southern California (USC) Medical Center again confirmed that C1028 gave a wet COF of 0.64 to a surface on which all pedestrians slipped when it was wet (Sotter, 2005).

A study published in 2007 compared the readings of nine slipmeters on three surfaces (wet and dry) to rank their relative traction against the results of human subjects. The results indicate that traditional dragsleds, manual or powered, do not reflect the rankings of the human subjects (Powers et al., 2007).

Despite these shortcomings and limitations, some vendors of flooring-related products misuse this test method in promotional, marketing, and technical information related to their "slip-resistant" products (see Exhibit 6.1, Slip-Resistant Floor Treatment Study 2000).

In 2007, despite the above issues, the standard was reinstated with a new precision and bias statement after being withdrawn for a year. Because the original supply of calibration tiles was depleted, a new tile was specified. The standard continues to permit the device to be used for wet testing, despite clear evidence that such testing is invalid.

C1028 is referenced by several other ASTM standards, including:

Under the jurisdiction of C18 Dimension Stone, subcommittee C18.08 Selection of Exterior Dimension Stone:
- *ASTM C1528-02 Standard Guide for Selection of Dimension Stone for Exterior Use* is for architects, engineers, specifiers, contractors, and material suppliers who design, select, specify, install, purchase, fabricate, or supply natural stone products for construction applications.

Under the jurisdiction of C03 Chemical-Resistant Nonmetallic materials, subcommittee C03.02 on Monolithics, Grouts and Polymer Concretes:
- *ASTM C722-04 Standard Specification for Chemical-Resistant Monolithic Floor Surfacings,* which addresses requirements for aggregate-filled, resin-based, monolithic surfacings for use over concrete floors in areas where chemical resistance and the protection of concrete are required.
- *ASTM C1486-00 Standard Practice for Testing Chemical-Resistant Broadcast and Slurry-Broadcast Resin Monolithic Floor Surfacings* addresses methods for preparing test specimens and testing procedures for broadcast or slurry-broadcast monolithic floor surfacings in areas where chemical resistance is required.

4.5 BRUNGRABER MARK I PORTABLE ARTICULATED STRUT SLIP TESTER (PAST)

While working for the National Bureau of Standards (NBS), which is now known as the National Institute for Standards and Testing (NIST), Dr. Brungraber developed the NBS-Brungraber in the 1970s (Brungraber, 1977). This slip tester was originally known as the "NBS Standard Static COF Tester" and, later, as the Mark I Slip Tester. The NBS-Brungraber (or Mark I) is approved for dry testing only as a Portable Articulated Strut Tester (PAST). Even so, the apparatus is specified for testing tub and show surfaces as ASTM F-462, which is performed with a water/soap solution (see Section 4.5.3). Similar in principle to the James Machine, the Mark I is also an articulated strut instrument approved only for dry testing (Figure 4.6). However, it is a portable device that can test actual floors; it also uses a graduated rod that provides a direct reading from the device.

FIGURE 4.6 The NBS-Brungraber, or Mark I, PAST.

4.5.1 OPERATION

A 10-lb weight is attached to the upper end of a pair of vertical rods, which are free to move through linear ball bushings in the carriage. The carriage is also free to move horizontally along a second pair of rods under the control of a spring. Connecting the lower ends of the vertical rods and the test foot is a 10-in. shaft, hinged at both ends. The 3 in. × 3 in. (76 mm × 76 mm) test foot (the same dimensions as the James Machine test foot) is generally made of standard leather.

The tester is placed with the test foot in contact with the walkway surface and the carriage to the left end of the horizontal rods, so that the shaft is vertical and the test foot is subjected only to a vertical force. As the carriage moves to the right under the control of the spring, the inclination of the shaft increases, resulting in an increasing lateral force on the test foot. When this force exceeds the vertical force times the static COF, a slip occurs, which activates a trigger mechanism that blocks further motion of an indicator rod that has been drawn along by the carriage by means of a weak magnetic force (Figure 4.7).

Some calculation is required to convert this to a slip resistance measurement. Because the test foot rests on the walkway surface, the instrument is still subject to sticktion, making it inappropriate for wet testing.

4.5.2 AVAILABILITY AND STANDARDS

The U.S. Government patented the NBS-Brungraber (U.S. Patent No. 3,975,940; issued August 24, 1976). Available for license to any U.S. entity, the only producer of the instrument has been Slip Test Inc., owned and operated by Dr. Brungraber. Although still in use, only about 100 units were ever sold, and Slip Test has indicated that it no longer manufactures the instrument.

FIGURE 4.7 The NBS-Brungraber graduated rod, from which readings are obtained.

There were two U.S. test methods for the operation of the Brungraber Mark I. Although still available for purchase from ASTM, *ASTM F1678-96 Standard Test Method for Using a Portable Articulated Strut Slip Tester (PAST)* (http://www.astm.org/DATABASE.CART/WITHDRAWN/F1678.htm) under the jurisdiction of F13.10, was withdrawn in 2005 because the test method did not contain a precision statement. The other is F462, which is still an active standard.

Dr. Brungraber's next invention, the Mark II, has gained wider acceptance.

4.5.3 F15 Consumer Products

Formed in 1973, the purpose of F15 is to develop safety and performance standards for consumer products, encompassing a broad range of subjects including lighters, playground equipment, infant carriers, toys, beds, candles, scooters, and bathtubs. Recognizing that its work may overlap with that of other ASTM committees, the scope of F15 requires that it coordinate with committees having mutual interest in the subject of a given standard.

In the early 1970s, the need for slip resistance standards for bathtubs became apparent. As one of a number of products that had been identified as hazardous by the Consumer Product Safety Commission (CPSC), bathtubs were ranked fourteenth on the CPSC Product Hazard Index. ASTM formed Committee F15 Consumer Product Safety to develop such standards. Initial research conducted from 1973 to 1975 indicated that bathtubs far exceed shower stalls in terms of frequency and severity of injury. Most injuries were slips and falls experienced entering or leaving the tub, or while changing between standing and sitting positions.

4.5.3.1 F462

F462 is the Standard Consumer Safety Specification for Slip-Resistant Bathing Facilities (http://www.astm.org/cgi-bin/SoftCart.exe/DATABASE.CART/REDLINE_PAGES/

F462.htm?L+mystore+ruwy0545) under the jurisdiction of technical committee F15 Consumer Products.

At the time F462 was written, the only viable instruments available were the HPS, pendulum, James Machine, and the newly developed PAST (Brungraber Mark I or NBS-Brungraber). Through the process of elimination, primarily on practical grounds (e.g., insufficient space to operate an HPS, excessive speed of the pendulum, and inappropriate size/weight of the laboratory-based James Machine), the Brungraber Mark I was selected as the instrument upon which to base the standard.

To establish a baseline and threshold for the test, 50 bathtub surfaces were tested, with individual results ranging from 0.003 to 0.417. As a result of this testing, the acceptable level of slip resistance was set as 0.04. The rationale was that the standard should be set at twice the value of the highest reading obtained on the untextured bathing surfaces tested.

The F462 standard calls for testing surfaces with the Brungraber Mark I PAST fitted with a Silastic 382 rubber test pad (medical grade), performed with soapy water (using Federal Specification P-S-624g or ASTM D799-74 specifications). Measurements are made in nine specific areas of the tub, and all nine areas must meet or exceed the threshold for the tub to pass. The bathtub meets the specification if slip resistance is .04 or greater, bearing no known relationship to a threshold of safety for human ambulation.

Standard F462 was first approved in 1979 and has been reapproved several times, most recently in 2007. Reapproval means that, except for editorial changes, the standard was renewed without changes.

A primary concern with this test method is that the soap scum builds up on the tub surface due to soap and shampoo interaction with water hardness and other contaminants; this results in a smooth coating over the slip-resistant surface. As a result, the bather is not standing on the textured surface that was tested, but on the soap scum film, which levels the texture significantly, reducing its slip resistance. Also, the consistency of the soapy water mixture is problematic. Even if a consistent product were readily available, the resulting mixture does not represent anything that occurs between a bather's foot and the tub surface while showering.

ASTM requires that any specification that includes a test method must contain all the requirements of a test method. The major portion of F462 is a detailed test method. And in the 30 years since it was first promulgated, the document still contains no Precision and Bias (P&B) statement. P&B is fundamental to validating a test method. Without it, there is no support that the protocol specified in the standard is any more accurate or appropriate than another.

4.6 BRUNGRABER MARK II PORTABLE INCLINABLE ARTICULATED STRUT SLIP TESTER (PIAST)

In the 1980s, subsequent to the development of the Mark I device, Dr. Brungraber invented the Mark II. A gravity-based articulated strut device designed to avoid the stiction problem, the Mark II enables users to test wet surfaces by eliminating the

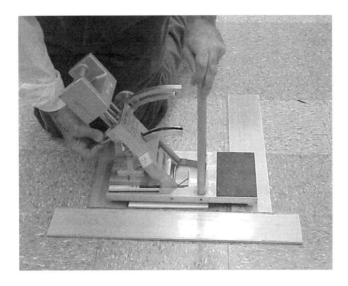

FIGURE 4.8 The Mark II PIAST.

residence time (or time delay) between the application of vertical and horizontal forces (Figure 4.8).

Similar to the Mark I, the Mark II is a portable device fitted with a 3 in. × 3 in. (76 mm × 76 mm) test foot, comparable to the test foot of the James Machine. It uses a 10-lb (4.5-kg) weight on an inclinable frame, with a test pad of Neolite Test Liner suspended above the walkway surface. For each trial, the angle is set to a progressively more horizontal position and the weight is released by means of a clip until a slip occurs. Readings can be taken directly from the instrument.

Due to its size and weight, as well as the dimensions of the test pad, the Mark II requires some expertise and effort to set up and operate on stairs. Because of its reliance on gravity for operation, it is unsuitable for measuring ramps in either the downhill or uphill direction. Instead, the Mark II must be used across the direction of travel.

In addition, the low slip resistance readings obtained in wet tests appear to be the result of the flat position of the test foot at initial contact, trapping water between the foot and the walkway surface. This creates a hydroplaning effect (see Figure 4.9).

Independent studies by several researchers (Medoff et al., 2003; Flynn and Underwood, 2000; Grieser et al., 2002) demonstrate that this effect can be overcome by using a grooved test foot, which provides a path for the displaced water. Although more study may be needed to optimize the specific design (i.e., depth, width, configuration of grooves, and directionality), using a grooved test foot on the PIAST has been shown to closely align results with that of the Variable Incidence Tribometer (VIT).

A study to determine the impact of groove count was performed, using test feet with 1, 3, 5, 7, 9, 11, 13, and 15 grooves, as well as a smooth test foot on polished granite and vinyl composition tile. Slip resistance was significantly higher for increasing groove counts under wet conditions. The results indicated that grooves effectively

FIGURE 4.9 A direct reading of slip resistance can be obtained from the PIAST's graduated protractor.

channel water film out from between the test foot and the walkway surface, reducing hydroplaning. The researchers concluded that, although smooth test feet may be useful when evaluating pedestrians using nontreaded shoes or well-worn soles, grooved test feet may provide a more realistic measure of the slip resistance when wearing the more common patterned footwear bottom (Joganich and McCuen, 2005).

In a study on the repeatability and bias of the VIT (see below) and the PIAST, horizontal/vertical force-ratio settings were compared to measured ratios detected by a force plate. When these detected ratios were assessed by standard statistical methods, both displayed a high degree of repeatability and a low degree of bias (Smith, 2003).

In a 1999 study, the validity, repeatability, precision, and consistency of three portable tribometers were evaluated, including the Portable Skid Resistance Tester (see Chapter 7, Overseas Standards), and the Brungraber Mark II. The study concluded that the Mark II was more valid but less consistent and precise than the pendulum (Grönqvist et al., 1999).

A study in 2007 investigated the effect of the size of the contact area on the measured friction in order to reduce the squeeze-film effect and the correlation between the measured friction and the perception rating of human participants. The results indicated that friction increased as contact area was reduced. The researchers concluded there is a need to reduce the pad size. Based on the results, they concluded that the proper size of the footwear pad might be around 1.5 in. to 2.5 in. It was observed that the movement of the footwear pad became somewhat unstable at impact when a smaller pad (e.g., one in. square) was used, and this problem could affect the validity of friction measurements (Chang et al., 2007).

The ASTM standard for this apparatus, F1677-05 *Standard Test Method for Using a Portable Inclineable Articulated Strut Slip Tester (PIAST)* (http://www.astm.org/cgi-bin/SoftCart.exe/DATABASE.CART/WITHDRAWN/F1677.htm?L+mystore+ruwy0545), was withdrawn in 2006 because it lacked a precision statement and because it was considered a proprietary device (i.e., patented and available through a single source). It is still available for purchase through ASTM. The Mark II device would likely be permitted under a test method under development by ASTM F13 (WK11411 *New Test Method for Obtaining Measurements with Portable Variable Angle Strut Slip Resistance Meters*) (see Section 4.7.2).

However, significant research has been performed on the repeatability and reproducibility of this apparatus. In 2008, ANSI Technical Report TR-A1264.3 Using Variable Angle Tribometers (VAT) for Measurement of the Slip Resistance of Walkway Surfaces was released, which provides detail on the operation, protocol, and studies available for this and the English XL (see below).

The Mark II is still available from Dr. Brungraber's company, Slip Test Inc. (1900 Fourth Avenue, Spring Lake, NJ 07762), which is also the only manufacturer. Two applicable patents exist: U.S. Patent No. 4,759,209 issued July 26, 1988 and U.S. Patent No. 4,798,080 issued January 17, 1989. Currently, the Mark II can be purchased for $4,500.

More recently, Dr. Brungraber has produced a variation of the Mark II, known predictably as the Mark III. His primary objective was to make the instrument significantly lighter. This was done by replacing the 10-lb (4.5-kg) weight with a spring providing equivalent pressure and other machining efficiencies. Dr. Brungraber is also replacing the clip and pull handle with a handwheel adjustment similar to the VIT.

4.7 ENGLISH XL VARIABLE INCIDENCE TRIBOMETER (VIT)

In the early 1990s, William English developed the English XL™, an articulated strut device similar in principle to the Mark II. The English XL is approved for dry and wet testing as a VIT. The English XL does not rely on gravity, but is powered by a small carbon dioxide cartridge at a set pressure (Figure 4.9).

The instrument is an aluminum frame onto which is attached a hinged aluminum mast (Figure 4.10). At the base of the mast is a spring and joint assembly onto which is attached a circular test foot with a diameter of 1.25 in. (32 mm). The angle of the mast can be adjusted by a hand wheel from vertical 90° to 45°. The pneumatic mast is powered by a carbon dioxide cartridge, under high pressure, through a control valve that actuates the cylinder until a slip occurs. The slip index can be read directly from the protractor mounted on the instrument. Although the manufacturer and standards for the instrument recommend the use of Neolite Test Liner for floor testing (see Section 4.9.2), any material can be mounted on the test foot.

The instrument has several features that make it suitable for field testing:

- The test foot cylinder contacts the walkway surface in a heel-first attitude of contact.
- The velocity of contact is consistent with strobe flash experiments of heel contact speed (Perkins, 1978).
- The size of the test foot is similar to the area of heel contact in human walking.
- The compressed gas delivers a uniform force and permits testing of inclined surfaces such as ramps without the impact of gravity.

- Similar to the PIAST, the application of vertical and horizontal forces is simultaneous, thus avoiding residence time and permitting reliable wet test results (Figure 4.11).

In a study by Powers et al. in 1999, the PIAST and VIT demonstrated remarkably low bias and good repeatability. Work by Chang and Leamon (1997) reported that the PIAST and VIT were both capable of distinguishing surface roughness on wet quarry tile.

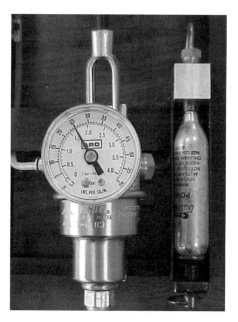

FIGURE 4.9 Using the regulator gauge, pressure is set at ±25 PSI.

FIGURE 4.10 The English XL VIT.

U.S. Tribometers

FIGURE 4.11 The VIT slipping on wet marble tile.

Under the corresponding ASTM test method F1679 the apparatus has undergone ruggedness testing, a procedure to measure the influence of potential variables on test results. In addition, several ASTM workshops have generated much data on the reliability and consistency of the VIT. Two studies done for the revision of the Occupational Safety and Health Administration (OSHA) Steel Erection Standard bear out the correlation of the instrument's output to the human perception of walking by comparing how the VIT and walkers' rank sets of materials of various degrees of slip resistance (see http://www.cosh.org/docs/d0100/d000037/d000037.html). Also see the OSHA report on slip resistance for structural steel in Exhibit 5.1. This and additional detail on the operation and output of this instrument is available in *ANSI Technical Report TR-A1264.3 Using Variable Angle Tribometers (VAT) for Measurement of the Slip Resistance of Walkway Surfaces.*

Although the VIT is capable of measuring steps, the stair fixture orients the instrument opposite the direction of travel. This has the potential to yield to higher-than-expected results due to the lesser amount of wear in this direction.

For the purposes of calibration, the manufacturer specifies the use of the current TCNA C1028 calibration tile (Figure 4.12) and specifies a wet test result of 0.20 on that surface. The 6-in.-square version can be obtained for $5.00 from TCNA, 100 Clemson Research Blvd., Anderson, SC 29625.

The VIT is currently still under patent (U.S. Patent No. 5,259,236,; issued November 3, 1993), and the only manufacturer is William English Inc. It is available for purchase at http://www.englishxl.com for $4,200 as of this writing.

ASTM published two standards for the operation of the VIT. *F1679-04 Standard Test Method for Using a Variable Incidence Tribometer (VIT)* (http://www.astm.org/cgi-bin/SoftCart.exe/DATABASE.CART/WITHDRAWN/F1679.htm?E+alertstore), under the jurisdiction of F13.10, was withdrawn in 2006 because of the proprietary nature of the device and the lack of precision and bias statement, which are the same reasons the Brungraber Mark II standard was withdrawn. This standard is still avail-

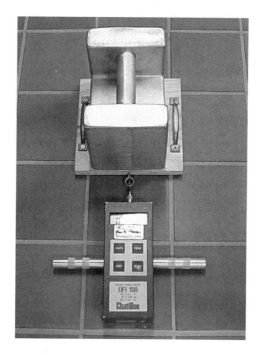

FIGURE 4.12 A C1028 apparatus fully assembled and ready to use.

able for purchase from ASTM. The other ASTM standard was D5859, discussed further below.

ASTM F1679 is referenced by the following standards under the jurisdiction of D07 Wood, subcommittee D07.02 on Lumber and Engineered Wood Products:

- *ASTM D7032-07 Standard Specification for Establishing Performance Ratings for Wood-Plastic Composite Deck Boards and Guardrail Systems (Guards or Handrails)* addresses procedures to establish a performance rating for Wood-Plastic Composite (WPC) deck boards. This specification also defines the procedures to establish a performance rating for WPC guards and handrails. The purpose of this specification is to establish the basis for code recognition of these products.
- *ASTM D7031-04 Standard Guide for Evaluating Mechanical and Physical Properties of Wood-Plastic Composite Products* addresses test methods appropriate for evaluating a wide range of performance properties for WPC products. It was developed from evaluations of experimental and currently manufactured products.

In addition, the English XL, like the Mark II, would be covered under a standard under development by F13, WK11411 *New Test Method for Obtaining Measurements with Portable Variable Angle Strut Slip Resistance Meters* (see Section 4.7.2).

4.7.1 D01 Paint and Related Coatings, Materials, and Applications

Established in 1902, D01 is responsible for over 670 standards, related to coating classification, sampling, preparation, application, analysis, quality assurance, and performance requirements.

4.7.1.2 D5859

D5859, the Standard Test Method for Determining the Traction of Footwear on Painted Surfaces Using the VIT *(http://www.astm.org/cgi-bin/SoftCart.exe/ DATABASE.CART/WITHDRAWN/D5859.htm?E+mystore)*, addresses a method of measuring the slip resistance of footwear (wet and dry) on painted walkway surfaces using the VIT. Although originally developed in D01, this document was subsequently transferred from D01 Paints to ASTM F13.10. Shortly afterward in 2004, F13 elected to withdraw this standard.

4.7.2 WK11411

Still under development as of this writing, Work Item WK11411 *Standard Test Method for Obtaining Measurements with Variable-Angle Tribometers* addresses the use of variable-angle tribometers (VAT) to obtain slip resistance readings. A VAT is a tribometer in which the test foot is set in motion before making contact with the surface, simultaneously applying horizontal and vertical forces to the surface. Two commonly used tribometers that qualify as a VAT are the Brungraber Mark II and the English XL. ASTM F13 (under the jurisdiction of F13.10 Traction) has plans to develop additional test methods by tribometer class (e.g., dragsled, pendulum).

4.8 TEST PAD MATERIAL

Test pad material is the surface used on slip resistance testers that makes contact with the floor surface to obtain the slip resistance measurement. Appropriate test pad material is essential to ensuring accurate, reliable, and valid results. Of the variety of test pad materials used, only Neolite Test Liner and leather have gained wide acceptance in the United States.

4.8.1 Leather

Prior to the use of Neolite, leather was the primary material used as a representative footwear bottom material. "Standard" leather is still used, which normally references government specification KK-L-165 Revision C(2)—Leather, Cattlehide, Vegetable Tanned and Chrome Retanned, Impregnated, and Soles, initially developed in May 1969 and amended June 1976 (http://www.fss.gsa.gov/pub/fedspecs/sort1e.cfm).

However, many concerns arise when using leather as a constant in slip resistance testing. R. B. Hunter recognized most of these concerns in his research of 1929. Some of these include:

- No matter how "standardized" you try to make it, leather is not a homogenous material. In fact, being an organic substance, each piece of leather could be considered a unique material.
- It has different properties at different levels of thickness. Sanding of leather often results in a material with frictional properties different from the pre-sanded surface.
- Leather is highly absorbent and highly sensitive to humidity. Once used for wet testing, its properties are permanently altered. This means that once leather material has been used in a wet test, it is no longer useful for dry testing.
- Leather is also not representative of heel material. Most heels are of a synthetic compound. Essentially, slips occur more on the rubber heels of leather-soled shoes.
- Leather can react differently depending on how worn the material has become. Because it reacts in a way unique to all other types of materials, it is only of value as a test material when it can be reasonably expected that most or all individuals in the area will be wearing leather soles.

Leather conforming to Federal Specification KK-L-165C (Type 1, Class 6) is, as of this writing, only known to be available through Parsons Tanning Co. (333 Skokie Blvd, Suite 105, Northbrook, IL 60062).

4.8.2 Neolite® Test Liner

In the early 1990s, ASTM Technical Committee F13.10 began a study to identify a more suitable material than leather for walkway surface slip resistance testing. It was found that the best available material was a test grade of Neolite, originally developed, patented, and trademarked by the Goodyear Tire & Rubber Company. It is now specified by several ASTM test methods.

Despite protests to the contrary, this type of Neolite was, at one time, used by the footwear industry as a heel material. Documents from the U.S. Trademark Electronic Search System (TESS) verify that Goodyear registered this material in 1953 as "soles and heels composed of an elastomer and a resin."

Neolite is the original styrene-butadiene rubber (SBR), of which many varieties are routinely used as footwear bottom material. Under the trade name Neolite, it was sold in sheet form and die-cut to sole shapes for adhesive attachment. Its special properties include flexibility and adequate wear resistance in thin substances, and it felt and behaved underfoot much like leather, not like rubber. Soling of this kind has now become widely known and used and is often referred to as resin-rubber.

To date, no other suitable material for floor friction testing has been identified for the following reasons:

- Neolite is generic, durable, and stable. The characteristics of the material do not change under normal conditions, regardless of wear or moisture.
- The special test (or scientific) grade of Neolite Test Liner used is manufactured with quality controls for hardness and consistency of physical properties. Neolite Test Liner is manufactured to meet a now withdrawn ASTM

standard for Neolite. It is also made to a specific recipe and tested for consistency of physical properties according to a former Rubber Manufacturers' Association (RMA) specification HS-3 for use as a friction pad material, specifying shore hardness of 93–96, specific gravity of 1.25 ± 0.02, and $1/8$ in. (3 mm) thickness.
- The traction properties of Neolite are in the median range of commonly used shoe bottom materials.
- Neolite has been proven reliable and repeatable over many years in service as a friction pad material, and as the material of choice for the Horizontal Pull Dynamometer Pullmeter (C1028), the HPS, the PIAST, and the VIT (Flynn and Underwood, 2000). It was the material of choice by the National Bureau of Standards (now the National Institute of Standards and Technology) and others in studies on the use of tribometers on wet surfaces.

Neolite Test Liner is manufactured and supplied primarily through Smithers Scientific Services, Inc. (425 W. Market Street, Akron, OH, 44303-2099), although most any manufacturer of footwear bottom material can produce it. The consensus of many slip resistance experts is that Neolite Test Liner is the best currently available material for testing of floor surfaces to establish slip resistance.

4.8.3 RUBBERS AND OTHER FOOTWEAR MATERIALS

A variety of rubber compounds have been used as a friction pad material. In most cases, these have been in relation to overseas test methods such as the pendulum tester and Tortus-type devices (see Chapter 7, Overseas Standards).

Most rubbers have a curing period during which their properties are unstable. They also have a finite shelf life, after which their properties relating to slip resistance change again. In addition, it is difficult to locate the source of a consistent, long-term formulation. These conditions make it somewhat impractical to use most of these types of materials.

Many rubbers are at the high end of slip resistance materials used for footwear, and can provide overly optimistic readings when assessing the slip resistance of flooring materials. In contrast, neoprene rubber, a specification of some U.S. government shoes, provides low traction on lubricated surfaces. The impact of wear on rubbers is another variable.

4.8.3.1 Neoprene

Neoprene is a synthetic rubber specified in some older government shoe specifications and as test pad material in some government traction specifications. It is poor friction material for slip resistance testers because it has proven unrealistically slippery on lubricated (e.g., wet) surfaces. For this reason, neoprene is rarely in more contemporary test methods.

4.8.3.2 Four S

Rubber and Plastics Research Association (RAPRA), a European consultancy, markets Four S (Standard Shoe Sole Simulating) Rubber, now also known as Standard

Pedestrian Hard Rubber in Europe. Although it is relatively insensitive to temperature, which minimizes that variable, it gives artificially high readings, is not consistently produced, and is costly. Although it is commonly used for some tribometers in Europe and Australia, it has not gained a significant following in the United States.

4.8.3.3 TRRL or TRL Rubber

RAPRA also promotes the United Kingdom's Transport and Roads Research Laboratory (TRRL) rubber for pendulum testers. It is also known as CEN Rubber because it is defined in European standard *EN 13036-3 Road and airfield surface characteristics—Test methods—Part 3: Measurement of pavement surface horizontal drainability,* which specifies a pendulum test.

Initially developed to simulate the rubber used in vehicle tires, it is now being used overseas for slip resistance testing. The rationale is that, as shoe sole materials have become softer, it may give a better indication of slip resistance under some conditions, particularly for areas in which people are barefoot. Unlike Four S Rubber, it is not insensitive to temperature. The UK SRG recommends that it be used at 20°C.

It is considered well suited for assessing and comparing very rough surfaces, such as road surfaces; however, this same quality makes it a poor choice for making finer discriminations on smooth or moderate rough flooring materials. In fact, studies have indicated that TRRL rubber consistently yields slip resistance results 20 points higher (more optimistically) than Four S Rubber.

4.8.4 ACTUAL FOOTWEAR BOTTOMS

It is a common misconception that material from the actual footwear of a claimant should be used as the test pad material in accident investigations. It is clear, however, that this would not provide an objective measurement of slip resistance, since the type of material and its wear/condition significantly affects the results of the readings.

In essence, it would be difficult to determine whether the floor surface or the sole of the footwear was primarily responsible for the slip. It is essential that a consistent material be used in order to make meaningful comparisons between floor surface readings.

4.9 USES FOR TRIBOMETERS

4.9.1 PROBLEM IDENTIFICATION—PREVENTION

In order to take corrective action, the location of problem areas must be known. In conjunction with a preliminary analysis of losses, slip resistance testing can assist in identifying such conditions before they result in accidents. Areas to be considered include high traffic areas, entrances and exits, and areas with a history of frequent and/or severe falls.

4.9.2 LITTLE OR NO PRIOR HISTORY

In some cases, loss history cannot be analyzed because this information is unknown or unavailable. Examples include new ventures, the use of new flooring materials, inadequate prior accident reporting and investigation procedures, and the acquisition of new properties or businesses. For these situations, slip resistance testing alone can be instrumental in identifying potential areas of concern.

4.9.3 Claims Defense/Documentation

Periodically performing and documenting slip resistance testing on surfaces that may be subject to slip and fall claims as part of an ongoing prevention program can be effective in minimizing incidents and their related costs.

4.9.4 Accident Investigation

Any factual investigation done today can be scrutinized through the legal process of discovery later. Because of this, it is essential that factual investigation support for claims, including slip resistance testing, be as thorough and accurate as possible. Prompt testing of the area and clear and concise documentation of the results can be critical in determining whether floor surface conditions contributed to the accident.

4.10 OPERATOR QUALIFICATIONS OF COMPETENCY

It is essential to ensure that designated staff is properly trained in slip resistance testing. Without being able to support the knowledge and ability of staff to operate a tribometer, there is little value in conducting testing at all. The best way to assure competency is to require certification. This serves not only to provide an increased measure of competence to operate the test apparatus according to a specified protocol, but it also can provide context by including information on the subject and factors involved in slip resistance testing. Certification should include:

1. Knowledge, which can be demonstrated by classroom participation or a written self-study program that includes a final examination
2. Practical application requiring demonstration of the proper use of the test device to obtain competent results

Presently, only the Variable Incidence Tribometer (VIT) or English XL is known to have a manufacturer-approved user certification program that also requires certification maintenance. The certification as of this writing is being administered by the manufacturer.

4.11 EQUIPMENT CALIBRATION AND MAINTENANCE

Most tribometers require some amount of care. Equipment must be calibrated, maintained, and stored per the manufacturer's specifications. This is essential to assuring and defending the integrity of test results.

Depending on the type of apparatus, a wide variety of components may require periodic or ongoing attention. Examples of tribometer maintenance needs include gauge calibration, joint and valve lubrication, apparatus leveling, tightening of bolts, broken welds, and detection and repair of damaged or bent parts.

Frequent user calibration is also recommended between the required manufacturer calibrations. This normally involves the use of a stable ceramic or similar tile, which the user tests upon return from the manufacturer and periodically to ensure results remain consistent.

It is also advisable to document all user and manufacturer calibration results, including the date, location, method, individual performing work, and test instrument/serial number.

4.12 PROGRESS ON THE ASTM "GOLD" STANDARD

Currently, there is no harmonization of pedestrian slip resistance test methods. The results of one tribometer cannot be correlated to the results of another. To overcome the issue of different tribometers providing different results, an effort that began to develop a unified standard for slip resistance reached a major milestone in 2008.

4.12.1 VALIDATION AND CALIBRATION OF WALKWAY TRIBOMETERS

In 2000, ASTM undertook a major effort to resolve these differences by means of a proposed approach that is based on the relative ranking of materials against a reference set of materials. In 2008, the first of a set of standards to accomplish this objective, *Standard Practice for Validation and Calibration of Walkway Tribometers using Reference Surfaces* was passed by ASTM F13.

Using the data obtained from a substantial research effort with the USC Musculoskeletal Biomechanics Research Laboratory (see Section 4.13), a set of reference materials was developed through human subject walking experiments. Besides having different slip resistance values falling within the range of the human perception of traction, the reference materials have slip resistance values (and, therefore, the rank order) that are constant regardless of the apparatus. In essence, a single "ruler" was established against which all walkway tribometers, regardless of their type or operating principle, would perform.

The four materials selected are common types of walkway surface material: black granite, porcelain, vinyl composition tile (VCT), and ceramic tile.

The primary audience for this document is the tribometer manufacturer, and it provides the information necessary to validate a tribometer against this set of established materials. The degree to which this approach is adopted by manufacturers remains to be seen. One potential issue to this approach is that ASTM does not oversee (i.e., certify or register) the accuracy of the application of their standards. This means that it is the responsibility of the manufacturer to evaluate their products properly, and it is the responsibility of the purchaser to verify the same. The standard also has application to users, because it provides specifications for calibration of tribometers.

U.S. Tribometers

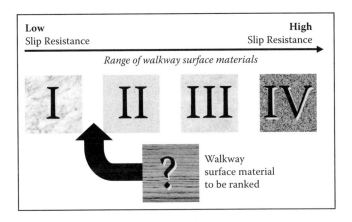

FIGURE 4.13 Primary ASTM "gold standard" methodology.

Tribometer Friction Test Results	Reference Material Pair
0.22	I
0.35	Test Result (I–II)
0.44	II
0.59	III
0.73	IV

FIGURE 4.14 ASTM "gold standard" ranking.

4.12.2 Field Testing with Walkway Tribometers

The second pending document is intended for use by the tribometer user. At the site of the evaluation, each of reference surfaces would be tested and recorded, ensuring that they are ranked in the proper order. Next, the walkway surface in question would be tested with the same test foot used against the reference materials (see Figure 4.13).

The walkway surface would be ranked against the set of reference materials. For example, assume that by using the four reference materials (ranging in values between 0.22 and 0.73), the results of the walkway surface in question are 0.35. That places the walkway surface of interest between two of the reference materials. The test results might be expressed by its relative ranking to the reference materials (e.g., 1 to 2), not the numerical results of the slipmeter (see Figure 4.14).

4.12.3 Thresholds

Historically, a single friction coefficient was considered to differentiate between slip-resistant and non-slip-resistant walkways. More recently, the slip resistance value is considered relative to the type of activities expected on the walkway surface. For example, dancers on a dance floor require less friction than do pedestrians, and workers pushing heavy loads across a floor require more friction. So, it is likely that no single threshold, such as the frequently cited 0.5 figure, is sufficient to appropriately distinguish slippery from tractive in all instances.

ASTM has taken the position that their role is to develop the "ruler," but not the thresholds. Thus, it is up to others (e.g., industry, government, other standards bodies) to use these standards in establishing appropriate levels of acceptability for flooring and ambulatory activities.

4.12.4 Benefits

Using the approach of a relative ranking of a floor surface/material against a uniform set of external calibration materials has strong potential for standardization of slip resistance results at the international level for several reasons. The approach:

- Is not hardware dependent, thereby eliminating the need to develop multiple standards and permitting the use of any tribometer, provided it can meet the performance criteria specified
- Standardizes a set of reference materials for testing
- Eliminates the variety of scales used to report slip resistance results, thereby reducing confusion and misinterpretation

Development of this set of standards is a slow-moving process for several reasons, most notably the amount of research required, the many entrenched interests involved, and the challenge of producing a standard from a basic concept to a practical application.

4.13 GROUNDBREAKING RESEARCH

Work continuing at the Division of Biokinesiology and Physical Therapy at the USC (http://pt.usc.edu/labs/mbrl/) promises to settle longstanding debates regarding thresholds of slip resistance (Figure 4.15). With access to the latest equipment and the expertise to apply it effectively, the project, headed by Dr. Chris Powers, Director of the Musculoskeletal Biomechanics Research Laboratory and chair of ASTM F-13 subcommittee F13.40 on Research, is beginning to relate slipmeter readings to human ambulation. This is being accomplished through a series of trials involving people walking across a set of force plates in conjunction with tribometer workshops on the same surfaces and under the same conditions. This approach may also provide the basis for slip resistance bias (e.g., an accepted reference value).

Walking subjects cross a floor area several times at a normal gait while wearing goggles that do not permit the subject to see the floor. The subject is unaware of the location of the force plates or that a force plate was lubricated, thus resulting in a natural and unbiased slip. The walker is harnessed to an overhead trolley to prevent injury, but even this is set up to measure the degree of vertical support used in recovery from the slip. Features of this project that make it unique from other research include:

- Powerful computer modeling is used to study slip events.
- Subjects are unaware of floor conditions, thus providing realistic and unbiased results.

FIGURE 4.15 Activity during one of the USC workshops.

- Slipmeter readings are taken close to the subject's point of slip immediately following each event.
- Seven digital video cameras are set up to capture the motion of key portions of the subject's movements, using infrared markers located on the walking subject that allow for analysis from a variety of angles.

The purpose of the 2004 workshop was to assess the viability of using human subject walking trials to create a reference against which tribometer readings could be compared. Human subject slip events during walking were used to objectively rank the slipperiness of three surfaces with and without a contaminant (six conditions). Nine tribometers were used to independently measure and rank surface slipperiness for all six conditions (Table 4.1).

Eighty-four subjects were randomly assigned to one of six floor surface conditions. Subjects first performed three to six nonslip walking trials followed by a trial in which the floor panel of interest was inserted. The ranking of surface slipperiness was based on the number of slip events observed in each group.

Nine tribometers were used to measure the COF of the six surface conditions. For each surface condition, the COF was measured four times and averaged. The ranking of floor surfaces for each tribometer was based on the values obtained.

Tribometer measurements were compared to the gait-based ranking. Based on the number of slips in the human subjects, the surface conditions were divided into three levels of slipperiness: not slippery, slippery, and very slippery. Six tribometers (Brungraber Mark II and Mark III, English XL, Wessex and Sigler Pendulums, and Tortus II) correctly ranked the surfaces. Three tribometers (C1028, Horizontal Pull Slipmeter, and Universal Walkway Tester) did not (Powers et al., 2005).

As of this writing, the results of the 2007 workshop are pending publication. This follow-up study used more "real-world" flooring materials.

TABLE 4.1
Tribometers Used in the Workshops Held at the USC Musculoskeletal Biomechanics Research Laboratory

2004 Workshop	2007 Workshop
Dragsled Machines	**Dragsled Machines**
Horizontal Pull Slipmeter	Horizontal Pull Slipmeter
C1028	C1028
Tortus II	Tortus II
Universal Walkway Tester	Tortus III
	BOT-3000
Variable Incidence Machines	**Articulated Strut Machines**
Brungraber Mark II	Brungraber Mark I
Brungraber Mark III	
English XL VIT	**Variable Incidence Machines**
Pendulum Machines	Brungraber Mark II
Sigler	Brungraber Mark III
Wessex	English XL VIT
	Pendulum Machines
	Sigler
	Wessex

There is concern that the USC research done to date is based on a single instrument of each type and that the results depended on that instrument being properly calibrated and operated. Also, the human subject group was made up only of young, able-bodied individuals and does not reflect the demographics of the general population, which can skew the baseline results against which tribometers are measured.

Regarding the 2007 workshop, there is also concern that the differences in slip resistance between flooring materials tested in the laboratory required significantly more precision than is needed in actual pedestrian slip resistance testing.

However, this work done is an excellent start. This is an ongoing project, and additional research is planned using more tribometers. For more information, see http://pt.usc.edu/labs/mbrl/research/slip.html.

EXHIBIT 4.1
Comparison Chart: U.S. Slip Resistance Test Methods

Apparatus	Tp	Yr	Av	W	U.S. Standard	P	Wt	L	Pad	OD	Cost ($)	More Information
James	AS	40s	Y	N	ASTM F489* ASTM D2047 UL 410	N	N/A	M	Leather, Silastic	H	13–23,000	http://www.cspa.org (plans for manual apparatus) http://www.michem.com/equipment/test-equipment/floor-care-testing-equipment http://www.astm.org (Committees F13.10, D21.06)
HPS	DS	60s	Y	N	ANSI/ASTM F609 ANSI A1264.2	Y	10	W	Neolite	M	1500	http://www.astm.org (Committee F13.10) http://www.trusty-step.com http://www.cscforce.com/slip.htm
Mark I (Past)	AS	70s	N	N	ANSI/ASTM F1678* ASTM F462 ANSI 1264.2	Y	Unk	W	Leather	L	N/A	Slip Test, Inc. (Sliptestin@aol.com) http://www.astm.org (Committees F13.10, F15.03) Patent 3,975,940—owned by the U.S. Government
C1028~	DS	80s	Y~	Y	ASTM C1028	Y	53	M	Neolite	H	300	http://www.astm.org (Committee C21.06)
Mark II (PIAST)	AS	80s	Y	Y	ANSI/ASTM F1677* NFPA 1901 ANSI A1264.2 OSHA 1926.754*	Y	Unk	W	Neolite	L	4000	Slip Test, Inc. (Sliptestin@aol.com) http://www.astm.org (Committee F13.10) Patents 4,759,209 and 4,798,080
English XL (VIT)	AS	90s	Y	Y	ASTM F1679*, D5859* NFPA 1901 ANSI A1264.2 OSHA 1926.754*	Y	4	W	Neolite	L	4700	http://www.englishxl.com http://www.astm.org (Committee F13.10) Patent 5,245,856

Notes: ~ There is no manufacturer for this device. It must be constructed. (1) manual propelled, (2) electrically propelled, (3) Jablonsky enhancement, and (4) computerized version (Michelman). *Withdrawn. Tp—AS: Articulated Strut, DS: Dragsled, P: Pendulum. Yr—Decade in which the apparatus was developed. Av—Commercially available. W—Approved by U.S. Standards for wet and dry testing. P—Portable. Wt—Approximate weight in pounds (field devices only). L—Listed for walkway surfaces or product merchantability. Pad—Primary test foot material. OD—Operator dependent (for velocity, force, angle, or complexity of setup)—Low, Moderate, High. Cost—Approximate Cost. Unk—Unknown.

EXHIBIT 4.2
Comparison Chart: Slip Resistance Test Methods Not Recognized in the United States

					Drag Sled Class Instruments			
Apparatus	A	Stnds	P	C	Pad	OD	Cost	More Information
					Motor Driven Drag Sled Instruments			
UFTM	N	None	Y	US	Various	MD	N/A	Rotational movement of two 1" diameter test feet, developed by NIOSH in the 1970s—never commercially produced
Topaka	N	None	Y	US	Paper		N/A	Williams Scientific Co., 1985
Kett 94 Ai	Y	None	Y	Jap	Brass		3700	Japanese Patent http://www.kett.co.jp/e/products/pro6/94i.html, http://www.kett.com/prod10.asp
In-tech Slip Meter	Y	None	Y	UK	Unknown		700	Allied Ltd—U.K. Patent applied for http://www.gripmeter.co.uk/index.html
Hoechst	N	None	N	Ger	Metal		N/A	
Tortus II, III	Y	AS/NZ	Y	UK	4S		3000	http://www.mastrad.com/
Sellmaier	Y	None	Y	Ger	Various		3500	Version of Tortus—http://www.pioneer-eclipse.com/products/other/other.html
BOT-3000	Y	None	Y	US	Various		5995	Version of Tortus—http://reganscientific.com/index.html
CEBTP Skidmeter	N	None	Y	Fr	Tire Rubber		N/A	A dragsled pull along two rails, controlled by a motor and gearbox
Welner	N	None	Y	US	Various		N/A	Patent 5,736,630 (April, 1998)
Majcherczyk	N	None	Y	Fr	Various		N/A	Patent 4,4081,989—owned by Centre Experimental de Recherches et d'Etudes du Batiment et des Travaux Publics
English UST	N	None	Y	US	Neolite		N/A	Prototype of a fully automated HPS, never commercially produced. Patent 4,8985,015 (January, 1990)

U.S. Tribometers

Apparatus	A	Stnds	P	C	Pad	HP	Cost	More Information
Hand Operated Drag Sled Instruments								
ASM 725	Y	None	Y	US	Neolite	Y	800	http://www.americanslipmeter.com
GMG 100	N	None	Y	Ger	Various	N/A	N/A	Manually pulled devices result in undue operator influence
Static Friction Tester	N	None	Y	US	Unk	N/A	N/A	Measurement Products Company (5-pound weight)
Schuster	Y	Non-US	Y	Ger	Unk	Unk	Unk	Unk
Model 80	Y	None	Y	US	Leather	Y	1000	Technical Products Company
PTI Drag Sled	N	None	Y	US	Various	N/A	N/A	Pennsylvania Transportation Institute (PTI) Tester was used in trials for ADA—never commercial produced, Bohdan T. Kulakowski, Ph.D., Penn State
Pazzaglia	N	None	Y	US	Various	N/A	N/A	Patent 5,245,856 (September, 1993)
Walker	N	None	Y	US	Various	N/A	N/A	Patent 2,225,140 (May, 1939)

Apparatus	A	Stnds	P	C	Pad	OD	Cost	More Information
Drag Sled Class Instruments								
Other Drag Sled Instruments								
Davies	N	None	Y	US	Various	CR	N/A	Patent 3,187,552 (June, 1965)
Surface Friction Analyzer	N	None	Y	US	Rubber	SP	N/A	Patent 3,828,605 (August, 1974), Fazekas—owned by Elias Productions

Notes: Tortus et al. (incl varieties such as Gabbrielli, FSC 2000, Tortus II), UFTM—Universal Friction Testing Machine. A—Commercially available. P—Portable. C—Country of origin. OD—Operator Dependent (HP—hand pulled, CR—cranked, and SP—spring driven are more operator dependent than MD—motor driven). Unk—Unknown. +—Dragsled class instruments are subject to stiction, and are invalid for wet testing.

Continued

EXHIBIT 4.2 (Continued)

Comparison Chart: Slip Resistance Test Methods Not Recognized in the United States

Apparatus	Type	A	Stnds	P	C	Pad	OD	Cost	More Information	Comments/Sensitivities
Other Apparatus										
Harrall FTM	Unk	N	None	Y	US	Paper	M	N/A	N/A	Patent 2,299,895 (October, 1942)
Hunter	AS	N	None	Y	US	Leather	H	N/A	N/A	Never commercially produced, see Chapter 3
Sigler	P	N	None	Y	US	TRL, 4S	H	N/A	N/A	Erratic readings on wet surfaces, no longer in production, velocity and mechanics do not relate to human ambulation
Frederik	Unk	N	None	Y	US	Various	M	N/A	N/A	Patent 2,955,455 (October, 1960)
BPST	P	Y	Non-US AS/NZ	Y	UK	4S	H	$5,000	http://www.munro-group.lco.uk/ http://www.wessexengineering.co.uk	Erratic readings on wet surfaces, velocity and mechanics do not relate to human ambulation, developed for road skid resistance
Stanley Wessex PFT (FIDO)	Unk	N	None	Y	Sw	Unk	H	N/A	http://www.tft.lth.se	Developed for highway skid resistance
Ramp Test	R	Y'	Non-US AS/NZ, DIN	N	Ger	Various	H	Unk	http://www.en.din.de	Incline walking not related to mechanics of level walking, highly subjective evaluation

U.S. Tribometers

Name	Type	P	Standard	A	C	Surface	OD	Price	URL	Notes
CSPA 0202	N/A	Y*	None	Y	US	Paper	N/A	N/A	http://www.cspa.org	Highly subjective assessment for comparison purposes only (see Chapter 5)
Slip Alert	RC	Y	None	Y	UK	4S	L	$5,750	http://www.slipalert.com/	A dynamic COF drag-sled
Kirchberg Rolling Slider	RC	N	None	N	UK	4S	L	N/A	N/A	Constructed by HSL according to the design described by Kirchberg et al.; has three rubber sliders arranged in a similar pattern to those on a GMG100 drag sled
SATRA PFT	Unk	N	None	Y	UK	4S	U	N/A	http://www.satra.co.uk/index.php/content/view/full/126	See Chapter 7 under SATRA
SATRA STM 603	Unk	Y	EN 13287	N	UK	Various	L	Unk	http://www.satra.co.uk/index.php/online_store/categories/test_equipment/slip_resistance/slip_resistance_tester	

Notes: Type: AS—Articulated Strut, DS—Dragsled, P—Pendulum, RC—Roller Coaster, R—Ramp. A—Commercially Available. P—Portable. C—Country of Origin. OD—Operator Dependent (for velocity, force, angle—or complexity of setup)—Low, Moderate, High, Unknown. Unk—Unknown. *—For more information on many of these instruments, see Chapter 7.

EXHIBIT 4.3
Neolite Patent

May 12, 1953   G. H. GATES   2,638,457
SHOE SOLE COMPOSITION
Filed Oct. 2, 1945

```
[MODIFIER]  [STYRENE]   [BUTADIENE]  [CATALYST]   [WATER]
                                                  [EMULSIFIER]
      ↓         ↓            ↓            ↓           ↓
    [STYRENE PHASE]  →   [EMULSION]  ←   [WATER PHASE]
                             ↓
                    [COPOLYMERIZATION]
                             ↓
                    [REACTION STOPPED] ← [ANTI-OXIDANT]
                             ↓
                          [LATEX]
                             ↓
                       [COAGULATION]
                             ↓
                         [WASHING]
                             ↓
                          [DRYING]
                             ↓
                [BUTADIENE-STYRENE COPOLYMER]
                             ↓
                  [BLENDING WITH A RUBBER]
                             ↓
                    [RESIN-RUBBER BLEND]
                             ↓
               [ADDITION OF CELLULOSIC FLOCK]
                             ↓
                      [COMPOUNDING]
                             ↓
                      [VULCANIZATE]
```

Inventor
George H. Gates

By

Attorney

EXHIBIT 4.3 (*Continued*)
Neolite Patent

Patented May 12, 1953

2,638,457

UNITED STATES PATENT OFFICE

2,638,457

SHOE SOLE COMPOSITION

George H. Gates, Cuyahoga Falls, Ohio, assignor to Wingfoot Corporation, Akron, Ohio, a corporation of Delaware

Application October 2, 1945, Serial No. 619,876

7 Claims. (Cl. 260—17.4)

This invention relates to a new vulcanized shoe sole composition in sheet form comprising a substantial and effective amount of an inorganic filler and a cellulosic flock material and as a binder therefor the vulcanizate resulting from the vulcanization of a mixture of rubbery copolymer and resinous copolymer of butadiene-1,3 and styrene.

The drawing is a flow chart showing the process of producing the blend and the vulcanizate of this invention.

In the manufacture of shoe soles, leather has been considered a more useful material than natural rubber, although rubber possesses qualities rendering it desirable for use in footwear. A material which possesses the desirable characteristics of leather, together with the desirable characteristics of rubber, and without the undesirable characteristics of either, would constitute a great advance in the manufacture of shoe soles and in large measure increase foot comfort.

It is well-known that leather is subject to deterioration by moisture because of the constant wetting and drying action that leather soles undergo which soon cause them to become stiff, thereby interfering with proper foot comfort. Rubber, on the other hand, is not so affected, but presents other disadvantages as a shoe sole; for example in the matter of the transmission of heat from the walking surface to the foot. Rubber does not possess sufficient stiffness for foot comfort unless excessively loaded with pigments. In an attempt to impart the necessary stiffness to rubber for shoe sole use, various pigments have been used, including carbon black, whiting, etc., but the proper degree of stiffness is not obtained until such a large amount of stiffener or pigment has been incorporated that the pigmented rubber becomes "dead" in the sense that it has little or no resiliency or springiness. The pigmented rubber is also too heavy for practical shoe sole use. Where carbon black is used as the stiffener, the highly pigmented rubber badly marks the surface walked upon.

Ordinarily, rubber, natural or synthetic, is too flabby or soft for shoe sole use without loading and too difficult to cement to the shoe uppers. Furthermore, when stitched, the sole tends to roughen out in a hill and valley effect due to the tension put upon the material by the stitches. Also, the stitches tend to cut through the material. Ultimately, sore feet may be caused by the creep characteristics of such a soft and flabby material. Again, shoe soles made of rubber cannot be made harder than about 75–80 Shore hardness by the use of ordinary rubber pigments because the resulting pigmented rubber has a tendency to flex-crack under higher pigment loadings.

It has now been discovered that the synthetic rubber from butadiene-1,3 and styrene, heretofore ordinarily unsuitable as a shoe sole material, may be made more desirable than leather for shoe sole use when the rubber is blended with a resinous diene-vinyl copolymer and compounded with a cellulosic flock material. When this rubber is blended and compounded with these materials, an unexpected and highly desirable result is produced in that the resulting material has an exceptional "leathery feel," together with the proper resilience and springiness.

The rubber component of the new composition is a rubbery copolymer resulting from the polymerization of a mixture containing as the essential copolymerizable monomers butadiene-1,3 and styrene or a substituted styrene and having an elongation of at least 300% and the ability to retract to substantially its original shape.

The resinous component of the blend may vary from a hard, brittle resin to a stiff, flexible material depending upon the ratio of butadiene-1,3 and vinyl aromatic monomer present in the mixture polymerized. This resinous characteristic is present in the product resulting from the reaction of a mixture of butadiene-1,3 and vinyl aromatic monomer when present in a ratio between about 30/70 and about 5/95; i. e., about 30 to 5 parts of butadiene-1,3 to about 70 to 95 parts of vinyl compound (e. g., styrene).

The vinyl aromatic monomer of the resin may also be referred to as being an alpha, beta unsaturated aliphatic substituted aromatic monomer, specific examples being styrene, the substituted styrenes, e. g., metachlorostyrene, p-chlorostyrene, isomeric dichlorostyrenes, alkyl styrenes, e. g., methyl styrene, isopropyl styrene, and vinyl naphthalene, etc. A particularly desirable class of these monomers is constituted by the vinyl aryl monomers, the desired member being styrene.

The following description illustrates generally the preferred conditions to be employed in producing the resin component. Generally the drawing shows that the copolymerization is carried out in the emulsion stage in which a styrene phase, including a modifier and styrene, is reacted with butadiene-1,3 in the presence of a water phase, including a catalyst, an emulsifier

EXHIBIT 4.3 (*Continued*)
Neolite Patent

2,638,457

3

and water. The copolymerization is continued at a temperature necessary to effect reaction of the styrene with butadiene until the desirable hydrocarbon conversion is reached. The resulting latex is then coagulated and the coagulum is washed and dried. Where it is desired to stop the reaction at a certain percent hydrocarbon conversion value, an antioxidant is introduced into the latex in sufficient amount to terminate the reaction and to protect the resulting coagulum against deterioration by oxidation.

A more specific example illustrates the copolymerization of a styrene phase, including 85 parts of styrene and 0.1 part of dodecyl mercaptan with 15 parts of butadiene in the presence of a water phase including 200 parts of water, 5 parts of sodium rosinate and 0.1 part of potassium persulfate at a temperature of 125° F. for a period of time to form a latex having a solids content of 32.5%, after which time 0.5% of phenyl beta-naphthylamine is introduced into the reaction. The latex is precipitated by the addition of a 3% commercial alum solution, the resulting coagulum being washed with water and dried to produce a resin having a softening point of 62° C.

Any suitable modifier may be added to the styrene phase. Suitable modifiers include those generally referred to as mercaptans containing at least six carbon atoms and particularly such mercaptans as isohexyl-, octadecyl-, and dodecyl mercaptan. Other desirable modifiers are the dialkyl polysulfides, the tertiary alkyl mercaptans, and the dialkyl xanthogen disulfides.

The modifier may be added in an amount between about 0.05% and about 3.0%, preferably in an amount between about 0.08% and about 2.0%, and it has been found particularly desirable to use about 0.1% in each case on the combined weight of the aryl and diene components used. The modifiers act to increase the solubility of the resulting resin in such solvents as benzene and toluene and also adjust the degree of stiffness of the resin. The greater the amount of modifier, the greater the solubility and the softer the resulting resin.

The water phase includes a catalyst, water and an emulsifier. Suitable catalysts that may be used are potassium persulfate, benzoyl peroxide, hydrogen peroxide, perborates, percarbonates. The catalyst may be used in an amount between about 0.1% and about 1%. Typical emulsifiers are those which may be generally referred to as the fatty acid soaps, e. g., sodium stearate and the rosin acid soaps, e. g., sodium rosinate, and the alkali metal salts of alkali sulfuric acid esters, e. g., sodium lauryl sulfate, and the alkali metal salts of alkyl aryl sulfonates, e. g., sodium dodecyl benzene sulfonate. The emulsifier may be present in amount between about 0.1% and about 5%.

Water is present in an amount based upon the total amount of monomers being reacted and may be used in a monomer/water ratio between about 100/60 to about 100/200. Generally, the copolymerization is carried out at a temperature between about 20° C. and about 70° C. for a period of time between about 4 hours and about 100 hours, depending upon the percent conversion desired, the charged monomer ratio, the catalyst used and the type of monomers present.

The butadiene-styrene rubber is blended with the resinous copolymer and compounded with a cellulosic flock material and with a substantial and effective amount of inorganic filling material and a vulcanizing agent and vulcanized to produce the shoe sole material of this invention.

4

A synthetic rubber which is admirably adapted for this blending and compounding is the one resulting from the copolymerization of butadiene-1,3 and styrene in which the butadiene is present in a predominant amount, and particularly where the butadiene-styrene ratio ranges from about 60/40 to about 85/15.

A desirable shoe sole material is produced when blending the resinous material with a rubber in a resin-to-rubber ratio from between about 5/95 to about 75/25.

A typical blending formula is one in which about 100 parts of the rubber is blended with between about 25 to about 100 parts of the resin and between about 10 and about 25 parts of a cellulosic flock material. This composition may then be further compounded with suitable accelerators, softeners, sulfur and fillers to produce a vulcanizable shoe material which may then be vulcanized and cured at a temperature between about 300° F. and about 340° F. for a period of time between about 10 minutes and about 1 hour to produce a shoe sole material having the desirable characteristics mentioned hereinbefore.

A representative compounding formula is as follows:

	Parts by weight
Rubber	100
Diene-vinyl resinous material	25–100
Cellulosic flock material	10–25
Sulfur	2–3
Organic accelerator (Tuads, Captax, etc.)	1–2
Zinc oxide	3–5
Filler (clay, whiting, silene, etc.)	60–100
Stearic acid	0.5–1
Softener (paraffin, pine tar, asphalt derivatives, etc.)	5–20

In the formula above, Tuads is tetramethyl thiuramdisulfide, Captax is mercaptobenzothiazole and silene is a hydrated calcium silicate.

The usual blending operation is conducted in a Banbury mill in which the rubber is first added, and after a few minutes breakdown the resin is added in a small amount, together with the sulfur or other desirable rubber accelerator and the zinc oxide, until the batch is heated to a temperature of about 200° F. and about 225° F. The flock material, softener, filler and stearic acid may then be added and the milling continued until a thoroughly mixed mass has been obtained.

This material may then be extruded into a sheet of desirable thickness for shoe sole use and cured in the usual manner to give a vulcanized material which then may be used in uncut form for shoe soles or may be precut into standard shoe sole sizes.

It has been observed that the presence of the cellulosic flock material is essential in producing a shoe sole composition having the desirable properties mentioned. For example, the presence of this flock material improves the non-skid properties of the composition on icy and wet surfaces, and materially reduces the tendency for the composition to "grow" or "spread," by which quoted terms is meant that the dimensional stability of the sole composition has been improved. The flock also improves the "feel" of the composition, by which quoted term is meant that the composition has a "feel" of leathery smoothness which is desirable in a shoe sole material. The flock also prevents flex-cracking of the blend alone. The flock may be added in an amount between about 10 parts to about 25 parts

EXHIBIT 4.3 (*Continued*)
Neolite Patent

2,638,457

per 100 parts of rubber used and it is preferred to use between about 15 parts and about 25 parts. Any cellulosic flock material may be used, including cotton and rayon flocks. However, it is preferred to use the rayon flock.

Suitable changes may be made in the details of the process without departing from the spirit or scope of the present invention, the proper limits of which are defined in the appended claims.

I claim:

1. A composition in sheet form comprising a filler including about 10 to about 25 parts of a cellulosic flock material together with a substantial and effective amount of mineral filler, and as a binder therefor the vulcanizate resulting from the vulcanization of (1) a rubbery copolymer of a major proportion of butadiene and a minor proportion of styrene, (2) 25 to 100 parts of a hard, thermoplastic resin obtained by copolymerizing a mixture of 70-95 weight percent of styrene and 30-5 weight percent of butadiene in aqueous emulsion in the presence of a substance which promotes the solubility of the resulting resin in benzene and selected from the group consisting of mercaptans containing at least six carbon atoms, dialkyl polysulfides, and dialkyl xanthogen polysulfides, and (3) sulfur as a vulcanizing ingredient for the binder, the parts being by weight per 100 parts by weight of rubbery copolymer.

2. A composition in sheet form comprising a filler including about 10 to about 25 parts of a cellulosic flock material together with a substantial and effective amount of mineral filler, and as a binder therefor the vulcanizate resulting from the vulcanization of (1) a rubbery copolymer of a major proportion of butadiene and a minor proportion of styrene, (2) 25 to 100 parts of a hard, thermoplastic resin obtained by copolymerizing a mixture of 70-95 weight percent of styrene and 30-5 weight percent of butadiene in aqueous emulsion in the presence of a substance which promotes the solubility of the resulting resin in benzene and being an alkyl mercaptan having 6 to 18 carbon atoms, and (3) sulfur as a vulcanizing ingredient for the binder, the parts being by weight per 100 parts by weight of rubbery copolymer.

3. A composition in sheet form comprising a filler including about 10 to about 25 parts of a cellulosic flock material together with 60 to 100 parts of mineral filler, and as a binder therefor the vulcanizate resulting from the vulcanization of (1) a rubbery copolymer of a major proportion of butadiene and a minor proportion of styrene, (2) 25 to 100 parts of a hard, thermoplastic resin obtained by copolymerizing a mixture of 70-95 weight percent of styrene and 30-5 weight percent of butadiene in aqueous emulsion in the presence of a substance which promotes the solubility of the resulting resin in benzene and selected from the group consisting of mercaptans containing at least six carbon atoms, dialkyl polysulfides, and dialkyl xanthogen polysulfides, and (3) sulfur as a vulcanizing ingredient for the binder, the parts being by weight per 100 parts by weight of rubbery copolymer.

4. A shoe sole comprising a filler including about 10 to about 25 parts of a cellulosic flock material together with a substantial and effective amount of mineral filler, and as a binder therefor the vulcanizate resulting from the vulcanization of (1) a rubbery copolymer of a major proportion of butadiene and a minor proportion of styrene, (2) 25 to 100 parts of a hard, thermoplastic resin obtained by copolymerizing a mixture of 70-95 weight percent of styrene and 30-5 weight percent of butadiene in aqueous emulsion in the presence of a substance which promotes the solubility of the resulting resin in benzene and selected from the group consisting of mercaptans containing at least six carbon atoms, dialkyl polysulfides, and dialkyl xanthogen polysulfides, and (3) sulfur as a vulcanizing ingredient for the binder, the parts being by weight per 100 parts by weight of rubbery copolymer.

5. A shoe sole comprising a filler including about 10 to about 25 parts of a cellulosic flock material together with a substantial and effective amount of mineral filler, and as a binder therefor the vulcanizate resulting from the vulcanization of (1) a rubbery copolymer of a major proportion of butadiene and a minor proportion of styrene, (2) 25 to 100 parts of a hard, thermoplastic resin obtained by copolymerizing a mixture of 70-95 weight percent of styrene and 30-5 weight percent of butadiene in aqueous emulsion in the presence of a substance which promotes the solubility of the resulting resin in benzene and being an alkyl mercaptan having 6 to 18 carbon atoms, and (3) sulfur as a vulcanizing ingredient for the binder, the parts being by weight per 100 parts by weight of rubbery copolymer.

6. A shoe sole comprising a filler including about 10 to about 25 parts of a cellulosic flock material together with 60 to 100 parts of mineral filler, and as a binder therefor the vulcanizate resulting from the vulcanization of (1) a rubbery copolymer of a major proportion of butadiene and a minor proportion of styrene, (2) 25 to 100 parts of a hard, thermoplastic resin obtained by copolymerizing a mixture of 70-95 weight percent of styrene and 30-5 weight percent of butadiene in aqueous emulsion in the presence of a substance which promotes the solubility of the resulting resin in benzene and selected from the group consisting of mercaptans containing at least six carbon atoms, dialkyl polysulfides, and dialkyl xanthogen polysulfides, and (3) sulfur as a vulcanizing ingredient for the binder, the parts being by weight per 100 parts by weight of rubbery copolymer.

7. A shoe sole composition in sheet form comprising a filler including 10 to 25 parts of a cellulosic flock material, together with a substantial and effective amount of mineral filler and, as a binder therefor the vulcanizate resulting from the vulcanization of a mixture of (1) a rubbery copolymer resulting from the polymerization of a mixture containing 60 to 85 weight percent of butadiene-1,3 and 40 to 15 weight percent of styrene, (2) 25 to 100 parts of a hard, thermoplastic resin obtained by copolymerizing a mixture of 70-95 weight percent of styrene and 30-5 weight percent of butadiene in aqueous emulsion in the presence of a substance which promotes the solubility of the resulting resin in benzene and being an alkyl mercaptan having 6 to 18 carbon atoms, (3) 2 to 3 parts of sulfur, (4) 1 to 2 parts of an organic accelerator, (5) 3 to 5 parts of zinc oxide, (6) 0.5 to 1 part of stearic acid, and (7) 5 to 20 parts of a softener, the parts being by weight and per 100 parts by weight of rubbery copolymer.

GEORGE H. GATES.

(References on following page)

EXHIBIT 4.3 (*Continued*)
Neolite Patent

2,638,457

7
References Cited in the file of this patent

UNITED STATES PATENTS

Number	Name	Date
2,029,371	Hickler	Feb. 4, 1936
2,039,529	Guinzburg	May 5, 1936
2,393,208	Waterman et al.	Jan. 15, 1946
2,414,803	D'Alelio	Jan. 28, 1947
2,419,202	D'Alelio	Apr. 22, 1947
2,452,999	Daly	Nov. 2, 1948
2,477,316	Sparks et al.	July 26, 1949
2,541,748	Daly	Feb. 13, 1951

8
FOREIGN PATENTS

Number	Country	Date
345,939	Great Britain	Mar. 16, 1931

OTHER REFERENCES

Rubber Age, November 1947, page 200.
Modern Plastics, February 1947, pp. 100–102.
Vanderbilt 1942, Rubber Handbook, pp. 140–149, published 1942 by R. T. Vanderbilt Co., N. Y.
India Rubber World, January 1945, page 422.
India Rubber World, February 1945, p. 590.

5 U.S. Standards and Guidelines

The body pays for a slip of the foot, and gold pays for a slip of the tongue.

Malaysian proverb

5.1 INTRODUCTION

Although ASTM is the most active in development of pedestrian safety related standards in the U.S., it is by no means the only one. Beyond the work of ASTM, a variety of federal government documents (e.g., Occupational Safety and Health Administration [OSHA], military, and federal specifications) and other consensus standards (e.g., NFPA International, American National Standards Institute [ANSI]) are available, in addition to work by private industry.

When referring to U.S.-based regulations, standards, and specifications, it is important to understand several aspects of these documents such as the perspective and biases of the developer, the stated and intended focus of the document (which oftentimes is quite narrow), and the method by which the document came into existence (see Exhibit 7.1, National Standards Bodies).

5.2 OCCUPATIONAL SAFETY AND HEALTH ADMINISTRATION (OSHA)

OSHA refers to four separate slip resistance standards, each promulgated at different times and with varying degrees of enforceability. This information can be reviewed on OSHA's website at http://www.osha.gov.

5.2.1 SECTION 1910.22 GENERAL REQUIREMENTS

Recommendation: Federal Register, April 1, 1990, Vol. 55, No. 69, p. 13408, 29 CFR, Walking and Working Surfaces and Personal Protective Equipment (Fall Protection Systems); Notice of Proposed Rulemaking—Section 1910.22 General Requirements (Appendix A to Subpart D Compliance Guidelines).

1. Surface conditions.
2. Slip resistance. A reasonable measure of slip resistance is static coefficient of friction (COF). A COF of 0.5, which is based upon studies by the University of Michigan and reported in "Work Surface Friction Definitions,

Laboratory and Field Measurements, and a Comprehensive Bibliography," is recommended as a guide to achieve proper slip resistance. A COF of 0.5 is not intended to be an absolute standard value. A higher COF may be necessary for certain work tasks, such as carrying objects, pushing or pulling objects, or walking up or down ramps.

Slip resistance can vary from surface to surface, or even on the same surface, depending upon surface conditions and employee footwear. Slip-resistant flooring material such as textured, serrated, or punched surfaces and steel grating may offer additional slip resistance. These types of floor surfaces should be installed in work areas that are generally slippery because of wet, oily, or dirty operations. Slip-resistant-type footwear may also be useful in reducing slipping hazards.

The OSHA 1910 "standard" for slip resistance is not a law or a standard; it is a proposed nonmandatory appendix item that has yet to be adopted as a standard. It specifies a slip resistance of 0.50 or higher for the workplace. Nonmandatory appendices neither add to nor detract from the obligations contained in the OSHA standards. OSHA clarified this provision in a 2003 standard interpretation letter (Fairfax, 2003). In part, it states:

> OSHA does not have any standards that mandate a particular COF for walking/working surfaces. While there are devices to measure the COF, no OSHA standard specifically requires that employers use or have them. As you may know, there is a nonmandatory appendix (Appendix A to Subpart D) in the Notice of Proposed Rulemaking for Walking Working Surfaces for general industry that discusses COF. Although the notice of proposed rulemaking was published on April 10, 1990, the final rule has not yet been issued.

Although OSHA inspectors have been known to cite this specification under the General Duty Clause, it is not a frequent occurrence. Applying this section can be difficult, because OSHA has yet to specify an apparatus and test protocol upon which to base a citation, effectively making it of questionable enforceability.

5.2.2 MANLIFTS 1910.68(c)(3)(v)

> Surfaces. The upper or working surfaces of the step shall be of a material having inherent nonslip characteristics (coefficient of friction not less than 0.5) or shall be covered completely by a nonslip tread securely fastened to it.

This reference specifies no apparatus and test protocol as a basis for a citation and, thus, is an arguably unenforceable regulation.

5.2.3 FIRE BRIGADES 1910.156(e)(2)(ii)

OSHA requirements for fire brigades in 1910.156(e)(2)(ii) call for Class 75 footwear that is water resistant for at least 5 in. above the bottom of the heel and equipped with slip-resistant outer soles.

Protective footwear shall be water-resistant for at least 5 inches (12.7 cm) above the bottom of the heel and shall be equipped with slip-resistant outer soles.

5.2.4 Appendix B to 1926 Subpart R—Steel Erection Regulatory (3) [Withdrawn]

Slip resistance of skeletal structural steel. Workers shall not be permitted to walk the top surface of any structural steel member installed after five years after effective date of final rule that has been coated with paint or similar material unless documentation or certification that the coating has achieved a minimum average slip resistance of 0.50 when measured with an English XL tribometer or equivalent tester on a wetted surface at a testing laboratory is provided. Such documentation or certification shall be based on the appropriate ASTM standard test method conducted by a laboratory capable of performing the test. The results shall be available at the site and to the steel erector. (Appendix B to this subpart references appropriate ASTM standard test methods that may be used to comply with this paragraph (c)(3)).

Appendix B to Subpart R—Acceptable Test Methods for Testing Slip-Resistance of Walking/Working Surfaces (1926.754(c)(3)). Non-Mandatory Guidelines for Complying with 1926.754(c)(3).

The following references provide acceptable test methods for complying with the requirements of 1926.754(c)(3).

- *Standard Test Method for Using a Portable Inclineable Articulated Strut Tester (PIAST)(ASTM F1677–96)*
- *Standard Test Method for Using a Variable Incidence Tribometer (VIT)(ASTM F1679–96)*

OSHA announced that the final Steel Erection Standard would be effective on January 18, 2002, and fully implemented within five years. However, in January 2006, OSHA revoked the provision addressing the slip resistance requirement in response to comments that indicated several test methods were unlikely to be completed by the July effective date.

The standard addressed the hazards identified as major causes of injuries and fatalities in the steel erection industry. The slip resistance provision was not intended to be the sole means of protecting workers from fall hazards, but to complement other requirements of the standard as part of an overall strategy for reducing fall-related injuries and fatalities.

The provision required that coated structural steel meet a specified level of slip resistance when measured using ASTM test methods. The ability to comply with the slip resistance provision depended upon: (1) the completion of industry protocols for slip testing equipment, and (2) the availability of suitable slip-resistant coatings. The lack of completed test methods delayed the development of suitable slip-resistant coatings. In addition, there was not adequate testing of coatings for durability, especially in corrosive environments.

A thorough discussion on the rationale and science that resulted in these specifications is located in the OSHA Structural Steel Standard Commentary in Exhibit 5.1.

5.3 AMERICANS WITH DISABILITIES ACT (ADA)

Public Law 101–336, 7/26/90. Federal Register Vol. 56, No. 144, Chapter A4.5.1, Friday, July 26, 1991 Rules & Regulations (http://www.usdoj.gov/crt/ada/adahom1.htm)

> A Federal court in Florida returned indictments against eight persons on charges of conspiracy to commit mail fraud and money laundering involving a multi-million dollar telemarketing scheme.
>
> Hundreds of people spent millions of dollars buying useless slip-resistant chemicals, which actually can make surfaces more dangerous. The "skid-resistant" adhesive chemicals would purportedly meet ADA standards for slip resistance. The investors, who paid between $10,000 and $14,000 to store the chemicals, were told that company salesmen would sell the chemicals to businesses in their area and they would make $5 or more on each gallon sold.
>
> After an investor purchased the chemicals, the sham company would fake small orders of about $100 each for a few months. By the time a victim realized the fraud, the defendants were hidden behind layers of sham corporations and could not be located. (U.S. Department of Justice, 1996)

Over twenty-seven million Americans report some difficulty in walking. Of these, eight million have a severe limitation; one-fifth of this population is elderly. Ambulatory persons with mobility impairments are particularly at risk of slipping and falling even on level surfaces (Access Board, Technical Bulletin #4, 2003).

The ADA provisions for slip resistance, like OSHA, were developed as a guideline, since the specification appears in an appendix. Because no test protocol, apparatus, or test foot material is specified, it is difficult to apply them. The ADA specified slip resistance of at least 0.60 for level surfaces and 0.80 for ramps, where accessible by persons with disabilities.

This is clearly stated in a 2003 Access Board bulletin shown below:

> It is impossible to correctly specify a slip-resistance rating without identifying the testing method, tester, and sensor material to be used in evaluating the specified product and equally invalid to compare values obtained through one methodology to those resulting from different testing protocols. (Access Board, 2003)

These thresholds were established based on a flawed two-phase research sponsored by the Architectural and Transportation Barriers Compliance Board (ATBCB). The first phase of the study involved five able-bodied individuals walking at fast speeds in dissimilar footwear over a clean, dry force plate and the use of a detergent-wetted NBS-Brungraber (PAST, or Mark I) with a Silastic 382 rubber test pad (similar to ASTM F462 for bathtub evaluation). In another phase of the study, nine mobility-disabled subjects walked across a force plate. Since this study had such a limited scope, it was inappropriate to arrive at any thresholds for the general population (Fendley and Marpet, 1996).

According to correspondence from the Assistant Attorney General John R. Dunne dated January 19, 1993:

> There are no enforceable standards for coefficients of friction in the regulations. The Appendix to the Guidelines, which is advisory only, discusses recommended coefficients

U.S. Standards and Guidelines   165

of friction in section A4.5.1 (page 35,678) . . . the recommended coefficients of friction are provided only as advisory guidance and not as regulatory requirements.

On November 16, 1999, the slip resistance specifications of 0.6 and 0.8 were removed. Although Americans with Disabilities Act Accessibility Guidelines (ADAAG) published new guidelines in 2004, they are as of this writing still under review and not yet incorporated into the ADA. The original comment period closed in May 2005. The Department of Justice (DOJ) plans to issue another notice and comment period before finalizing adoption of new standards. The DOJ indicates that it could take up to a "couple years" to complete this process. Until this time, the original ADA standards remain in effect. The 2004 ADAAG guidelines do not cite an acceptable method for assessing walkway slip resistance even though they indicate that the surface should be slip resistant. (Access Board, 2003)

It is important to understand that ADA requirements were promulgated for the benefit of physical impaired individuals, making their relevance to the ambulation of able-bodied persons arguable.

5.3.1 ADAAG 4.5 Ground and Floor Surfaces/A4.5.1

A4.5—Ground and floor surfaces along accessible routes and in accessible rooms and spaces including floors, walks, ramps, stairs, and curb ramps, shall be stable, firm, slip-resistant, and shall comply with 4.5.

A4.5.1—People who have difficulty walking or maintaining balance or who use crutches, canes, or walkers, as well as those with restricted gaits, are particularly sensitive to slipping and tripping hazards. For such people, a stable and regular surface is necessary for safe walking, particularly on stairs. Wheelchairs can be propelled most easily on surfaces that are hard, stable, and regular. Soft, loose surfaces such as shag carpet, loose sand or gravel, wet clay, and irregular surfaces such as cobblestones can significantly impede wheelchair movement.

Slip resistance is based on the frictional force necessary to keep a shoe heel or crutch tip from slipping on a walking surface under conditions likely to be found on the surface. Whereas the dynamic coefficient of friction during walking varies in a complex and nonuniform way, the static COF, which can be measured in several ways, provides a close approximation of the slip resistance of a surface. Contrary to popular belief, some slippage is necessary to walking, especially for persons with restricted gaits; a truly nonslip surface could not be negotiated.

OSHA recommends that walking surfaces have a static COF of 0.5. A research project sponsored by the Access Board (ATBCB) conducted tests with persons with disabilities and concluded that a higher COF was needed by such persons. A static COF of 0.6 is recommended for accessible routes and 0.8 for ramps.

It is recognized that the COF varies considerably due to the presence of contaminants, water, floor finishes, and other factors not under the control of the designer or builder and not subject to design and construction guidelines, and that compliance would be difficult to measure on the building site. Nevertheless, many common building materials suitable for flooring are now labeled with information on the static COF. Although it may not be possible to compare one product directly with another, or to guarantee a constant measure, builders and designers are encouraged to specify materials with appropriate values. As more products include information on slip resistance, improved uniformity in measurement and specification is likely. The ATBCB's advisory guidelines on slip-resistant surfaces provide additional information on this subject.

5.4 ACCESS BOARD RECOMMENDATIONS

The ATBCB (http://www.access-board.gov/adaag/about/bulletins/text/surfaces.txt) of the U.S. Department of Justice was created to ensure federal agency compliance with the Architectural Barriers Act (ABA). The ATBCB adopted the ADA recommendations and has stated several specifications as simply a guideline, not a requirement or a standard.

In a 2003 bulletin, the ATBCB clearly stated the following:

> Researchers' recommendations for a static coefficient of friction for surfaces along an accessible route, when measured by the NBS-Brungraber machine using a silastic sensor shoe, were approximately 0.6 for a level surface and 0.8 for ramps. These values are included in the advisory material in the Appendix to ADAAG, but are not in any way mandatory. (Access Board, 2003)

5.5 FEDERAL SPECIFICATIONS

5.5.1 RR-G-1602D

RR-G-1602D is the Federal Specification for Grating, Metal, Other than Bar Type (Floor, Except for Naval Vessels). Originally published in 1970 (current version dated February 29, 1996), this specification is a procedure for testing the slip-resistant characteristics of metal gratings used in floor surfaces within military and General Services Administration (GSA) applications. Threshold values range from approximately 0.34 to 0.49 depending on the material (e.g., aluminum, steel), condition (e.g., dry, mud, ice, grease, detergent), and test foot material (e.g., leather, boot rubber, shoe rubber, Neolite®, and Hypolon) where such flooring is subject to foot traffic. The test method specified is a homemade equivalent to a C1028 instrument, using a 3-in. diameter sole material, a dead weight of 175 lbs, and a *"continuous recorder to record the resistance offered by the grating to the material."* Section 3.7 provides thresholds of "anti-slip," and Section 4.4.3 specifies the test materials and method. These specifications can be obtained at http://www.everyspec.com/FED+SPECS/RR-G-1602D-10643.

5.6 U.S. MILITARY SPECIFICATIONS (NAVY)

The following specifications specify the use of a horizontal pull slip tester.

5.6.1 MIL-D-23003A(SH)

This military specification refers to Deck Covering Compound, Nonslip, Rollable.

5.6.2 MIL-D-24483A

This military specification refers to Nonslip Flight Deck Compound.

5.6.3 MIL-D-0016680C (Ships) and MIL-D-18873B

These military specifications are for *Deck Covering Magnesia Aggregate Mixture*. These specifications both provide static coefficient of friction (SCOF) and dynamic coefficient of friction (DCOF) thresholds of 0.3 to 0.6 (against dry leather and rubber, respectively) and 0.4 to 0.6 (against wet leather and rubber). They specify the use of either generic inclined plane or dragsled-type instruments.

5.6.4 MIL-D-3134J

This military specification for Deck Covering Materials (1988), outlines procedures for testing under dry, wet (with 4% salt), and oily (SAE 10W) conditions using leather and rubber test pads. Although the test method options are the same as 18873B (see Section 5.6.3), the thresholds vary. It can be obtained at http://www.wbdg.org/ccb/FEDMIL/d3134j.pdf.

5.6.5 MIL-D-17951C (Ships)

This military specification written in 1975 for Deck Covering, Lightweight, Nonslip, Silicon Carbide Particle Coated Fabric, Film, or Composite, and Sealing Compound, *supersedes* the 1961 version. Thresholds are similar to those of 18873B (see Section 5.6.3).

5.6.6 MIL-W-5044C

This military specification, written in 1970 for Walkway Compound, Nonslip, and Walkway Matting, Nonslip, supersedes the 1964 version. Describing a homemade dragsled device, it specifies a variety of thresholds, depending on the type of surface, the condition (e.g., dry, water, oil), and the type of slider (e.g., leather, rubber). Thresholds range from 0.45 to 1.00.

5.7 ASTM INTERNATIONAL (FORMERLY AMERICAN SOCIETY FOR TESTING AND MATERIALS)

Organized in 1898, ASTM is one of the largest voluntary standards development organizations in the world. ASTM is a nonprofit organization with more than 32,000 members from more than 100 countries. Due to this international scope, the organization was renamed ASTM International in 2001 (http://www.astm.org).

ASTM develops standards in 130 areas and covers subjects such as metals, paints, plastics, textiles, construction, energy, the environment, consumer products, medical devices, electronics, and many others. More than 11,000 ASTM standards are published annually, all of which are "full consensus" documents (see Exhibit 7.2, Approaches to Development of Standards).

Standards development begins when members of a committee identify a need, or other interested parties approach the committee. From there, task group members prepare a draft standard which is reviewed by its parent subcommittee through a letter ballot. After the subcommittee approves the document, it is submitted concurrently to the main committee and the general membership of ASTM. Members are provided the opportunity to vote on each standard. Negative votes cast during the balloting process, which must include a written explanation of the voters' objections, must be fully considered prior to submittal to the next level in the process. Final approval of a standard depends on concurrence by the ASTM COS (Committee on Standards) that proper procedures were followed and due process was afforded.

The number of voting producers on a committee cannot exceed the combined number of voting nonproducers (e.g., users, ultimate consumers, and those having general interest); this is intended to prevent any one party from dominating any aspect of the process. If a member feels the standards development process has been unfair, appeal options are available within the Society.

ASTM has been the most active U.S. standards-making organization in the development of slip resistance-related standards.

Note: ASTM standards cited herein are subject to reapproval and revision. As such, it is important to consult the most current edition of these standards.

5.7.1 Technical Committee ASTM F-13

In the 1970s, ASTM Committee E-17 Skid Resistance included subcommittee E17.26 Methods of Measuring Pedestrian Friction, which did a substantial amount of research into slip resistance measurement, including a symposium on Pedestrian Friction in 1977 (and the subsequent Special Technical Publication known as STP 649).

Long since disbanded, the work of subcommittee E17.26 continued with ASTM Committee F-13. The title of the ASTM F-13 technical committee was Safety and Traction for Footwear. Established in 1973 and originally created to develop standards relating only to footwear, its name was misleading because its scope also included safety and traction for walkway surfaces, as well as practices related to the prevention of slips and falls. The committee changed its name to Pedestrian/Walkway Safety and Footwear in February 2003 to more closely align it with the scope shown below:

> ... to develop consensus standard test methods, guidelines, practices, definitions, criteria, and nomenclature for pedestrian safety, for the fit and function of footwear in relation to interfaces, walking surfaces and devices used for the evaluation of pedestrian traction. (ASTM F-13)

F-13, one of ASTM's larger committees at approximately 300 members, is divided into several subcommittees, one of which is F13.10 Traction, which has jurisdiction over methods of measuring pedestrian slip resistance.

Other standards relating to the measurement of pedestrian slip resistance are the responsibility of other ASTM committees, although these are more focused on test methods intended to validate the merchantability of lines of products instead of evaluation of in service walkway surfaces. Until recently, no single focal point of slip resistance standards existed, even within ASTM. The C21 Ceramic Tile task group was free to develop a standard that it felt was needed within the industry, and the D21 Polishes committee did the same. Over the span of many years, the committees closely guarded these standards which eventually permeated their respective industries. At the same time, as our knowledge and technology have progressed, the usefulness of these standards has diminished.

To streamline the process of standardization in this area, ASTM formally designated Technical Committee F-13 as the only source for the development of slip resistance test methods. The rationale was that F-13 was well positioned in terms of scope and expertise to develop generic test methods for slip resistance. Other committees with specialized needs (e.g., ceramic, resilient, polish, paint) could then reference the appropriate F13 standard and craft documents that included supplementary information (e.g., specimen preparation, thresholds, limits) for their application.

In addition to slip meter standards (see Chapter 4, U.S. Tribometers), F-13 has a number of other standards relating to pedestrian safety.

5.7.1.1 ASTM F695

ASTM F695-01 is the *S*tandard Practice for the Evaluation Test Data Obtained by Using the HPS or the James Machine for Measurement of Static Slip Resistance of Footwear Sole, Heel, or Related Materials.

Under the jurisdiction of F13.50 (Walkway Surface Practices subcommittee of F13), this document provides a method for a comparative ranking of test results performed using test methods in F489 and F609. It is also referenced by ASTM G115-04 Standard Guide for Measuring and Reporting Friction Coefficients, which is under the jurisdiction of G02 Wear and Erosion, residing in subcommittee G02.50 Friction.

5.7.1.2 ASTM F1240

ASTM F1240-01 is a Guide for Categorizing Results of Footwear Slip Resistance Measurements on Walkway Surfaces with an Interface of Various Foreign Substances.

The purpose of this standard, also under F13.50, is to assist in selecting appropriate footwear in areas where foreign materials can contribute to slips and falls. There are plans to combine these into a single document in the near future due to the similarity of F1240 and F695. The work to begin updating and combining these with F695 began in 2002; this effort is still underway.

5.7.1.3 ASTM F1637

ASTM F1637-02 is the Practice for Safe Walking Surfaces standard.

This standard addresses common walkway surface defects and design considerations, including where slip-resistant surfaces should be provided. This document is the responsibility of F13.50. Although brief, this document contains useful guidelines on walkway surface design (see Chapter 1, Physical Evaluation).

5.7.1.4 ASTM F1646

ASTM F1646-07 is the standard for Terminology Relating to Safety and Traction for Footwear.

This standard, a document of F13.91 (the terminology subcommittee of F-13), defines commonly used terms relating to slip resistance and footwear.

5.7.1.5 ASTM F1694

ASTM F1694-96 (2004) is the Standard Guide for Composing Walkway Surface Investigation, Evaluation and Incident Report Forms for Slips, Stumbles, Trips and Falls.

Under the jurisdiction of F13.50 and updated in 2008, this standard provides recommendations for recording walkway surface investigation, evaluation, and incident report data pertaining to slips, trips, stumbles, and falls. It is intended to provide guidance in the development of custom reporting systems. This guide provides suggested content appropriate for inclusion into a questionnaire or report. Exhibits include a field investigation guide for slips, stumbles, trips, fall incidents, a sample walkway evaluation form, and a sample incident report form.

5.7.1.6 ASTM F802

ASTM F802-83 (2003) is the Standard Guide for Selection of Certain Walkway Surfaces When Considering Footwear Traction.

Under the jurisdiction of F13.50, this guide is intended to assist in the selection of walkway surfaces where the presence of foreign materials may increase the potential for a slip or fall.

5.7.1.7 ASTM F2048

ASTM F2048-00 is the Standard Practice for Reporting Slip Resistance Test Results.

First published in 2000, this F13.10 standard provides guidance on how to document the results of slip resistance testing. F2048 includes a sample format which can be used to collect this information.

5.8 NFPA INTERNATIONAL (FORMERLY NATIONAL FIRE PROTECTION ASSOCIATION)

NFPA International (NFPA) is a nonprofit organization founded in 1896 (http://www.nfpa.org). With more than 75,000 members, NFPA's mission is to reduce fire and other hazards by developing and advocating scientifically based consensus codes and standards. NFPA maintains more than 300 consensus codes and standards in 250 technical committees with more than 6,000 members, many of which are used as a basis for legislation and regulation nationally and internationally.

NFPA operates on a full consensus basis, meaning that committee balance (to prevent one interest from dictating the committee's agenda), transparency (public review of draft documents), and consensus are required elements in the process.

5.8.1 NFPA 1901

NFPA 1901 is the Standard for Automotive Fire Apparatus. This standard (Section 13-7.3) recognizes the use of the variable incidence tribometer (VIT, or English XL™) for surfaces used for stepping, standing, or walking on new mobile fire apparatus. The minimum average slip resistance requirements are shown below:

- 0.68 for exterior surfaces (wet) when measured per ASTM F1679 (English XL VIT). It was subsequently amended to also permit the use of the portable inclinable articulated strut slip tester (PIAST, or Brungraber Mark II), with thresholds of 0.52 (wet) when measured per ASTM F1677.
- 0.58 for interior surfaces (dry) when measured per ASTM F1679 and 0.47 (dry) when measured per ASTM F1677.

5.8.2 NFPA 101/5000

The 2006 edition of NFPA 101 (and 5000, NFPA's building code), updated every three years, is as vague as the other building codes with regard to slip resistance. However, the appendices of each of these documents (101-A7.1.6.4 and 500-A12–1.6.4) specify that ASTM F1679 (VIT) and ASTM F1677 (PIAST) are acceptable methods for measuring the slip resistance of walkway surfaces and stair treads.

These standards require that walkway surfaces be slip resistant under "foreseeable conditions." The appendices also make it clear that foreseeable conditions are defined as those which are likely to be present given the operations or activities normally conducted in a given area. An example of a foreseeable condition is of a swimming pool deck/area which is expected to be wet.

5.9 AMERICAN NATIONAL STANDARDS INSTITUTE (ANSI)

ANSI, a nonprofit organization, has been the administrator and coordinator of U.S. voluntary standards for more than 90 years. The ANSI organization does not itself develop standards. It facilitates development by conferring the authority to develop standards to qualified organizations. More than 175 entities are accredited by ANSI to develop consensus standards. ANSI conducts scheduled audits to assure the standards process is being observed by accredited organizations. Currently, ANSI includes more than 1,000 members and has developed more than 11,000 standards.

ANSI is a founding member of and the official U.S. representative to the International Organization for Standardization (ISO) (see Chapter 7, Overseas Standards). ANSI participates in 78% of all ISO technical committees and is responsible for accrediting U.S. Technical Advisory Groups (U.S. TAGs), which develop and communicate U.S. positions on activities and ballots of the international technical committee. ANSI is also a member of the Pan American Standards Commission (COPANT).

ANSI standards are to be withdrawn if not revised or reaffirmed within five years unless an extension has been granted. ANSI standards that are not revised or reaffirmed within ten years are automatically withdrawn.

5.9.1 ANSI A1264.2-2006

A1264.2-2006 is the Standard for the Provision of Slip Resistance on Walking/Working Surfaces.

The American Society of Safety Engineers (ASSE) acts as secretariat for this ANSI standard, which was officially adopted in August 2001 and was updated in 2007. It differs from other slip resistance standards (such as those from ASTM International and Underwriters Laboratories) in that it is oriented to workplaces instead of public places, and it specifies a numeric threshold of safety for walking. It is the only standard aside from the James Machine standard for polished surfaces (ASTM D2047) to set a minimum value.

The goal of the standard is to provide criteria to businesses in order to reduce the potential for employee slips and falls. Although it can be considered a state-of-the-art standard, revisions to this standard will continue to be necessary as advances in the field are made. The standard provides for the minimum performance requirements necessary for increased safety on walking/working surfaces in the workplace. Since many workplaces are also subject to pedestrian foot traffic from the public, the standard can also deliver similar benefits in reducing the potential for public liability.

In some instances, the standard refers the user to other ANSI and ASTM documents for further details. Due to the broad scope and many specifications provided or referenced in A1264.2, support from a knowledgeable and experienced safety consultant can provide constructive and valuable assistance. Three basic areas are addressed by the standard: provisions for reducing hazards, test procedures and equipment, and a slip resistance guideline.

Provisions for reducing hazards
- Footwear applications and considerations, including traction properties and considerations for the work environment
- Floor mats and runners, including location, installation, inspection and maintenance, storage and care, and cleaning and trade-out
- Housekeeping and maintenance procedures, including requirements of training, staff supervision, and floor monitoring
- Pre- and post-incident warnings, including signage/symbols, and placement
- Controlled access, including barricades and containing hazards
- Selection of walkway surface materials, including discussion of floor finishing and treatment options
- Snow and ice control and removal

Test procedures and equipment
- ASTM International test methods
- Dry surfaces equipment
- Test equipment for wet or contaminated surfaces

Slip resistance guideline (dry)
- Acceptable test methods
- Test foot material specifications
- A threshold of safety guideline for dry walkway surfaces of 0.5

U.S. Standards and Guidelines   173

5.9.2   ANSI A1264.3-2007

Released in 2008, *ANSI/ASSE TR-A1264.3-2007 Using Variable Angle Tribometers (VAT) for Measurement of the Slip Resistance of Walkway Surfaces* was developed by the A1264.2 subcommittee of the A1264 ASC Safety Standards for Floor and Wall Openings, Railings, and Toeboards and Fixed General Industrial Stairs. It was developed to provide guidance for the performance of slip resistance testing of in-place walkway surfaces, such as sidewalks, catwalks in factories, grocery store floors, and floors in commercial kitchens.

The report covers the technical aspects, research, legislation, standards activities and operation of the two widely used VATs commercially available for testing of walkway surface slip resistance, the Brungraber Mark II and the English XL (see Chapter 4, U.S. Tribometers).

5.9.3   ANSI A137.1-1988

This ANSI standard, Specifications for Ceramic Tile, is a guide for manufacturing or specifying four types of tile. It references ASTM C1028 as the method for determining the slip resistance of glazed wall, ceramic mosaic, quarry, and paver tile. This standard is currently only available through the Tile Council of North America (TCNA, see Section 5.13.2). Their ANSI 108 Committee is secretariat for this standard, which as of this writing has a 2007 draft under review. Per the A137.1 2007 draft, even though minimum and maximum requirements are set forth for the SCOF of various materials, they are for practical purposes regardless of the test results.

A review of the October 2007 roster of the ANSI A108 Committee showed the vast majority of members are tile manufacturers, distributors, contractors, and associations representing manufacturers, with minimal representation from actual users or general interest groups. There are many irregularities in the classification of members. A review of voting members indicates that 55% are manufacturers, distributors, or associations representing both manufacturers and distributors. ANSI regulations requires that no more than 50% of a technical committee should be from the manufacturing interest group, to reduce the potential for "sweetheart" standards being developed that are easily met and serve primarily to enhance perception in the marketplace. By establishing classifications in which distributors are classified as users, it becomes easier to obtain a voting majority and gain control of a committee.

5.10   UNDERWRITERS LABORATORIES

Established in 1894, Underwriters Laboratories Inc. (UL) is an independent, non-profit product safety testing and certification organization (see http://www.ul.com and http://ulstandardsinfonet.ul.com [standards search]). UL maintains more than 1,200 standards.

Until recently, the UL standards process was closed; only those specifically selected and invited to participate were involved. In an effort to be recognized as a true consensus standards-making organization, UL has recently elected to adopt ANSI standards development procedures.

UL now forms Standards Technical Panels (STPs), consisting of knowledgeable and interested parties with open meetings and drafts made available for public comment prior to publishing. Essentially, standards are developed by UL using the ANSI canvass method instead of the more comprehensive full-consensus method (see Exhibit 7.2, Approaches to Development of Standards). UL is the largest ANSI-accredited standards developer using the canvass method. As a result of using this new approach, UL documents are eligible for ANSI approval. Presently, the UL method of consensus does not appear to afford complete transparency or a suitable means of universal public comment, because documents proposed for approval must be purchased from UL in order to be reviewed.

5.10.1 UL 410

UL 410 is the standard for Slip Resistance of Floor Surface Materials (as Measured by the James Machine).

Many products claim to be classified or certified by UL for slip resistance. Although UL does not "certify" products with respect to slip resistance, it does "classify" products as such. Around 1970, UL began the classification of some products instead of listing them as a way for UL to increase revenue through fees.

UL applies the treatment to a surface and tests it to UL 410 using the James Machine. To be permitted to use the UL classification marking, test results must meet or exceed the specified slip resistance threshold, and the manufacturer must pay a fee. In 1953, UL stopped releasing actual test results to applicants. UL currently issues documentation only as to whether or not the results have exceeded the threshold. The average slip resistance threshold is set at a minimum of 0.5 for these products. The slip resistance for an individual surface is set at a minimum of 0.45.

A classification mark appears on products that UL has evaluated for specific properties, a limited range of hazards, or suitability for use under limited or special conditions. In the case of slip resistance, the product will display this notice:

CLASSIFIED BY UNDERWRITERS LABORATORIES INC.® AS TO SLIP RESISTANCE ONLY.

This is a product standard that applies to the following types of floor surfaces and treatments:

- Floor covering materials made of wood or composite materials
- Water-based floor treatment materials
- Fillers, sealers, varnishes, and similar floor treatment materials
- Detergent materials
- Abrasive-grit-bearing floor treatment materials
- Floor treatment materials other than water-base
- Sweeping compound materials
- Walkway construction materials used as floor plates, ramps, and stair treads that are made of natural stone, composite materials, abrasive-grit surface materials, and metal

U.S. Standards and Guidelines 175

The scope of UL 410 is significantly broader than ASTM D2047, in that it is intended for use for virtually all types of walkway surfaces and finishes. It is important to understand that the intent of this standard was to provide for consistent laboratory testing under ideal conditions. According to James himself, the instrument was designed only as a method to compare the results of a treated surface material to that of the same material untreated. While James discussed the 0.5 threshold, it did not appear to be his intent to use the instrument or the threshold as a means to categorically differentiate a slip-resistant surface from a non-slip-resistant surface.

In its current form, UL 410 has numerous drawbacks that impede its viability as a valid test method (see Chapter 4, U.S. Tribometers). They include:

- No provisions are recommended for calibration of the instrument. This is an essential step for the manually propelled version of the instrument, the only and most operator-dependent version covered under this document.
- The document permits wet and contaminated testing, although the literature repeatedly demonstrates that the apparatus is subject to stiction.
- Unlike ASTM, UL documents require no testing to validate the protocol or apparatus. Thus, there is no precision and bias statement (P&B) (nor to date, has one been planned) to establish the accuracy or repeatability of the test method.
- The document calls for the use of leather, a highly variable substance inappropriate for wet or contaminated testing. Leather is not a homogenous material; each piece is unique. Sanded leather often results in a change in frictional properties. Once used for wet testing, its properties are permanently altered. Leather can react differently depending on how worn the material has become. UL 410 does not even specify standard leather (government specification KK-L-165), allowing the introduction of yet another variable which is unaccounted.
- With one exception, the substrates, or floor materials on which floor treatments are tested, are not standardized; this adds yet another variable.

The following limitations must also be kept in mind when considering a UL classification per slip resistance:

- The James Machine is a laboratory device designed to test under ideal conditions. The test cannot test in-service flooring and is suited to new product materials only, thus making its primary and most appropriate use for the merchantability of new flooring materials. Use of this standard should be restricted only to quality control testing, to determine dry or static coefficient of friction rather than slip resistance.
- The James Machine was not designed for wet testing or for materials other than those with smooth, untextured surfaces.
- There are no known published studies or other scientific documentation supporting the validity of the James Machine using the prescribed combination of materials, contaminants, and test feet. Despite repeated requests over many years, neither the research data of Sidney James, nor that of

flooring manufacturers has been made public to support the 0.5 COF and its correlation with pedestrian safety.
- The scope of UL 410 includes the testing of floor treatments, duplicating an existing ANSI standard: ANSI/ASTM D-2047.
- Calculations in the design do not account for the added weight of the substantial struts, thereby introducing an uncorrected variable.
- Given that the substrate, the tile used to test the treatment, is often 0.5 to start with in its untreated condition, the treatment need only be neutral and does not have to degrade the slip resistance in order to pass the test.

Among others, UL made two notable changes in their standards development procedures:

- A majority of the committee is no longer required to return an affirmative vote.
- No quorum is needed at committee meetings. Approval can be obtained by a simple majority vote of STP members in attendance.

In 2003, UL unsuccessfully balloted UL410 to become an ANSI standard. A detailed outline of the issues can be reviewed online in a document from the American Society of Safety Engineers (ASSE) in their September 2003 response at http://www.asse.org/publications/standards/sansi/sansi26.php.

5.11 MODEL BUILDING CODES

Primary specifications relating to slip resistance and dimensional criteria for pedestrian safety are contained in Section 10—Means of Egress (see Chapter 1, Physical Evaluation). Whereas each of these codes requires that walkway surfaces and stair treads be slip resistant, none provide requirements or even guidance on acceptable thresholds or methods of assessing slip resistance.

5.12 OBSOLETE STANDARDS

5.12.1 FEDERAL TEST METHOD STANDARD 501A, METHOD 7121

This is the standard for Floor Covering, Resilient, Non-Textile; Sample and Testing.

First published in June 1966 and withdrawn many years ago, this government test method specified the Sigler pendulum device to measure dynamic COF for nearly 20 years. The scope of the document was the measurement of DCOF for resilient nontextile floor coverings with relatively smooth surfaces using rubber (Standard Reference Compound No. 3 in method 14111 of Federal Test Method Standard No. 601, for rubber test heel and leather [Federal Specification KK-L-165]) (http://www.fss.gsa.gov/pub/fedspecs/sort2.cfm).

5.12.2 U.S. GENERAL SERVICES ADMINISTRATION SPECIFICATION PF-430C(1)

This is the Finish, Floor, Water Emulsion standard.

This government specification, in place for many years, required that floor finishes meet a 0.5 static COF requirement using the James Machine. It was canceled in July 1999 (http://www.fss.gsa.gov/pub/fedspecs/sort5c.cfm).

5.12.3 ASTM D4518-91

ASTM D4518-91 is the Standard Test Method for Measuring Static Friction of Coating Surfaces. This standard was under the jurisdiction of D01 Paints. Originally published in 1985, this standard has since been withdrawn. It covered two separate methods: an Inclined Plane Test and the Horizontal Pull Test.

5.12.4 ASTM D-21 GRAY PAGES

In the early 1980s, ASTM Technical Committee D21 was unable to pass several test methods for alternatives to the James Machine. The methods were subsequently published in the gray pages of the Book of Standards (V 15.04):

- Proposal P 125 Test Method for Static and Dynamic Coefficient of Friction of Polish-Coated Floor Surfaces as Measured by the Horizontal Slip Tester (see ASTM F609)
- Proposal P 126 Test Method for Static Coefficient of Friction of Polish-Coated Floor Surfaces as Measured by the NBS-Brungraber Articulated Arm Tester (see ASTM F1678)
- Proposal P 127 Test Method for Dynamic Coefficient of Friction of Polish-Coated Floor Surfaces as Measured by the NBS-Sigler Pendulum Impact Tester (see Chapter 7, Overseas Standards)
- Proposal P 128 Test Method for Static Coefficient of Friction of Polish-Coated Floor Surfaces as Measured by the Topaka Slip Tester—A dragsled class of tribometer, this apparatus uses a 110-volt motor to pull a 5 in. × 8 in. (13 cm × 20 cm) canvas bag containing five pounds of lead shot across a walkway surface at 3.3 in. (8 cm) per second. Bond paper (25% rag content) is specified as test foot material instead of a footwear-bottom-related material. Similar to the HPS (F609), the weight is connected to the motor by a nylon cord and is fitted with three ½-inch diameter "rubber buttons" that contact the paper. No standard exists in the U.S. or abroad, although the instrument is now reportedly sold by a vendor of floor finish systems and is promoted as a valid test method.

5.13 U.S.-BASED INDUSTRY ASSOCIATIONS AND ORGANIZATIONS INVOLVED WITH SLIP RESISTANCE

5.13.1 CERAMIC TILE INSTITUTE OF AMERICA (CTIOA)

Based in Culver City, California, the stated mission of the CTIOA (http://www.ctioa.org/index.cfm?pi=FSI), is to "promote appropriate and expanded use of ceramic tile and natural stone through education." The CTIOA is an industry group representing the interests of the ceramic tile industry. Membership in this small organization

TABLE 5.1
CTIOA-Endorsed Slip Resistance Thresholds

Instrument	Test Foot	Condition	Locations	Threshold
Tortus	4S rubber	Wet	Level floors	0.50
			Bathtubs, showers, pool decks	0.70
Pendulum	4S rubber	Wet	Level floors	35 BPN
	TRRL rubber		Showers, pool decks	

consists of three categories: tile contractors (13), tile manufacturers and distributors (18), and associates (29). Some tile manufacturers and slip resistant-floor treatment vendors refer to meeting or exceeding requirements as required by CTIOA.

Historically, CTIOA has specified the use of ASTM C1028, which is a test method only and does not specify a level of safety. C1028 cannot be exceeded, because it does not specify safe or unsafe levels of slip resistance. However, CTIOA adopted the James Machine guideline of 0.50 even though no known correlation exists between these instruments.

The CTIOA maintains a Ceramic Tile Industry Technical Committee which is unaffiliated with any approved standards development organization. Part of the committee's mission is to identify needed representation on ASTM and ANSI committees that involve ceramic tile or its installation. Several years ago, CTIOA released a series of position papers endorsing the use of non-U.S. slip resistance test methods: the ramp, the Tortus, and the pendulum (see Chapter 7, Overseas Standards), in lieu of the C1028. CTIOA endorses the thresholds listed in Table 5.1. No thresholds are indicated for inclined surfaces such as ramps.

Although no published peer-reviewed research is cited, the CTIOA primarily references draft ISO standard 10545-17, a series of personal communications, and its own paper to support this position.

The CTIOA also offers floor safety testing surfaces, including testing for coefficient of friction to ASTM C-1028, surface roughness, Tortus testing (dry and wet), pendulum testing (dry and wet), and expert witnessing.

In an undated bulletin (http://www.ctioa.org/reports/cof19.html), the CTIOA endorsed the use of a single commercial product, Non-Slip 21 treatment for slippery floors. Their justification only cited that the science staff of a "major Canadian university" conducted a "confidential investigation" of floor treatments. Based on the fact that Non-Slip 21 was the only product that did not have "detrimental effects" on the tested surfaces, they elected to endorse the product publicly. There was no further discussion of the ability of the product to improve traction or reduce slips and falls, only that it did not damage the flooring.

For more information on these test methods, see the Chapter 7, Overseas Standards.

5.13.2 TILE COUNCIL OF NORTH AMERICA (TCNA)

The TCNA (http://www.tileusa.com/), based in Anderson, South Carolina, is a trade association representing manufacturers of ceramic tile, tile installation materials, tile equipment, raw materials, and other tile-related products. According to the TCNA, the Tile Council, established in 1945 as the Tile Council of America (TCA), was created, "with the sole purpose of expanding the ceramic tile market in the United States." In 2003, the TCA became the TCNA to reflect its membership expansion throughout North America.

Based on an industry forum and consensus, the TCA publishes literature on nationally recommended ceramic tile installation specification guidelines and ANSI standards on ceramic tile and installation systems. For more than 58 years, the TCA has worked to expand the market of American-made ceramic tile and related installation products, tools, and raw materials. The TCA lab conducts ANSI, ISO, and ASTM testing to assess a product or system's ability to meet high performance criteria. TCA members include:

- Ceramic tile and accessory manufacturers
- Art/studio tilemakers
- Raw material and equipment suppliers
- Associate installation materials
- Affiliated products, including providers of specialized products, tools or equipment, treatment or maintenance products, or metal or resin tiles

The TCNA provides consulting and product testing services for ceramic tile. The TCNA Laboratory, a wholly owned subsidiary of the TCNA, offers testing for tile, stones, and installation materials per ASTM, ANSI, and ISO standards. They use ASTM C-1028 for coefficient of friction testing, as well as the BOT-3000 apparatus for static and dynamic coefficient of friction testing. No test method or protocol for the BOT-3000 is specified.

The TCA is secretariat for a number of ANSI standards for ceramic tile under ANSI Committee 108, including ANSI A137.1-1988.

5.13.3 NATIONAL FLOOR SAFETY INSTITUTE (NFSI)

After considerable lobbying before several ASTM technical committees, from the mid 1990s through 2006, the NFSI was unable to garner support to develop standards based on various incarnations of the Tortus (see Chapter 7, Overseas Standards). The NFSI was started in 1997 by the principal of Traction Plus, a company that markets a variety of floor treatments, a proprietary tribometer (a predecessor of the BOT 3000), slip-resistant footwear, wet floor signs, and other related products.

NFSI proceeded to develop its own standards, product certification, and walkway auditing program outside of a recognized consensus standards process. With membership in the organization, access to their database of information on slips and falls is provided, reportedly obtained from governmental databases, private industry loss data, and insurance company loss data. No assessment of the validity and usefulness of this data have been done by an independent authority.

In 2006, NFSI obtained accreditation to be an ANSI standards developer. It formed the ANSI B101 Committee and several subcommittees, and began the process of adopting existing NFSI standards (as previously written by the NFSI founder and an unnamed subcontractor) as ANSI standards.

There are ongoing objections from the ASTM Technical Committee F-13 and the ASSE on behalf of the ANSI A1264 Committee on the basis that the work of NFSI's ANSI committee is duplicative of longstanding ANSI standards promulgated by both organizations. ANSI regulations do not permit duplication.

Other concerns regarding the activities of NFSI include the following:

- The failure to make public any studies or other research upon which the standards are based.
- When first balloted, one document included substantial portions taken verbatim from two in force ANSI standards (ANSI/ASTM F-1637 and ANSI A1264.2) without permission.
- In violation of ANSI and NFSI's own regulations, which state that all meetings are open to interested parties, the author was barred access as an observer to any NFSI ANSI subcommittees.

5.13.3.1 NFSI Position on Slip Resistance Testing

One issue of concern involves the NFSI position on slip resistance test methods.

There is only one manufacturer for the device upon which NFSI standards are based, and there is no compelling reason for an exception to the ANSI policy on proprietary devices. Regan Scientific Instruments (formerly UWT, LP) markets the BOT-3000 in the U.S. The BOT-3000, formerly the Universal Walkway Tester (UWT), is yet another Tortus variation.

The following raises questions regarding the objectivity of specifying a proprietary device by default:

- NFSI and Traction Plus (the company in which the NFSI executive director and committee secretary have financial interests) have endorsed this instrument.
- It is the only test method permitted to obtain NFSI Product Certification.
- It is the only device for which NFSI has a Walkway Auditor Certification Program.

The scope of the interlaboratory study (ILS) report completed on the BOT-3000 was limited to dry testing, yet an NFSI document in development specifies the use of this instrument for wet testing. Although the report claims that the ILS performed for the BOT-3000 met the requirements of ASTM E-691, it fails to do so in numerous areas. The study:

- For portable test equipment, ASTM E691 describes the laboratory as the person and equipment doing the test. There were four BOTs used, so there were four laboratories, not the eight people claimed.

- E691 describes the material as the substance being tested. Two substances were tested, not the four claimed in the study, in which each instrument was called a "material."
- E691 specifies that materials to be tested reflect a broad range of results, yet the two materials used, formica and Teflon, have similar expected COF values.

This means that four "laboratories" and two similar "materials" were tested, which is insufficient to meet the criteria of E691. In addition, all testing was performed dry.

As of this writing, work on these and other standards are still under way.

5.13.4 Consumer Specialty Products Association (CSPA)

Headquartered in Washington, D.C., and formerly the Chemical Specialty Manufacturers Association, the Consumer Specialty Products Association (CSPA) (http://www.cspa.org) is the industry trade group representing the makers of formulated products for home and commercial use. Formed in 1914, CSPA members total more than 220 and include companies engaged in the manufacture, formulation, distribution, and sale of consumer specialty products for household, institutional, and industrial use. CSPA member companies account for more than $14 billion in annual revenues.

The CSPA Polishes & Floor Maintenance Division represents more than 80% of manufacturers, marketers, and suppliers of waxes, polishes, floor finishes, and a wide range of floor maintenance products. Its stated mission includes: ". . . influencing the direction and scope of regulations and legislation pertaining to floor care products," and is a ". . . liaison with national and international organizations to monitor regulatory and technological issues."

Under the CSPA Polishes and Floor Maintenance division is the Scientific and Regulatory Affairs Committee, a subcommittee of which is the "Walkway Safety Subcommittee." ASTM D3758, Standard Practice for Evaluation of Spray-Buff Products on Test Floors, refers to CSMA Bulletin Number 245–70, Comparative Determination Slip Resistance of Floor Polishes. This test method, established in 1970, is a subjective relative ranking of slip resistance, comparing coatings to another of accepted value. In 2002, this test method was redesignated as CSPA 0202. The test calls for the individual conducting the test to slide a clean sheet of writing paper 8½ in. × 11 in. (21 cm × 28 cm) or longer ("such as conventional office stationary") across the floor with a shod foot, and rate the surface from one ("much less slip resistance") to five ("much greater slip resistance"). The method concludes by stating the obvious: "No estimation of the precision or accuracy has been developed for this test method."

In a non sequitur, the bulletin also states that CSPA considers any floor polish tested by ASTM D2047 that results in a static COF of at least 0.5 to meet the requirements of the proposed FTC Rule 5. This would make it permissible to promote the product as slip resistant (see Chapter 3, Principles of Slip Resistance).

It is important to remember that this test method was not developed by a consensus of a balance of interested parties, but by an industry trade group with a mission to promote its members' products. CSPA is also a supplier, in some cases the only known supplier of the following:

- Official Vinyl Composition Tile (OVCT), required as a calibration surface for testing under ASTM D2047, and now available under the name TM1, Vinyl Composition Tiles for Wax Testing
- Leather conforming to Federal Specification KK-L-665C, also required for the ASTM D2047 test method
- James Machine charts, a basic requirement for testing under any standard using this apparatus

5.13.5 RESILIENT FLOOR COVERING INSTITUTE (RFCI)

The total membership of RFCI consists of six of the largest manufacturers of resilient flooring. RFCI (http://www.rfci.com) is an industry trade association of North American manufacturers. RFCI was established to support the interests of the resilient floor covering industry. The stated objectives of RFCI include the following:

- Promoting the use of resilient floor covering as a product category
- Monitoring and responding to federal, state, and local legislation and regulations that affect the industry and its products

Most standards-making activities that interest RFCI are under the jurisdiction of ASTM Technical Committee F06, Resilient Floor Coverings. However, the issue of slip resistance test methods developed in ASTM F13 is also considered important enough to prompt substantial RFCI involvement. RFCI has conducted research regarding slip resistance test methods, but their work is not made public. As with other similar organizations, it must be understood that work done by RFCI is not performed by a balance of interested parties, but by an industry trade group with a mission to promote its members' products.

5.13.6 FOOTWEAR INDUSTRIES OF AMERICA (FIA)

The FIA became part of the American Apparel and Footwear Association (AAFA) (www.americanapparelandfootwear.org) in 2000, when it merged with American Apparel and Manufacturers Association. A national trade association representing apparel, footwear, sewn products companies, and their suppliers, the mission of AAFA is to promote and seek to "enhance members companies' competitiveness, productivity and profitability in the global marketplace."

The purpose of the FIA Slip Resistant Footwear Committee is to study and evaluate the slip-resistant qualities of footwear soling materials by testing with the four most recognized testing machines under varied conditions and surfaces. Task force groups are studying the following segments of slip resistance:

- Outsole compound and design
- Environment and conditions

U.S. Standards and Guidelines   183

- Testing methods
- Statistical analysis and data research

Members of the FIA automatically become members of the world's largest footwear research and development organization, the SATRA Footwear Technology Centre based in Kettering, England. SATRA has a contractual arrangement with FIA to provide technical information and products and services (see Chapter 7, Overseas Standards).

5.13.7 Contact Group on Slips, Trips, and Falls (CGSTF)

The CGSTF (http://www.safeworkresearch.com) is intended to provide an international forum for discussion on scientific matters related to slips, trips, and falls accidents. It is a source of information, news, and discussion on scientific aspects of slips, trips, and falls. Of particular value is the list of consensus standards and peer-reviewed publications by year on this topic.

5.13.8 International Ergonomics Association (IEA)

In 2007, the IEA created a technical committee on Slips, Trips, and Falls (STF). The goals of the committee are to:

- Facilitate communication among members around the world
- Help organize international STF symposia at major international conferences
- Coordinate activities among regional organizations and individuals
- Promote collaboration among members and organizations

The committee addresses injuries due to falls from heights, falls on the same level, trips, and loss of balance in occupational and leisure settings. Members include researchers, safety professionals, and practitioners. In addition, the committee maintains a distribution list of interested nonmembers.

The first conference sponsored by this committee was held in August 2007. The goals of the conference were to provide a technical forum on slip, trip, and fall accidents for the safety researchers and consultants around the world. This forum would allow them to present their latest findings and to exchange research ideas. This conference covered all aspects related to the problem from the research perspective to practical experience such as biomechanics, slip resistance measurements, standardization, pedestrian fall prevention, flooring and stairway design and development, design of footwear, anti-skid devices, and housekeeping.

5.13.9 National Safety Council (NSC)

The NSC (http://www.nsc.org) was founded in 1913 and chartered by Congress in 1953. Its mission is "to educate and influence society to adopt safety, health and environmental policies, practices and procedures that prevent and mitigate human suffering and economic losses arising from preventable causes." The Council is a nonprofit, nongovernmental, international public service organization dedicated to improving the safety, health, and environmental well-being of all people. As a nongovernment organization, the NSC does not legislate or regulate.

The boards and their officers and committees, aided by some 2,000 more volunteers, determine policies, operating procedures, and programs to be developed and carried out by the council's professional staff of more than 300. Board members represent business, labor, council chapters, government, community groups, trade and professional associations, schools, and individuals. Members of the NSC include 18,600 companies of all sizes from a broad spectrum of industries, representing 33,300 locations and 8.5 million employees around the world.

The NSC maintains more than 20 publications, programs, posters, and other materials related to slip and fall prevention, most of which consist of basic information.

5.13.10 AMERICAN ACADEMY OF FORENSIC SCIENCES (AAFS)

With more than 5,000 members, the AAFS (http://www.aafs.org/) consists of ten sections representing a wide range of forensic specialties. The Engineering Sciences Section deals with ergonomic and human biomechanical related issues, including slips and falls.

The AAFS publishes the Journal of Forensic Sciences, an internationally recognized scientific journal. AAFS also offers newsletters, an annual scientific meeting, and seminars. AAFS holds its annual scientific meeting each February, where more than 500 scientific papers, seminars, workshops, and other special events are presented. Professionals gather to present the most current information, research, and updates in this expanding field.

Through its journal and the annual meeting, which occasionally has a track of sessions devoted to slips and falls, AAFS has been the forum for many peer-reviewed research projects related to slip resistance and fall prevention.

EXHIBIT 5.1
OSHA Report on Slip Resistance for Structural Steel

5214 FEDERAL REGISTER/VOL. 66, NO. 12/THURSDAY, JANUARY 18, 2001/RULES AND REGULATIONS (PP. 5214–5218)

Paragraph (c)(3) will reduce the risk of steel erection workers slipping on coated steel members installed three years after the effective date of this standard. At that time, it will prohibit employees from walking on the top surface of any structural steel member that has been coated with paint or similar material, unless the coating has achieved a minimum average slip resistance of 0.50 when wet on an English XL tribometer, or the equivalent measurement on another device. This paragraph does not require that the particular coated member be tested. Rather, it requires the test to be done on a sample of the paint formulation produced by the paint manufacturer. The testing laboratory must use an acceptable ASTM method, and an English XL tribometer or equivalent tester must be used on a wetted surface, and the laboratory must be capable of employing this method. The test results must be available at the site and to the steel erector. Appendix B lists two appropriate ASTM standard test methods that may be used to comply with the paragraph. If other ASTM methods are approved, they too are allowed under this provision.

The final paragraph differs from the proposal in two significant respects. Proposed paragraph (c)(3) would have prohibited employees from walking on the top surface of any structural steel member with a finish coat that decreased the coefficient of friction (CoF) from that of the uncoated steel.

The final text sets a specific slip resistance for the coated surface, when tested wet. In addition, proposed paragraph (c)(3) stated that the paragraph applied to coated steel installed at the effective date of the standard, rather than, as in the final, three years later.

THE HAZARD

Based on SENRAC's (Steel Erection Negotiations Rule Advisory Committee) discussions and the rulemaking record, OSHA finds that working on steel surfaces coated with paint or other protective coatings presents slip and fall hazards to employees and that this standard must reduce this hazard using feasible means. SENRAC described the hazards as the use of paint or coatings on steel for structures exposed to highly corrosive materials (such as those used in mills and chemical plants) or exposed to varying weather conditions (such as stadiums). In the proposal, OSHA set out SENRAC's concerns as follows:

The Committee found that a major cause of falls in the steel erection industry is the presence of slippery walking, working, and climbing surfaces in steel erection operations when fall protection is not used. The problem initially arises from the application of protective coatings on structural steel used, for example, in the construction of mills, chemical plants, and other structures exposed to highly corrosive materials as well as in the construction of stadiums or other structures exposed to varying weather conditions. It is usually impractical to leave the steel uncoated and then to paint the entire structure in the field after erection. Unfortunately, steel coated with paints or protective coatings can be extremely slippery. When there is moisture, snow, or ice on coated steel, the hazard is increased . . . (63 FR 43467).

As discussed below regarding § 1926.760, accident data in this record demonstrate that falls from elevations of 30 feet or less resulted in many ironworker injuries and fatalities. In addition, the Agency recognizes that slips on the same level also lead to many injuries. We believe that provisions to reduce the slip potential of surfaces walked on by steel erection workers are clearly needed. OSHA and SENRAC examined the factors involved in slippery surfaces and determined that the most effective and feasible approach is to increase slip resistance and allow employees to walk on only those coated surfaces which meet a threshold for acceptable slip resistance. Much of the discussion in this rulemaking involves issues regarding which slip-resistant threshold to set; whether it is feasible to measure it; and whether compliance with such a provision is technically and economically feasible.

Commenters affirmed the existence of a serious hazard from coated surfaces; many asserted that slick or slippery paint is very dangerous (Exs. 13–49, 13–66, 13–95, 13–345, 13–348, and 13–355B). Most of these commenters (Ex.13–66 and a group of 124 ironworkers in Ex. 13–355B) added that slippery paint is the worst condition they run into on structural steel, and they asked that the paint be made safe. Other ironworkers (Ex. 13–355B) asserted that epoxy paint was hazardous to erectors. All together, 230 of these ironworkers commented in support of a provision to make painted steel less slippery. A comment from a structural steel fabricator (Ex. 13–228) stated that they agreed that "painted [steel], moist or wet, is slipperier."

In contrast to the comments asserting that coated surfaces present a slipping hazard, a comment from an engineer for a state government agency (Ex. 13–359) stated that slippery surfaces were attributable to a variety of causes, such as weather conditions, which can reduce traction on coated or uncoated surfaces (Ex. 13–359). He added that there was no basis for the requirements that addressed a CoF in subpart R "since there are no accepted methods for determining friction at the job site and tests would not be relevant to site conditions." In addition, the American Iron and Steel Institute Steel Coalition submitted a consultant's report asserting that it is not really necessary to know a CoF in evaluating pedestrian traction, and that it is important to rate the traction under various relevant conditions (Ex. 13–307A, pp. 24–25).

In response to the first concern that slippery surfaces are attributable to a variety of causes, OSHA points out that requiring less slippery coatings in no way suggests that employers should ignore other unsafe conditions. The general construction standard for training § 1926.21 requires employers to "instruct each employee in the recognition and avoidance of unsafe conditions . . ." This includes slipping hazards due to factors such as moisture from weather conditions and unsafe footwear. OSHA agrees however, with its expert witnesses, William English, David Underwood and Keith Vidal, who stated in their report, that "contaminants" (including rain water, condensation and ice) and shoe bottom construction are important factors, but are

U.S. Standards and Guidelines

not as easily controlled as surface coatings (Ex. 17, p. 2). Also, the rule will require wet testing, thus accounting for most weather-related slip hazards.

In response to the second concern that it is not really necessary to know a CoF in evaluating traction, the final rule text does not set a required CoF—the 0.50 measurement is a slip resistance measurement for the walking surface. While related to CoF (a ratio of forces), the 0.50 referred to in the final rule is a measurement on a tester that is designed to mimic (to some extent) the dynamic forces involved in walking on a surface. While different types of shoe material (and different amounts of wear) affect the amount of traction experienced by the worker, the record shows that it is not feasible to establish a requirement that would account for all the factors that relate to the CoF. Nor would it be feasible to measure slip resistance at the site under the numerous and ever-changing "relevant conditions." The English reports and testimony of English, Underwood and Vidal (as discussed below) shows that setting a requirement for the walking surface (when wet) will improve traction.

A commenter suggested that OSHA focus on ironworkers' footwear rather than specifying a slip resistance for the paint (Ex. 13–307A, pp. 2–5). The Agency finds that this type of approach would not work as a substitute for addressing the slip resistance of the paint because ironworkers' footwear typically become contaminated with mud, gravel, and other substances that would alter the slip resistance characteristics of the sole material (Exs. 203X, p. 213 and 204X, p. 292).

Other commenters recommended that only uncoated surfaces be allowed to be erected (Exs. 13–41, 13–138 through 13–142, 13–234, and 13–341). The record does not demonstrate that uncoated steel is necessary for employee safety since surface coatings can provide equivalent or greater protection against falls. Also, SJI [Steel Joint Institute] identified several significant problems with requiring the steel to be uncoated when erected. Among these would be increased costs associated with painting the steel in the field after it was erected, which it estimated would amount to $450 to $800 million, and a slowing of the construction process by two to four weeks (Ex. 204X; p.17).

USE OF THE TERM "'FINISH COAT"

The final rule specifies the acceptable slip resistance of structural steel "coated with paint or similar material," whereas the proposal limited the provision to steel which had been "finish-coated." This change clarifies that the provision applies to the surface of the coated structural steel when the steel is erected. OSHA believes that the rulemaking record demonstrates that the hazard posed by slippery coated steel is present irrespective of whether the coat is part of a multi-coat system. In addition, we note that both the English I study (Ex. 9–64) commissioned by SENRAC and the English II study (Ex. 17) commissioned by OSHA, which tested slippery coated surfaces, evaluated coatings that were not necessarily "finish" coats. According to Paul Guevin, an OSHA expert witness, the English II study looked at three types of slip-resistant primers: Alkyd paints without additives; zinc-rich primers, and alkyds or other resin-based primers with polyolefin (Ex. 18, p. 2). The modification to "coating" also responds to concerns that it would be difficult to determine which paints are "finish" coats. Thus, the reworded provision now clearly applies to steel members coated with standard shop primers where the shop primer is the uppermost coat when the steel is erected.

A number of commenters asked OSHA to clarify and/or define the term "finish coat" (Exs. 13–182, 13–209, 13–228, 13–363, and 13–367). One of these commenters (Ex. 13–182) opined that finish-coated means painting after erection, which they indicated was done in many situations. A fabricator (Ex. 13–228) commented that a finish coat is the final coat of a multi-coat paint system, whether it was applied in the shop or the field is immaterial. Another commenter (Ex. 13–367, p. 16) noted that "it is frequently not possible to determine if an applied coating is a single coat or a multi-coat system." The American Institute of Steel Construction (AISC) speculated (Ex. 13–209, pp. 31–32) that SENRAC's use of "finish-coat" was an attempt to address certain epoxies and polyurethanes, which are typically the second and third coats found in multi-coat paint systems, but that "[t]he scope of the proposed rule could be twisted to apply to all paints, not merely that small segment of the market that may present a problem." OSHA disagrees with this characterization of the provision's intended application. By deleting the term "finish coat," OSHA clarifies that the provision applies to coated steel on which employees must walk, regardless of whether the coating will remain the last coat of paint after the steel erection is over, and regardless of the chemical composition of the coating.

BENCHMARK SLIP-RESISTANCE CRITERION

The final standard requires that coated steel must score at a minimum average slip resistance of 0.50 as measured on an English XL tribometer or equivalent reading on another tester. Proposed § 1926.754(c)(3) would have required that the structural steel surface be no more slippery than bare, uncoated steel. OSHA stated in the proposal that SENRAC, after reviewing various industry presentations, "concluded that it could not determine a minimum value for slip-resistance or CoF, given all the variables to be considered, nor could it agree on an acceptable testing method" (63 FR 43468).

After reviewing the entire record, OSHA has determined that it is necessary to set a specific slip resistance value for coated steel. No other regulatory approach to reducing the risk of slipping is as appropriate. The record supports using the English XL value of 0.50 (or the equivalent) as the cutoff for acceptable coated steel surfaces on which employees may walk. The record demonstrates that acceptable testing methods will be available when the provision goes into effect.

The English II report noted that a level of 0.50 was reasonably safe and has been recognized for many years:

> The noncontroversial 0.50 threshold of safety that has been recognized in the safety engineering literature and case law for 50 years would provide a vast enhancement of footwear traction that would produce a significant improvement in the safety of ironworkers working at high elevations. (Ex. 17, p.12)

In post-hearing comments (Ex. 64), Mr. Guevin explained that when the Federal Trade Commission published a proposed rule for floor polishes in 1953 it determined a minimum of 0.50 when measured on a James machine to be a safe value (Ex. 64,

U.S. Standards and Guidelines 189

pp.3–4). In his testimony at the hearing (Ex. 200X; p.120), Dr. Underwood added that he understood that 0.50 came from rounding up a CoF of 0.35 to give a small margin of safety for walking slowly in a normal way. He indicated that the CoF of 0.35 came from determining a ratio of an average hip height of 3 feet (0.91 m) and a common distance of 2 feet (0.61 m) per step taken in a normal stride.

The English II study indicates that the recommendation of 0.50 on the English XL scale was based on the previously established benchmark of 0.50 CoF (Ex. 17, p.12). We find that the information and testimony from the rulemaking record show that 0.50 on the English XL scale is an appropriate minimum value to designate slip-resistant surfaces when measured under wet conditions using the ASTM methods referenced in Appendix B to this subpart.

As noted above, OSHA is changing the proposed benchmark for acceptable slip resistance, from bare steel, to a specific slip resistance value for the coated steel. Thus, there is no need for employers, paint companies or fabricators to measure the slip resistance of bare steel for purposes of complying with this standard. Some participants objected to using the slip resistance of bare steel as the benchmark. OSHA believes that the revised provision addresses these concerns. A comment from a builder's association (Ex. 13–121) stated that "it is next to impossible to provide CoF equal to original steel after coating it." The Steel Coalition wrote that the proposal's reference to a test for a comparative coefficient of friction in § 1926.754(c)(3) would not be practical or meaningful, and that coatings with a high slip resistance score would be considered unacceptable when compared to original steel with a higher score (Ex. 13–307, pp. 35–36). The American Institute of Steel Construction (AISC) (Ex. 13–209, p. 36) stated that "[t]he benchmark of bare steel is ambiguous." AISC explained that using bare, uncoated steel as a benchmark was problematic because it was impossible to find a single uniform steel surface with which to make comparisons—"there is no such thing as a uniform piece of bare steel" (Ibid, p. 30). The AISC also objected on the grounds that each piece of steel would have to be tested, before and after it was coated (Ibid, p. 30).

The Society for Protective Coatings (SSPC) (Ex. 13–367, p. 16) stated that ". . . data from the English study [English I study] shows that a pristine millscale steel surface received one of the poorest ratings by ironworkers and by the English machine. Therefore, it is extremely risky to make an assumption about slip resistance based on whether the steel is coated or uncoated."

During the hearing, Mr. English testified that he did not support the benchmark of original or bare steel:

> First of all, . . . pristine bare steel is pretty rare. Secondly, . . . the baseline would be variable. Thirdly, we find that pristine bare steel, it's slippery . . . And as a practical matter, it rarely occurs as a problem at erection sites. (Ex. 200X; pp.115, 128–129)

Some comments supported using bare steel as the benchmark of acceptable slip resistance. Journeymen ironworkers (54 individuals, Ex.13–207C) signed statements saying that they backed limiting coatings to the equivalent of bare steel. However they did not provide information concerning the feasibility or adequacy of relying on "bare steel."

In sum, the record supports OSHA's decision that bare steel is not an appropriate benchmark. We agree with the commenters who stated that there is considerable variability in bare steel surfaces due to both manufacturing specifications and extent of oxidation, that variability would also pose substantial problems in implementing the requirement, and that some bare steel is unacceptably slippery.

TEST METHODS

The final rule requires that beginning three years after the effective date of the rest of the standard, employees may not walk on coated steel unless the coating has been tested and found to meet the threshold 0.50 using an appropriate ASTM test method. Appendix B specifies two methods now approved by ASTM. The record shows that these methods are sufficiently accurate and yield sufficiently reproducible results for use in testing coatings to determine their compliance with the specified 0.50 measurement.

Evidence in the record shows that testing using the VIT (English XL) according to ASTM F1679–96 will provide reproducible and accurate results of the slip resistance of coated steel: the authors of the English II study stated that the VIT has achieved satisfactory precision and bias according to ASTM E691–92 Standard Practice for Conducting an Interlaboratory Study to Determine the Precision of a Test Method. The report of their testing showed that highly consistent results were produced from repeating the VIT tests, and that there was substantial correlation between the ironworker rankings with VIT rankings.

Also, the final rule's designation of approved ASTM testing methods as appropriate to determine compliance with a performance criterion is consistent with other OSHA standards. For example, in OSHA's standard for nationally recognized testing laboratories, an "ASTM test standard used for evaluation of products or materials" falls under the term "appropriate test standard" (as set out in the introductory text to paragraph (c) of that section, § 1910.7).

Various participants, however, claimed that the two ASTM testing methods lack precision and bias statements, which in their view render those standards "meaningless" (see e.g., Dr. Kyed's testimony Ex. 204X; p. 262 and Ex. 13–367; pp. 3–4). However, various witnesses (including one who offered the position above) stated that precision and bias statements often lagged behind a new approval by ASTM of a test method. "Test methods can be temporarily issued without these statements, but they must eventually comply with this requirement. Generally, it's a 5-year period" (Ex. 204X; p.262). Dr. Mary McKnight from the National Institute for Standards and Technology (NIST), testifying with a panel from the Society for Protective Coating (SSPC) [formerly the Steel Structures Painting Council], agreed that ". . . within 5 years, there will be a group of laboratories that become proficient in running the test method and who will participate in a round robin study. At the end of this process, ASTM includes a number describing statistical significance of different responses, with a 95-percent repeatability limit and/or confidence level" (Ex. 205X; pp. 56–68). In post-hearing comments (Ex. 71, p. 4), Mr. English stated that the ASTM F1679 precision and bias study has been approved by letter ballot, and at a recent meeting of the F13.10 Traction subcommittee, two-thirds of those present voted to find all negatives nonpersuasive.

OSHA concludes that the rulemaking record demonstrates that the methods identified in Appendix B are sufficiently reliable in evaluating the slip resistance of coated steel. The record also shows that this reliability is likely to be confirmed by the ASTM precision and bias statement process within the 5-year period this provision will be delayed.

In post-hearing comments, the major industry groups who objected to OSHA's designating ASTM methods stated that "several of their organizations actively participate in research and development efforts involving the validation and adoption of a testing machine and test methodology appropriate to coated structural steel" and recommended that OSHA delay the effective date for 3 years to allow further expert evaluation (Exs. 63, p. 7 and 75, p. 4). These groups also wanted this additional time to determine if implementation of the provision was feasible.

Although the ASTM methods are the best available, OSHA acknowledges that the ASTM methods lack a protocol for representative samples of steel and their preparation. The Agency anticipates that either these parallel issues will be addressed by ASTM within the time frame before paragraph (c)(3) becomes final (5 years after the effective date of the final rule) or alternative steps can be taken to ensure accounting for these parameters.

AVAILABILITY OF PAINTS TO MEET THE SLIP RESISTANCE BENCHMARK

The final standard delays the effective date of the slip-resistant coating provision for 5 years from the date the rest of the standard becomes effective. This is a change from the proposal, which would not have delayed the effective date. OSHA finds that although some slip-resistant coatings suitable for use in the steel erection industry are now available, widespread distribution and use of suitable coatings will take additional time. We have chosen a 5-year delay in agreement with the post-hearing requests of the major organizations commenting on this issue. These organizations submitted their comments as the Unified Steel Construction Consensus Group (USCCG) (Ex. 63), a group that consists of eight large organizations as signatories. The USCCG explained that their membership represents design, engineering, fabrication, manufacturing, and field installation components of the steel construction industry. (The following organizations were listed as signatories: The Steel Joist Institute; Steel Erectors Association of America; National Council of Structural Engineers Associations; National Institute of Steel Detailing; Council of American Structural Engineers; American Institute of Steel Construction; Metal Building Manufacturers Association; and the Society for Protective Coatings.) They stated that the rulemaking record was uncertain about the extent adequate coatings were now available, and that developing, testing and distributing appropriate slip-resistant coatings for the industry would take time. Also, during the rulemaking, many paint formulators and steel fabricators stated that they do not now use the specific paints tested in the English II study. (For example, see Ronner at Ex. 204X, pp. 15 and 108–109; and Appleman at Ex. 205X, pp. 139 and 157–158.) In addition, some formulators and fabricators and their representatives stated that there is a lack of information about whether the paints/coatings in use can meet the standard's slip-resistant threshold. (For example, see Ex. 13–367, pp. 7 and 17; Ex. 13–307, pp. 38–39; Ex. 13–209, pp. 36–37; and Ex. 206X, pp. 34–35.)

OSHA finds that there is some uncertainty as to the extent to which there are adequately slip-resistant coatings currently available that would meet the industry's needs. In view of the fact that there are many such coatings presently on the market (see Ex. 17, pp. 3 and 10–11; Ex. 18, pp. 1–2; Ex. 200X, pp. 54, 62–63, 70, 137–139, and 168–169; Ex. 204X, pp.193–194; Ex. 205X, pp. 139 and 157–158) and the technology for developing additional coatings is in place (see Ex. 205X, pp. 51, 93–94, 99–102, 139, 151–152, 157–158, 167–168 and 217–219; Ex. 63, pp. 3 and 7; and Ex. 64, pp. 2–3), it is reasonable to expect that the 5-year delay will provide enough time for the industry to develop coatings that comply with the final rule.

OSHA agrees that the record evidence on the availability of slip-resistant paint which meets the standard is conflicting. The witnesses who conducted the English I study commissioned by SENRAC (Ex. 9–64), and the English II study commissioned by OSHA (Ex. 17), testified that one reason for conducting these studies was to determine whether slip-resistant paint was widely available for use by the steel erection industry. They contended that slip-resistant paints are available. They surveyed fabricators first, to identify coatings actually in use for steel erection, tested these coatings in their studies, and found that most of them passed the tests for slip resistance (Ex.18, pp. 1–2). In post-hearing comments (Ex. 71, p. 4), Mr. English stated that "paints now being applied on something over 80% of the fabricated steel products in the U.S. can be easily made to comply with the proposed specification with no complications to application methodology, coatability, corrosion or UV resistance or any of the 'problems' raised by . . . those opposed to this standard." He added that the paints that do not already comply could be brought into compliance with "the simple addition of the plastic powder . . ." Another witness (Ex. 205X; pp. 220–221) acknowledged that zinc-rich primers that are currently being used "extensively" had good slip-resistant qualities. However, he also stated that they are not generally used by the industry (Ibid; pp. 139 and 157–158).

Various other rulemaking participants told OSHA that the coatings used in the English studies represented only a small percentage of coatings used in steel erection. According to a telephone survey of 180 fabricators conducted by Mr. Ronner for the Steel Joist Institute (SJI) (Ex. 28), only 14 (7%) used the paints tested in the English II study (Ex. 204X; p. 15), and that although slip-resistant coatings are now used for various military applications such as helicopter flight decks and aircraft carriers, they are not generally used by the steel erection industry (Ex. 205X, pp. 139 and 157–158). The SSPC commented that slip resistance has not been a design factor for coatings used on structural steel and that slip-resistant paints have not generally been tested for durability (Ex. 13–367, p. 7). A representative of the SJI (Ex. 204X, p. 13) testified that the zinc-rich primers, paint with polyolefin beads and some alkyd-based primers used in the English II study, are for spray applications only and are not recommended for dip operations. He added that steel joists typically are coated by dipping them in dip tanks (Ex. 204X; p. 13), and that the industry could not spray on paints due to state and Federal environmental restrictions. These commenters assert that there is no basis for assuming that the same slip resistance would be achieved if the paints were dipped, and that there are technical problems with applying some of

U.S. Standards and Guidelines 193

the slip-resistant paints by dipping. (See, for example, Mr. Ronner's testimony, Ex. 204X; p.13, and Mr. Appleman's testimony, Ex.205X; p. 93.) Both Mr. Guevin and Mr. English acknowledged that they do not know if the same slip results reported in the English II study for the paints with beads would be obtained if that paint had been applied by dipping (Ex. 200X; pp. 62–63).

Promising approaches to providing slip-resistant coatings for the steel erection industry were identified during the rulemaking. As explained in the English II study (Ex. 17, p. 11) and as Mr. Guevin (Ex. 200X, p. 56) stated by ICI Devoe in Western Canada developed a slip-resistant 3-coat system, using "DevBeads," an additive of polyolefin beads. However, various participants questioned whether grit particles such as polyolefin beads could be added to paints and primers in steel erection. For example, George Widas (OSHA expert witness who peer reviewed the English II study) questioned whether such coatings would retain their corrosion protection (Ex. 204X; p. 240); Mr. Sunderman of KTA Tator, Inc., questioned whether polyolefins would be degraded by ultraviolet light (Ex. 206X, p. 34–35). Mr. Sunderman also challenged the notion that specific properties of paint can be modified "randomly" without affecting the balance of properties, and without extensive testing and evaluation (Ibid, p. 35–36).

Several participants stated such that slip-resistant coatings could be developed for use in steel erection, but that time would be needed to do this. Robert Kogler, a research engineer, explained that testing corrosion control materials takes several years, and they still rely very heavily on long-term exposure data, but are coming up with accelerated testing that gives us reasonable data (Ex. 205X; p. 74, to same effect, see testimony of Dr. Appleman Ex. 205X; p. 51).

On a related issue OSHA finds that obtaining documentation or certification that coated steel meets this requirement also is feasible. However, paint manufacturers told OSHA in their post-hearing comments that they will work with interested parties to formulate, test and evaluate coatings to meet the standard's criteria. (See Exs. 63, p. 7 and 75, p. 4 and 205X; p. 218.) Mr. Guevin testified that based on his experience with contacting paint manufacturers to obtain slip-resistant coating for the English II study, and his knowledge of typical paint technical bulletins issued by manufacturers setting out specifications, tests conducted, and results, companies would readily certify if their coatings meet OSHA slip-index requirements in accordance with the recognized ASTM Method (Ex. 200X; p. 168). Thus, OSHA does not agree with a project manager for a steel fabricator (Ex. 13–300) who commented that the requirement was "not viable" because paint manufacturers will not provide documentation out of concerns for liability.

In sum, OSHA finds that although there are slip-resistant coatings in use for structural steel in limited specialized applications, most of them have not been adequately tested to determine whether they comply with the standard and meet the performance needs of other kinds of structures. The coatings industry has committed to develop, test and distribute coatings that comply with this standard in a reasonable time frame. OSHA believes that the hazard of slipping on coated steel is significant; that the paint and fabrication industries feasibly can produce and use coated steel that complies with this provision within the time frame stated in the regulatory text; and in any event, there are now coatings on the market that meet the standard that can be used to some extent even before the widespread production of new slip-resistant coatings. The need for this provision is amply supported in the record. We believe that by issuing a delay of the effective date of this provision the needs of the industries affected by this provision will be met and the long-term safety concerns of the workers who must walk on these surfaces will also be met.

6 Flooring and Floor Maintenance

Organizational Responsibilities

Component	Potential Responsibility
Specifying flooring materials	Architect, safety
Selecting floor treatments	Safety, purchasing
Applying floor treatments	Housekeeping, contractors
Specifying floor cleaners/equipment	Safety, housekeeping, purchasing
Specifying floor cleaning regimens	Housekeeping, safety
Performing floor cleaning	Housekeeping, contractors
Maintaining floor care equipment	Maintenance, contractors
Outsourcing floor care	Management, safety, operations

As you grow older, you stand for more and fall for less.

Proverb

6.1 INTRODUCTION

Successful slip and fall prevention efforts must include a strong understanding of flooring. This includes the types and construction of flooring materials, treatments, and cleaning products. In addition, this includes the proper methods of cleaning and treating floors, and selection, use, and maintenance of floor equipment and related supplies.

An understanding of how and why wear, cleaning, flooring materials, and finishes can affect traction will assist in weighing these significant variables when coming to conclusions about actions that may be needed to improve the traction of walkway surfaces.

6.2 THE THRESHOLD OF SAFETY

There is much misinformation about thresholds of safety for slip resistance. Contrary to published charts, no generally accepted safe, dangerous, or very dangerous thresholds have been established by any recognized independent U.S. authority. The one commonly mentioned level of acceptable slip resistance is 0.5.

It can be argued that the 0.5 threshold has been ratified by ASTM (D2047), UL (UL 410), and ANSI (ANSI 1264.2) (see Chapter 5, U.S. Standards and Guidelines). The 0.5 threshold is based on establishing the merchantability of new flooring-related materials and finishes using the James Machine (see Chapter 3, Principles of Slip Resistance and Chapter 4, U.S. Tribometers). In the United States, it is widely referred to in relation to static coefficient of friction (SCOF), while overseas 0.4 is the historically significant number in connection with dynamic coefficient of friction (DCOF).

Currently, the most significant value of slip resistance testing is its ability to determine the extent of an exposure (e.g., potential for pedestrian falls) and as a tool to establish levels of slip resistance by comparative methods. For example:

- Is this floor material or finish more or less slip resistant than others?
- How significantly is slip resistance affected by moisture or other contaminants?
- Is slip resistance stable or deteriorating due to wear?
- How do cleaning methods affect slip resistance?

For a history of the 0.5 threshold, see Chapter 3, Principles of Slip Resistance.

6.3 IDENTIFYING TYPES OF FLOORING AND THEIR PROPERTIES

The most compelling reason to properly identify types of flooring materials is that each involves different cleaning and maintenance requirements. Knowing the surface composition is also essential to selecting safe and effective floor treatments and finishes.

However, it can often be a challenge to properly identify the composition of flooring materials. Flooring materials can now be produced to closely resemble another. For example, a sheet vinyl floor may be colored and textured to appear as ceramic tile. The wide variety of colors, textures, patterns, and styles add to the confusion. Even the descriptive names of flooring materials can be misleading. Further complicating the identification process are opaque finishes that can be applied to materials which mask their true identity. Finally, there may be no recollection or record of the flooring material which was installed, especially if it was completed several years prior.

In one case, a material named as a type of granite was purchased and installed in an establishment serving alcohol because granite is not subject to damage due to alcohol or the acidic content of beverages. Later, it was discovered that the material was in fact an acid-etchable black limestone.

Being able to identify the type of material is helpful, but having more information readily available is always valuable. Considering that even if the class (e.g., resilient, nonresilient) and category (e.g., marble, granite, limestone) of a floor are known, the properties of slip resistance cannot necessarily be inferred with certainty. Slip resistance can vary based on the manufacturer due to the processes used, extent of quality control, and product line. In the case of inorganic materials such as natural stone, material properties can also vary by the source (e.g., country/region, quarry, and vein). In one database of natural stone materials (http://www.graniteland.com), 266 different types of marble and 524 different types of granite are available. Furthermore, the properties of individual pieces of tile can vary from one another.

Even with these drawbacks, an understanding of how to identify flooring materials does yield benefits in assuring that cleaning and finish requirements are appropriate, thus maintaining the desired appearance and maximizing the life of the floor.

Flooring and Floor Maintenance

6.3.1 Resilient Flooring

Depending on the texture, resilient flooring is generally highly tractive under dry condition, but often has low slip resistance when wet. Most resilient flooring is produced and sold in role or tile form. By definition, resilient flooring is flexible and pliant material. The following list further defines resilient flooring:

- Linoleum is a flexible floor covering made from oxidized linseed oil or a combination of drying oils, wood flour, and/or ground cork, resins, and pigment.
- Asphalt in 9-in. square tile form is made with fillers, pigments, and fiber (usually asbestos) with a binder. Its porous surface becomes brittle over time and is subject to fading and bleeding. It is commonly found in educational and religious institutions.
- Rubber is made from synthetic rubber, fillers, and pigments. It is popular because of its sound-deadening qualities. It has a resilient, dense surface and often does not require a seal or finish application. It can be sensitive to solvents and high alkaline cleaners, so neutral cleaners and a soft brush should be used when cleaning.
- Vinyl is made from vinyl resins (varieties of plastics) and is frequently used with other agents to make vinyl asbestos tile (VAT), vinyl enhanced tile (VET), and vinyl composition tiles (VCT). Vinyl tile remains the predominant floor type in the United States; however, each type of vinyl floor has different cleaning needs.
    - *Solid vinyl tile (SVT)*—SVT is most expensive to purchase because it has the highest vinyl content, but it is also the most durable and easiest to maintain. Usually, the higher the vinyl content, the more resistant it is to wear, abrasion, and indentation. It also keeps its shape better over time so that cracks and gaps do not form in the floor, trapping grease and dirt. Generally SVT floors do not require a floor finish. It can usually be maintained by dust/damp mopping and using a high-speed burnisher, minimizing maintenance costs.
    - *Vinyl composition tile (VCT)*—VCT floors account for approximately 80% of commercial floors in the United States and require the most maintenance. They require up to five coats of floor finish, frequent buffing/burnishing, and refinishing at least twice annually depending on foot traffic and weather conditions. A high-quality finish applied over a floor sealer can reduce the number of coats and refinishing cycles. Since VCT is not as flexible as other vinyl floors, they are more prone to cracking, scratching, and gaps forming between tiles. Cylindrical floor machines often work best with VCT floors, because they penetrate cracks and gaps, removing dirt and grease, which keeps the floor looking new longer.
    - *Vinyl enhanced tile (VET)*—VET has 25% more vinyl than VCT and is more resilient to wear and tear, abrasions, and indentations than VCT. They are also more flexible; gaps and cracks do not as readily developed. VET floors usually need only two coats of finish, and studies indicate that the high-gloss shine will remain longer if maintained with a cylindrical machine.

6.3.2 Nonresilient Flooring

The slip resistance of nonresilient flooring materials, which are characterized by their hard, inflexible properties, varies greatly. Compared with resilient flooring, a wider variety of nonresilient flooring is available.

Nonresilient tile is generally available in any of three states: virgin fired condition, glazed, and slip-resistant implants. Virgin tile, such as quarry tile, is generally a slip-resistant surface. Glazed tile is generally too slippery for heavy pedestrian areas. Glazed tiles can be manufactured with slip-resistant implants, but often the glazing negates the benefits of implants. Normally, virgin tile is installed and then sealed, which gives it a ceramic look and low slip resistance.

- Ceramic tile is a fired mixture of clays and is often glazed. Depending on the texture and finish, which can be high gloss, matt, or abrasive, ceramic tile can perform well or poorly under wet conditions.
- Quarry tile is a broad classification for baked red clay, set in concrete (grout), in 6- or 9-in. (15- or 18-cm) squares. It is usually deep red in color and left unglazed. A highly durable surface, quarry tile can perform well in dry and wet conditions when unsealed and untreated. Since quarry tile is commonly used in commercial kitchens, it is often subject to oil/grease contamination. If not cleaned properly, its natural slip-resistant properties can quickly degrade. Even when well maintained, wear takes its toll, reducing its slip resistance properties. Although quarry tile should not be treated, some choose to do so to protect the grout (see Chapter 9, Food Service Operations).
- Concrete slab is a mixture of cement, sand, crushed stone, and water. It is installed in paste form, and once it is set, the surface becomes hard, durable, and porous. The degree of slip resistance obtained from concrete depends largely on the method of finish (e.g., smooth trowel, rough brushed, textured, stamped) and the finish (e.g., paint with grit, paint without). Concrete tile is also available in varied forms. Depending on how it is finished, its appearance can be similar to natural stone materials.
- Terrazzo is an aggregated mix of granite and polished marble chips in poured Portland cement, leveled off to a smooth finish. Frequently used in public buildings, terrazzo displays slip resistance properties similar to marble. Terrazzo, which has a very porous surface, was listed by the National Bureau of Standards (now the National Institute of Standards and Technology) as a high-risk material for stairway treads and has low slip resistance under wet conditions. Some terrazzo is made with nonslip materials added, such as alundum grit. Since adding this material does not alter the appearance of the surface, documentation and testing are the only methods of verification. Terrazzo floors should not be waxed.
- Related to terrazzo, agglomerates are also marble chips, but are bound in resin and marble dust instead of cement.
- Marble is a crystallized calcium carbonate, a limestone with properties altered by intense natural pressure and heat. Highly polished marble is also quite tractive when dry. In wet conditions (untreated with a slip-resistant

finish), it becomes hazardous. Marble is often used for more elegant and costly installations including luxurious bathrooms in high-end hotels, of all places, and is one of the most common types of stone flooring. The slip resistance of marble varies depending on the origin, cut, and wear patterns of the marble floor. Polished marble generally has a relatively low coefficient of friction (COF) value. In some cases, marble can become worn to a higher, safer value. Marble located near a street entrance can be roughened by the grinding effect of debris on shoes.

- Other natural stones. Similar to marble, other natural stone surfaces are generally tractive in dry conditions, but can become hazardous under wet conditions.
  - Granite is an extremely hard rock surface. Naturally occurring in a variety of colors, it is frequently used in lobbies. Its appearance is improved with seal and finish, which degrades its slip resistance qualities. Fired granite is generally gray, dull, and relatively rough. This type of granite is often used for outdoor applications, including steps and walkways.
  - Slate is a blue or dark gray rock in flagstone-type floors found primarily in lobbies and outdoors. It does not polish naturally.
  - Limestone, such as travertine, can appear in a wide variety of textures, colors, and designs.
  - Sandstone includes bluestone, a relatively dense sandstone that is aptly named for its tones of blue, green, and purple, and flagstone, a hard sandstone easily to split.

6.3.3 OTHER TYPES OF FLOORING

- Wood, when properly sealed with a urethane or other permanent sealer to prevent moisture penetration, can be coated with one or two coats of floor finish and maintained like vinyl surfaces. Many woods in stained condition can be used as slip-resistant floors; however, the application of some sealants, or oiling floors, generally degrades slip resistance.
- Carpet (see Chapter 1 Physical Evaluation) is made of a variety of textiles with heavy jute backing and natural or synthetic fiber piles including wool, polyester, nylon, and other materials. Carpet is generally considered a slip-resistant floor covering, although soaked carpet with a small weave could become slippery.
- Astro turf is often used in outdoor applications, especially at casinos.
- Glass floors and stair treads are growing in popularity because of their aesthetic appeal. Since pedestrian safety is critical, the glazing must provide an acceptable level of slip resistance. Under the jurisdiction of ASTM Technical Committee E06 Performance of Buildings, subcommittee E06.56 on Performance of Railing Systems and Glass for Floors and Stairs, development of a related standard is underway. Work item *WK9258 New Practice for Practice for Design and Performance of Supported Glass Walkways* was started in 2005. This practice addresses elements related to load-bearing glass walkways, including performance, design, and safe behavior. It

addresses the characteristics unique to glass. Issues common to all walkways, such as slip resistance, are addressed in existing referenced standards.
- Processes designed to roughen the top surface of the glass to provide slip resistance include sandblasting, acid-etching, ceramic frit, and embossing. It is important to note that sandblasting can reduce the strength of the glass by as much as 50%. Therefore, glass flooring should never be sand blasted without a complete engineering analysis (Glass Association of North America, 2007). The key is applying the dots and lines with enough density and roughness. Most appear to meet a COF of 0.50 under wet conditions, although some have insufficient density of the slip-resistant spots, making it possible for a heel to slide between rows of dots.

One manufacturer, Eckelt Glas, uses a ceramic slip-resistant coating to the upper sheet of glass, which contains a hard abrasive material and is screen-printed onto the top-sheet of glass prior to the heat-strengthening process. During the process the ceramic coat is cured, and the abrasive frit beds down, fusing into the glass giving a permanently fixed finish.

Glass floors should be regularly inspected for damage, as impact from hard objects can crack the upper surface. Any damaged glass should be replaced as soon as possible. Cleaners and polishes may change the coefficient of friction and should be avoided (Glass Association of North America, 2007).

6.4 FLOOR FINISHES AND THEIR PROPERTIES

In many cases, floor replacement is not a cost-effective alternative, so other methods of improving the slip resistance qualities of walkway surfaces need to be considered. In general, the goal is to make the surface as slip resistant as possible. This is best accomplished by finding methods to increase surface roughness, while balancing the aesthetic needs of the business.

It is important to note that, while some may last for years, no known slip-resistant floor treatment is maintenance free. Each requires some upkeep, and most require reapplication at some point. Also, the degree to which these products are effective varies dramatically. Whereas some treatments demonstrate a significant increase in slip resistance, others can actually decrease it (see Exhibit 6.1, Slip-Resistant Treatment Study). Finally, floor appearance is usually affected by treatments. The desirable high gloss look can be dulled. This is another reason that testing prospective treatments on a sample, or a small, low-traffic area, prior to full installation is advisable.

6.4.1 THE RELATIONSHIP OF SHINE TO SLIP

In a series of four studies, the visual cues walkers use to predict slippery ground surfaces were investigated. In the first study, 91% of participants responded that they use shine to identify upcoming slippery surface. Three of these studies confirmed this reliance on shine to predict slip. Participants viewed surfaces varying in gloss, paint color, and viewing distance under various lighting conditions. Shine and slip ratings and functional walking judgments were related to the level of gloss and to

COF. However, judgments were strongly affected by surface color, viewing distance, and lighting conditions, all of which are extraneous factors that do not affect the surface COF.

Although walkers rely on shine to predict slippery ground, results suggest that shine is not a reliable visual cue for friction (Joh et al., 2006).

6.4.2 Conventional Floor Finishes

Polishes and waxes serve several purposes including:

- Making the floor easier to clean (dirt does not embed as easily)
- Providing durability to extend the life of flooring
- Providing a protective coating
- Increasing gloss to improve appearance

A multitude of chemicals go into polishes and waxes. They must be balanced, because each chemical influences a different property including gloss, durability, slip resistance, water resistance, black mark, salt, and detergent resistance. Specific ingredients, some of which can counteract another, affect each chemical.

Waxes are no longer made only from natural waxes. Synthetics are used more often. However, waxes contain little wax of either type, but are mostly of polymers with a small amount of wax for truth in advertising.

The more coats of wax or polish are applied, the higher the gloss. On average, one gallon provides one coat over an area of 2,000 sq ft. This provides a coating 1/10,000 of a millimeter thick, a fraction of the thickness of a coat of paint. Coating fills in imperfections, including peaks and valleys that would otherwise serve as runoff channels for water and other contaminants, causing them to smooth out.

Industry experts have stated that changes in formulation cannot be made to waxes in order to make them more slip resistant under wet conditions. The only way to do this would be to make the waxes rougher (e.g., by applying a grit), which would make them substantially less durable, and thus impractical.

In addition to floor surfaces, finishes also have an impact on the slip resistance of a given surface. Many floor finishes are suspended polymers (plastics) that become interlocked when the floor dries, becoming analogous to a sheet of plastic. The slip resistance that a finish is intended to provide can often be found on the container of floor finish used, under "Performance Specifications" (stated in terms of SCOF).

This rating must be taken with a grain of salt because results are usually achieved by using ASTM D2047 (see Chapter 4, U.S. Tribometers). The results are ultimately affected by how closely the application instructions are followed. Contaminants, cleaning, and maintenance can quickly erode the slip resistance of many conventional floor finishes. Factors include insufficient dilution of cleaning compounds, use of inappropriate detergents or floor finishing compounds, inadequate rinsing or burnishing, and settled or tracked-in dirt. Cleaner residue left on a floor will mix with the newly applied floor finish, destroying much of its water resistance. The only way to verify the actual slip resistance of the treated surface is to conduct testing.

6.4.3 "SLIP-RESISTANT" FLOOR TREATMENTS

6.4.3.1 Particle Embedding

Slip-resistant coatings, consisting of an adhesive base material with slip-resistant material (usually grit) may be used in some areas to increase traction. Some resist the idea of a slip-resistant treatment, fearing that the surface will become more difficult to clean. Engineering for slip resistance involves choosing the right size and shape particle to be embedded in the coating film and the proper spacing of the particles.

Choose the appropriate particle for your application. Effective particles for increasing slip resistance have an angular configuration and increase surface roughness. These particles must extend through the water or contamination on the surface and effectively engage the shoe bottom. Round particles such as glass beads or polypropylene are generally not effective.

Three frequently used abrasive grits are aluminum oxide, silicon carbide, and quartz. Aluminum oxide wears well and does not break easily. Silicon carbide has similarly effective wear qualities, but tends to be brittle and can become less effective when subject to high traffic.

Some synthetic acrylic particles have the necessary angularity, but because they lack the hardness of minerals, they are easily worn away in heavy-traffic areas. These acrylic particles are lightweight and can be suspended into the coating material. For heavier use areas, bleached aluminum oxide, colored quartz, or silica sand should be used. Aluminum oxide is generally considered the best slip-resistant particle because of its superior hardness.

Choose the correct particle size and distribution for your application. In general, larger particles spaced fairly close together provide the best slip resistance. As long as there is adequate spacing between the particles, cleaning remains easy. Smaller particles provide less texture and slip resistance but may be appropriate for certain interior applications.

Be sure that 30% to 50% of the particle is embedded in the resin matrix. If not enough of the particle is embedded in the resin, it may become dislodged during use. If not removed, these loose particles can produce a ball bearing effect and result in a more slippery surface. Conversely, if too much resin is applied for the size of the particle, an inadequate amount of the particle will protrude through the film to be effective. Often, slip-resistant particles are embedded into the primer material and then covered with a top coat (Architectural and Transportation Barriers Compliance Board, 2002).

6.4.3.2 Surface Grooving and Texturing

Shallow grooves and textures can be ground into existing concrete, asphalt, slate, marble, granite, and hard mineral tile floors to increase traction. In grooving, diamond saw blades produce precision cuts at specified depths, widths, and spacing. Grooves help drain liquids off the walkway to improve slip resistance. The direction of the grooves controls the direction of drainage. Saw blades are positioned close together for texturing and produce a surface with a ribbed appearance.

Unless cleaned properly, grooving and etching may not be an effective long-term solution. By cutting into the surface, this process can accelerate leaching of

Flooring and Floor Maintenance

contaminants such as grease and oil into the surface. Even alkaline-based cleaners and degreasers leave deposits and residues in grooves that can gradually reduce surface slip resistance, with the potential for making slip resistance worse than if no treatment had been applied at all. Cleaning grooved surfaces can be a challenge. Grooving is more commonly used for hard, exterior surfaces such as brick and stone.

6.4.3.3 Etching

Etching treatments contain diluted hydrofluoric acid or another acidic solution which creates micropores in the floor surface, resulting in roughening the surface to increase slip resistance. The treatment is applied, and a water rinse is used to stop the etching process. The degree of slip resistance depends on the strength of the acid solution applied to the surface and the length of time it is applied. Due to the hazardous nature of the chemicals involved and the importance of proper application, most etching treatments require installation by trained professionals.

Typically, the stronger the solution and higher the slip resistance, the less glossy the surface. Some experimentation with samples may be required to obtain the correct balance of gloss and slip resistance.

Etching can be effective on smooth, hard walkway surfaces such as stone/hard mineral surfaces and agglomerates (resin-based rock products). Vendors may have different formulations for different types of floor surfaces.

Etching can also result in surface conditions worse than the original. Deepened pores can fill with contaminants that, if not properly cleaned, can cause the surface to become more slippery. The etchant manufacturer directions for maintenance and cleaning procedures must be observed (see Section 6.6), and the slip resistance of the surface should be monitored regularly.

Since this treatment involves the removal of some amount of the floor surface material, it can decrease the service life of flooring. Etching treatments can last for years. Vendors have guarantees ranging from one to ten years; however, proper cleaning must be performed to avoid accumulation of contaminants. In most cases, any commercially available cleaner/degreaser would be sufficient as long as the surface is adequately rinsed.

6.4.3.4 Other Slip-Resistant Floor Treatments

Rubber or plastic coatings, including urethane polymers, epoxy, acrylic, and vinyl ester resins, adhere to the existing substrate or floor surface and provide a higher coefficient of friction than the existing surface. Light-duty types are suitable for wood surfaces, while heavy-duty types provide a nonslippery coating over wood, metal, or concrete. They may also contain rubber or plastic particles to give a textured surface.

Other treatments are designed to create a raised texture with the use of abrasive crystals bonded to the surface. Topical coatings such as these can be so dense as to inhibit adherence to the surface, requiring more frequent maintenance and reapplication than other types of treatments. Also, it should be noted that some coatings do not properly adhere to polished stone, porcelain, or similar surfaces.

Water-based floor coatings contain no volatile organic solvents. They are nonflammable and are environmentally safer. These coatings may be applied by brush, roller, spray, or trowel and generally retain nonslip properties in wet or oil conditions. Since water-based coatings contain no organic solvents, they are useful where ventilation is limited or an explosion hazard exists.

Some treatments are a hybrid of coating and etching. Hydrofluoric acid can be used to soften the grit to improve adherence to the floor surface.

If the floor coating does not create enough thickness, anti-slip properties will not be fully effective regardless of the quality of maintenance.

In general, coatings and other similar treatments can deteriorate more quickly than etching and grooving due to wear and cleaning. An advantage to a topical treatment is that it is often reversible, whereas etching and grooving are not.

6.5 THE IMPACT OF WEAR

Flooring materials in service will wear. Continued use will generally "polish" the surface, gradually reducing its slip resistance. This is due to the wearing down or smoothing of the asperities in the surface or the micropeaks in the material responsible for its roughness.

Directionality of surface roughness is also a factor to be considered in assessing slip resistance. Asperities become smoother in the direction of travel as the peaks are worn down and rounded. As such, a surface may demonstrate greater slip resistance when traversing the area in the direction perpendicular to normal travel. Likewise, surfaces with a directional pattern or grain may exhibit different slip resistance properties depending on the direction of travel.

6.6 FLOOR CLEANING

USING THE WRONG CLEANING PRODUCT

An equipment manufacturing company moved into a new office that had linoleum floors, where slipping accidents immediately began to occur on one floor at an alarming rate. The cleaning product vendor could not determine the problem, because their product was not known to cause such problems. Eventually the vendor learned from the cleaners that regular washing up liquid was being used to clean the floors instead of the specified cleaning product, because it was stored downstairs and inconvenient for her to get it. Once the floor was cleaned with the correct product, slip resistance improved immediately. As part of the corrective action, the floor cleaner is now stored on each floor. (http://www.hse.gov.uk/slips/experience/cleaningproduct.htm)

The effectiveness of floor cleaning can mean the difference between safe and slippery. When an area experiences a high frequency of falls, facilities often react by searching for a nonslip floor treatment without considering the actual cause of slippery floors, which may be contamination. Failing to properly clean and maintain floors makes it unlikely that a slip-resistant treatment will be effective in the long term. Cleaning is an especially effective issue to explore when slip resistance is marginal. In many cases, improvements in cleaning products and/or methods make a difference.

Studies have shown that revenues at stores with clean and shiny floors rose between 7% and 10% over stores with dingy floors. This means that for a superstore with $100 million in annual sales, an additional $7 to $10 million in sales would be at stake.

The financial losses due to dirty floors are difficult to measure because customers rarely inform store management of unacceptable conditions; however, they generally do inform their friends. Research shows that someone with bad news is likely to tell seven times as many people as with good news. When operating costs as cited above are combined with any loss of revenue, the impact on the bottom line is significant (Castle Rock Industries, 2007).

6.6.1 FLOOR CLEANING PRODUCTS

The quality and quantity of the following components help determine the effectiveness of a cleaner:

- Solvency—Ability to dissolve the material it contacts.
- Emulsification—Ability to break down a liquid into particles and suspend in liquid.
- Surfaction—Ability to lower surface tension and speed up solvency and emulsification.
- Surface tension—This is the tendency of water to bead up which retards surface wetting and inhibits cleaning. Surfactants are classified by their ionic (electrical charge) properties in water: anionic (negative charge), non-ionic (no charge), cationic (positive charge), and amphoteric (positive or negative charge). Soaps are anionic surfactants, made from fats and oils (or their fatty acids) by chemically treating them with a strong alkali.

The effectiveness of soaps lessens when used in hard water. Water hardness is the result of mineral salts that react with soap to form soap film.

6.6.1.1 Basic Types of Floor Cleaners

Alkaline cleaners are designed to react with fats and oils, converting them into soap. This process is called saponification. To remove this slippery residue, the floor must be thoroughly rinsed with clean, hot water. It cannot be permitted to dry; otherwise, polymerization can occur.

Acidic cleaners use a process called oxide reduction instead of saponification to remove contaminants. By using this process, polymerization cannot occur.

Neutral cleaners are typically used on glossy floor finishes, or those that can be dulled by the abrasive qualities of alkaline or acidic-based cleaners. These are usually relatively free-rinsing.

High-pH (alkaline) cleaners may be needed to remove grease stains, while acidic chemicals may be needed for mineral buildup from hard water or lime deposits. For daily maintenance, neutral cleaning chemicals are generally recommended. Acids and high-pH cleaners can scar hard floor surfaces such as marble, terrazzo, and other natural stones.

6.6.1.2 Six Categories of Floor Cleaners

A look at the MSDS of 350 floor cleaners allowed one researcher to classify 96% of all cleaners into one of six categories as shown below (Quirion, 2004).

- *NA:* neutral anionic
- *NN:* neutral nonionic
- *DA:* degreaser anionic
- *C:* cationic
- *DG:* degreasers based on glycol ethers
- *DL:* degreasers based on limonene.

Main ingredients of the six major floor cleaner categories are shown below:

- The *surfactants* help disperse dirt and fat in the cleaning solution.
- The *cosolvents* facilitate the penetration of the cleaning solution through the fat and dirt.
- The *alkaline salts* cut fat into smaller molecules that are easier to remove.
- Strongly alkaline floor cleaners that contain an appreciable amount of cosolvent are often designed as *degreasers.*
- Floor cleaners with a lower alkalinity and no or very little cosolvent are often referred to as *neutral.*
- The ionic nature of the surfactant, *anionic, cationic* or *nonionic,* is also an important distinction for the determination of the six categories.
- Two types of cosolvent are often added to degreasers: *glycol ethers* are water soluble, and *limonene,* extracted from citrus fruits, is not.
- Alkaline salts, such as *hydroxides,* are being replaced increasingly by *metasilicates* to generate the high pH of degreasers.

6.6.2 FLOOR CLEANING ISSUES AND METHODS

During preliminary field studies, it was observed that in most cases workers never received training on how to clean floors. It was also noted that workers and employers were eager to obtain more information on adequate cleaning methods (Quirion, 2004).

When a mop solution (water containing detergent) is applied, it emulsifies the soil and releases it from the surface. While suspended in the solution, the loosened soil should be easily removed by the mop, transferred to the rinse bucket, and discarded. If pickup is incomplete, the soil-laden solution settles into low spots on the floor, including grout lines and in the textured valleys of floor surfaces. The water evaporates, leaving a residue of detergent and soil particles that attaches firmly to the surface. Over time, if this cycle of incomplete cleaning is repeated, soil and detergent buildup becomes substantial and visible, especially when wet. Continuing such "slop mopping" can result in spreading this buildup to the remainder of the floor area.

The natural surface porosity of a porous floor surface gives the floor its slip-resistant characteristic. Over time, these pores become blocked, causing the floor to be more slippery. Organic and inorganic causes of pore blockage must be removed to achieve a safer floor. Traditional cleaners and degreasers can adequately remove fats, oils, grease, and solid deposits, but not inorganic stains.

This issue is different from sealed floors. Traditional cleaners may succeed in degreasing sealed floors; however, to achieve safer conditions, the floors must be effectively cleaned and provided with a slip-reducing agent.

Soil impregnation occurs when soil becomes trapped in the finish. To remove this embedded dirt, the wax/polish coat must be removed and reapplied. Burnishing can further embed the dirt. For this reason, the floor needs to be cleaned well in the first place using a dry mop, wet mop, applying cleaner, and reapply finish.

CASE STUDY: New Floor Required a Different Cleaning Regime, but Did Anyone Tell the Cleaners?

Following a number of slipping accidents, hospital management decided to replace an old kitchen floor. They decided on an epoxy based floor with a slip-resistant surface. Shortly after, the flooring supplier was contacted for assistance. The floor was stained in various areas and did not appear clean. The cleaners reported that the new floor was difficult to clean, was becoming slippery in parts, and was damaging their mops.

The supplier discovered that the cleaning instructions appropriate to the new floor were not being observed. Upon using a stiff brush and a bucket of warm water with the specified concentration of cleaner, the stains disappeared and the original floor color returned.

http://www.hse.gov.uk/slips/experience/tell-the-cleaners.htm (HSE)

6.6.2.1 Floor Stripping

Inadequate stripping and/or rinsing can leave a residue that interferes with the film formation process. Using a floor "neutralizer" can leave the floor coated with residual acid which reacts with the alkaline components in the polish causing instability and improper film formation (Interpolymer Corporation). General instructions for proper floor stripping are shown below:

- Sweep with a treated dust mop or vacuum the floor.
- Remove gum and other foreign material with a putty knife.
- Follow instructions for using the stripping solution and pour into the bucket. Start in the farthest corner from the entrance. Use cool tap water, because hot water can loosen tile adhesive and cause the solution to dry too fast.
- Dip mop head into stripping solution and wring slightly. Fan out the mop head on the floor and apply solution at the edges where there is the most buildup.

- Apply the solution in a 6 to 7 ft. (2 m) arc or side-to-side movement to cover the area between the edges. Cover only a 100–125 sq ft. (9.3–11.6 m²) area at a time.
- Let the solution soak for 4 to 5 minutes.
- Strip the area covered with solution, overlapping machine strokes using the floor machine with a stripping pad.
- Use the mop and empty bucket or wet/dry vacuum to pick up dirty solution. Do not allow solution to dry.
- Rinse area using fresh water and mop head which is slightly wrung out.
- Clean mop head with water after the first floor rinsing.
- Rinse twice more with fresh water. In the final rinse, add neutralizer per instructions. A neutralized floor provides a better bond of the finish to the tile.
- Allow floor to dry, and then restrip any high gloss spots.
- Allow floor to dry for at least one hour after final rinse.
- Check floor to be sure it is ready for finish by wiping the hand across it. If white powder appears, the floor has not been rinsed properly and must be rinsed again. When the white powder no longer appears, the floor is ready for finishing (ETC of Henderson).

6.6.2.2 Wet Mopping Floors

General instructions for proper wet mopping are as follows:

- Dust mop floor.
- Start at the farthest corner and work backward toward the door.
- Follow instructions for using the cleaning solution and pour into bucket until it is ¾ full. Fill the second bucket ¾ full with fresh water.
- Dip mop head into cleaning solution and wring mop head out slightly.
- When wet mopping a hallway, mop the floor along the edge of the baseboard first. Place mop head at the baseboard 3 to 4 ft. (1 m) away from the corner and mop toward the corner.
- Mop the open floor area by moving the mop side to side in a figure-eight motion. Overlap each stroke as you move back. Mop head should pass 1 ft. in front of the shoes. Hold the mop at a 15° angle from vertical.
- Each time both sides of the mop head are soiled, rinse mop head in fresh water bucket and wring. Change water as needed.
- Use only fresh water when using cleaning solutions requiring rinsing (ETC of Henderson).

6.6.2.3 Floor Buffing/Polishing

General instructions for proper floor buffing/polishing are shown below:

- Dust mop area.
- Damp mop area if necessary.
- Using a floor machine with a buffing pad, buff the floor area by:

Flooring and Floor Maintenance

- Starting along the baseboards at the farthest corner from the entrance into the room.
- Moving backward toward the entrance about 10 ft. at a time, buff the room using a side-to-side motion, overlapping the strokes.
- After buffing the entire floor area, use a clean, treated dust mop to pick up dust left from the buffing operation (ETC of Henderson).

6.6.2.4 Floor Care Equipment Maintenance

The quality of the treatment and cleaning of floor surfaces is dependent not only on the chemicals and procedures in use, but also the maintenance of the floor care equipment and supplies.

The following are some general guidelines and best practices for these tools:

Dust mops
- Do not use on wet, oily floors.
- Hang the mop with the head facing down without touching the floor when not in use.
- Change the mop head when soiled.
- Be sure that the mop block or frame is the proper size for the mop head.

Floor machines
- Rest the machine on its wheels, not on its brush or pad driver.
- Clean the machine and the electrical cord after each use.
- Inspect the cord regularly for fraying or loosening.
- Wear rubber overshoes and rubber gloves when operating on a wet floor.
- Consistently check nuts, bolts, and screws that may become loose.
- Perform motor adjusting or repair using qualified personnel.

Floor machine pads
- Install and center pad on the machine. An uncentered floor pad will wear unevenly and make the machine more difficult to control.
- Avoid bumping into objects when using the floor machine. If a pad in motion strikes objects, the pad can tear or snag.
- Use the correct cleaning method for each type of pad, because different types are made of different materials.
- Clean natural fiber pads by using the center die-cut piece from the pad or a medium bristle brush and brush away the accumulation from the pad.
    - For a polyester or nylon pad use one of these methods:
        - Soak the pad in stripping solution until dirt is softened or loosened, and then rinse using a water hose.
        - Wash under high-pressure water.
        - Launder in lukewarm water.
- Hang pads in a storage area to dry (ETC of Henderson).

Mop bucket and wringer
- Buckets should be checked before and after each use for cleanliness. Even a slightly dirty bucket will contaminate fresh water or solutions.
- Keep screws and bolts tightened.
- Keep casters properly lubricated and in good condition. Replace when necessary.
- Clean both the bucket and wringer daily. Do not lengthen the handle because too much force on the wringer can break it.
- Ensure the bucket and wringer are large enough for the mop that is being used.
- Remove loose mop strands and other articles caught in the wringer.
- Use a liner when applying a sealer or finish. Upon completion, dispose of the liner and clean buckets and place them in storage upside down.

Push brooms
- Rotate the brush frequently so as not to unduly wear one side.
- Do not lean heavily on the handle.
- Do not let the brush stand on the fibers, because this will bend them out of shape and impair its use.
- Use the broom only for the purposes intended.
- Use a large enough brush for the task at hand.
- Comb the brush at least weekly to ensure it remains in good condition.

Wet mops
- Most synthetic mop heads contain a fiber coating which makes them less absorbent. New cotton mop heads do not absorb as well initially because the fibers are still coated with natural oils. Before using a new mop, wash it in soap and tepid water, and rinse well.
- Mop heads are made of cotton, rayon, or material blends. Cotton mop heads are satisfactory for most floor care procedures with the exception of finishing/waxing. Since rayon mop heads leave little lint, it is a good choice for floor finishing and waxing. Since cotton holds water better, it is an appropriate selection for drying floors.
- Additional guidelines include the following:
  - Use a clean mop head when scrubbing, sealing, or applying a finish. Use old mop heads for applying stripping solutions.
  - Clean and wring (rinse well or launder) mop head after each use.
  - Launder mop heads in a synthetic mesh laundry bag.
  - Use a dedicated mop head for each procedure (e.g., scrubbing).
  - Do not twist or squeeze the mop too hard because this will break the strands.
  - Hang mop head to air dry after each use. Hang each without letting wet mops touch other equipment, walls, or floors when in storage. Use mop with care on splintered floors or floors with projecting nails, so as to not catch and tear the strands.
  - Economy of effort dictates that worn mop heads should be replaced.

Flooring and Floor Maintenance

- Do not leave mop head in chemicals or cleaning solutions even for short periods of time.
- Do not bleach mop head or use it with a solution with bleach.
• Do not wash in water over 160° F or dry in temperature above 150° F.

CASE STUDY: Cleaner Identifies Problem with Cleaning Equipment

A cleaner in a food factory noticed that the scrubber-drier was not removing the greasy contamination from the floor, making the floor slippery.

The issue was reported to the cleaning supervisor who determined that the wrong concentration of detergent was being used. It had become practice to use one capful of detergent in the scrubber-drier, far below the manufacturer's recommendation.

The supervisor also observed maintenance issues with the unit. The squeegee was in a poor condition, so it did not effectively remove water from the floor. The scrubber-drier was repaired, preventative maintenance implemented, and training provided to the cleaners. Once the recommended detergent concentration was used, there was a rapid improvement.

> Source: HSE Information Sheet—Slips and Trips:
> The Importance of Floor Cleaning

6.6.2.4.1 *Preventative Maintenance of Floor Machines*

Following is a guideline for performing preventative maintenance on floor machines.

Appearance
- Machine body is clean.
- Bump wheels and splashguards are operational.

Water system
- Solution tank is clean.
- Filter screens are clean.
- Water valves are operational.
- Water hoses are in good repair.

Recovery system vacuum
- Vacuum heads are clean.
- Recovery tank is clean.
- Machine hood is raised.
- Tank gaskets are dry.
- Drain system is operational.
- Squeegee is clean and assembly is operational.

Drive system wheels
- Front and rear wheels are greased.
- Brush pressure arm is greased.
- Pivot points are oiled.

- Brushes are clean, dry, and in good repair.
- Pad holders are operational, and the pad is the proper size.

Electrical system
- Switches are operational.
- Wiring is in good repair.
- Vacuum and drive motors are clean.
- Batteries
  - Water level is correct. Check weekly and refill as necessary using distilled water but do not overfill.
  - Posts and tops are clean.
  - Charger and connectors are operational.
  - Batteries are charged.
- Batteries are stored in dry, cool places, but above 32°F.

6.6.3 Floor Finish Indicators and Maintenance Issues

Indicators of a floor finish in poor condition include the following:

- Spills cannot easily be removed without leaving a stain.
- The floor has a yellow, gray, or patchy appearance (indicating wax buildup).
- The floor finish is flaking or chipping.
- The floor has a visibly dull finish with dark wear spots.
- Bare floor shows through the finish/sealer.
- When burnishing, the floor pad quickly becomes dirty.

Some maintenance issues that should be considered when establishing an effective floor maintenance program include the following:

- The polish film can get cut, which makes it difficult to properly clean as the dirt extends deeper into the film. Cleaning should exceed the standard procedure to effectively remove dirt embedded into a cut.
- Scuffmarks abrade the coating, leaving a rough surface. The more polish/wax that is applied, the less scuffing occurs, because the polish/wax coating absorbs the energy of the heel strike.
- High-speed burnishing removes some amount of the coating by abrading some of the film. Continuous burnishing without reapplying polish can result in thin or no coating left.
- Spray buffing is often used in institutional settings (e.g., schools and hospitals) to soften the finish; however, this process also removes layers of material.

6.6.4 Outsourcing Floor Care

While some companies maintain in-house staff and equipment for floor cleaning and treatment, most companies approach floor care in one of three ways:

- They own their own equipment and use in-store employees for labor.

Flooring and Floor Maintenance

- They own the equipment and outsource the labor to a building service contractor (BSC).
- They outsource the labor and the equipment to a BSC.

While equipment accounts for only 10% of a floor care budget, this only involves the acquisition of equipment compared to the costs of labor. However, the true cost of floor care equipment can be as high as 20%, particularly when it is not properly maintained.

6.6.4.1 Outsourcing Labor and Equipment

A common practice is to outsource all responsibility for floor care to a BSC. By doing so, the company loses negotiating leverage and ultimately, control of results. The average BSC relationship only lasts 2.5 to 3 years, because over the course of a contract, the quality of the floor care declines until the facility terminates the relationship and looks to hire a new contractor.

A major cause of eroding quality is the BSC's high employee turnover, which on average is 100% to 300% annually. Employees are not on board long enough to receive adequate training or to learn to properly operate and maintain the equipment. While this generally does not affect machine operation, since most auto scrubbers and burnishers are one-button machines, it does affect machine maintenance.

Floor care equipment requires daily maintenance such as replacing brushes and squeegees and checking battery water levels. Without adequate daily maintenance, machine performance declines rapidly, which can ultimately lead to equipment failure. According to one study, half of the facilities of one major retailer have at least one piece of floor care equipment that is broken or needs major maintenance work on a given day.

Poor machine performance and failures increase employee turnover. When machines underperform, floors do not shine and facility management becomes dissatisfied. When machines break down, contract laborers must resort to cleaning floors manually using mops and buckets. On average, a machine is out of service for approximately five to seven days. Mopping floors for an extended period, and then scrubbing and shining the resulting dingy floors make it difficult to achieve suitable quality and often causes workers to seek work elsewhere.

This cycle of poor machine performance and high turnover results in a floor care program prone to failure and eventual termination of the contract. The BSC suffers losses in repair costs, downtime, and lost productivity. Costs to the facilities increase each time they must change contractors, because it costs less to maintain a relationship than to begin a new one.

For the facility, a new contract means the BSC must acquire new equipment to service the account. The facility must then pay for a new fleet of equipment approximately every three years. The BSC will understandably purchase lower-quality equipment to reduce costs, further shortening the cycle of equipment downtime and increasing employee turnover.

6.6.4.2 Owning Equipment and Outsourcing Labor

To address these equipment concerns, some facilities purchase floor care machines and outsource the labor. This provides more control over equipment selection and more leverage with the BSC. Owning equipment brings unexpected costs. The initial investment in the hardware can be significant and is often amortized over five or more years, which can extend beyond the useful life of the machines, particularly when they are not properly maintained.

A machine fleet also requires additional manpower, to perform tasks such as arranging for repair. Many companies have accumulated a variety of machine models from different manufacturers. This requires reaching repair agreements with different suppliers and contacting the correct manufacturer for a given machine. An alternative is to work with service centers that can repair any machine. Unfortunately, most do not have national coverage. This may require the company to employ an equipment service group to manage contracts and dispatch manufacturers or a network of local service centers.

Another major factor is equipment abuse, which can account for up to 30% of all repair charges. If BSC employees abuse equipment, which can be verified by the service company, the company should charge applicable costs to the BSC. Some companies spend nearly a thousand dollars a month in manpower, IT infrastructure, and resources just to manage these issues. Failure to implement such a control process can result in paying hundreds of thousands of dollars in repairs that should legitimately be assigned to the contract laborer.

6.6.4.3 Owning the Equipment and Using In-House Labor

By having its own staff perform floor cleaning, companies reduce equipment abuse and eliminate charge-backs. However, the same capital investment must be made, a service support network must still be established, and condition of equipment must still be monitored. Studies have found that companies pay more to own and maintain their fleet than they realize. Onsite equipment surveys have discovered the following:

- Machines that could be repaired are simply removed from service and stored.
- Machines that should have been retired continue to be repaired.
- The costs for spare parts (accounting for 60% of a repair budget) are excessive.

One large national retailer discovered that every store had at least one piece of equipment standing idle. The retailer also realized it was paying list price for spares obtained locally, despite contracts that specified discounts. The retailer estimated at least 10% of the budget was spent to repair machines that should have been retired. Most retailers do not have such a clear picture of their floor care fleet. Since they are generally unaware of idle machines that could be repaired, they often overpurchase machines.

Equipment outsourcing is more than simply leasing versus buying. Many manufacturers offer lease programs that move the capital expense to a fixed monthly operating cost. Best practices for managing equipment include the following:

- *Assess equipment.* Begin with assessing your existing fleet of floor care equipment. Few companies have an accurate database of this machinery. For each facility, identify machines that should be repaired, machines that

Flooring and Floor Maintenance

need preventative maintenance, and those to be retired. Use asset tags to track equipment.

- *Equipment refurbishing.* If possible, locate a nationwide, vendor-independent service network that can repair machines from any manufacturer. Refurbish machines that need more extensive repairs. Negotiate with the service network to discount the cost of spare parts. With spare parts equaling 60% of the repair budget, these savings alone can be significant.
- *Maintenance for existing fleet.* Secure a scheduled maintenance program for existing machines. This can help to gain additional years of useful life from them, meaning machines will not have to be retired before they are fully amortized. This also mitigates potential training issues for BSC employees, because even routine maintenance is performed by the service vendor. With floor care machines up and running, facilities do not require additional scrub and shines that are often needed when equipment is broken. Working equipment also means less labor turnover and, again, higher productivity.
- *Scheduled equipment replacement.* Target a three- to four-year phase-in of new equipment, depending on the condition of the existing fleet. New equipment can be acquired through a capital lease or an operating lease. At the end of this period, the entire fleet will have been replaced with the latest floor care technology, providing another boost in productivity.
- *Long-term maintenance.* When responsibility is assigned to a service network for fleet maintenance, the machines will likely perform beyond the expected three-year life cycle (Castle Rock Industries, 2007).

Online Resource: CM B2B Trade Group

NTP Media is a publisher of information for the water treatment, car care, and facility maintenance industries. The company's CM B2B Trade Group™ offers products specifically for professionals in the carpet care, custodial services, and building maintenance markets. CM B2B Trade Group products include the following:

- *CM/Cleaning & Maintenance Management®* magazine, founded in 1963.
- *CM/Cleanfax®* magazine is targeted to the carpet cleaning and disaster restoration sector.
- CM e-News Daily™, an e-newsletter with more than 20,000 opt-in subscribers.
- Cleaning Management Institute®, a membership society for training and education of building maintenance professionals.

CM/Cleaning & Maintenance Management Online:
http://www.cmmonline.com/index.asp
CM/CleanfaxOnline:
http://www.cleanfax.com

6.7 CARPET MAINTENANCE

The frequency and associated costs of carpet replacement can be reduced with proper care by as much as five years from an average of six years to 13 years. Carpet replacement does not just involve the cost of the carpet and labor, but the disassembly/reassembly of work stations, recabling, and moving other equipment and fixtures out and back in.

According to State Farm research, carpet cleaning contractors budget only 8 to 10 cents per sq ft. on average. Based on the increased service life of carpet obtained through proper maintenance, it was estimated that anything less than 31 centers per sq ft. would result in a reduction of costs (Newman, 2003).

6.8 ASSESSMENT OF FLOOR TREATMENT/ CLEANING PRODUCTS AND METHODS

Slip resistance testing can provide a means of directly comparing the relative effectiveness of current or proposed floor treatments, cleaning products, and cleaning methods upon application and over time.

A systematic approach to selecting an effective floor treatment or evaluating a cleaning regimen is required. Finding the right product is often a process of elimination. Working with a professional will enable you to document the steps that support your flooring recommendation. A sound approach should include the following steps:

1. Develop a list of potential floor treatment products based on the type and use of flooring, and the appropriateness of the treatment for your application.
2. Review application and maintenance requirements. Methods vary by the type of treatment; some are simple and safe to apply, while others require professionals. Also, you must be prepared to meet care and cleaning requirements and the maintenance needs (e.g., frequency of application) in terms of time, manpower, and other resources.
3. Patch test the resulting short list of products on your flooring to further narrow the field. This is best done by a safety professional trained and experienced in the use of the tribometer selected.
4. Conduct a 30-day trial of those products that performed the best out of the box. Monitor the results by remeasuring the slip resistance, identify changes in the frequency of slip/fall incidents, and listen to feedback from staff. This will narrow the field to a small handful of products which should then be tested for an extended period of 90 days or more.

The bottom line in selecting a floor treatment is a controlled evaluation. Vendors offering a multifaceted set of solutions can be problematic. For example, a vendor might package shoes, floor treatments, signs, and other program elements. Given that *so many changes were made simultaneously,* it is difficult to isolate the aspect or aspects of the program that had a significant impact from those that were superfluous. You need to make one change at a time to accurately assess the impact of that change.

6.9 FLOOR MAINTENANCE CERTIFICATION

The only known certification programs related to floor maintenance are sponsored by industry groups or companies involved in the manufacturing of floor products.

6.9.1 IICRC Hard Surface Floor Maintenance Specialist (FCT)

The Institute of Inspection, Cleaning and Restoration Certification (IICRC) is the certification registry and standard setting organization for the inspection, cleaning, and restoration industry (http://www.advancedse.com.au/index.asp?page=training_IICRC). The IICRC was established in 1972 to provide a way in which professionals could receive training and become certified. It is the largest nonprofit registry of inspectors, cleaners, and restorers serving the United States, Canada, Great Britain, and Australia. It is owned and controlled by trade associations.

The IICRC's purpose is to set industry standards, establish required course work, grant accreditation of schools and instructors, administer uniform testing, and maintain a registry of certified professionals. It certifies that individuals meet prescribed levels of proficiency based on training, education, and experience. IICRC certification must be renewed annually by earning IICRC approved continuing education credits (CECs).

6.9.2 Certified Floor Safety Technician (CFST)

Founded in 1999, SureGrip Floor Safety Solutions markets products and implements floor safety programs. SureGrip sponsors the Certified Floor Safety Technician® educational program (http://www.suregripfloors.com/cfst.php) with the International Board of Environmental Health and Safety, Inc. (IBOEHS) North America. This course, endorsed by the IBOEHS, includes in-depth studies in floor safety, the physics of slips and trips, floor safety regulation, traction enhancement, maintaining a safe floor environment, and marketing floor safety.

6.9.3 Rochester Midland Corporation

A specialty chemicals manufacturer since 1888, the Green Housekeeping section (Greener Parks) of the Institutional Division offers evaluation of cleaning practices and equipment, and training and certification for personnel on maintaining resilient, wood, synthetic, concrete, and terrazzo floors.

6.10 FLOOR TREATMENT STUDY

In an extensive study, ESIS®, Inc. (ESIS), evaluated 10 floor treatments which clearly demonstrated the wide range of effectiveness of treatments marketed as slip resistant. There is a host of reasons why product names have not been released which are summarized below.

Unfortunately, there is no "silver bullet" for slip resistance. Each set of conditions and variables is different. The study determined only whether or not the treatments worked out of the box, on new (unworn and uncontaminated) marble and ceramic tiles. This is not necessarily an indication of how these products perform in the presence of contaminants, under worn conditions, or on other types of flooring. The study attempted to prove that, even under identical conditions, floor treatments vary. When significant variables are added such as wear, contaminants, and types and grades of flooring, it is unrealistic to reach anything but broad conclusions about these products.

Other concerns include application and maintenance. Some of the treatments require professional application and use caustic chemicals. Some require more frequent cleaning and reapplication than others.

All these issues come down to the fact that each situation needs to be evaluated on its own. There is no "one size fits all" situation. What will work for one facility may not be appropriate for another. A treatment that works effectively in wet conditions may be impractical or ineffective in greasy situations. A treatment that delivers superior slip resistance on marble may have no discernable effect on vinyl. Even a treatment highly effective on one grade, type, or manufacture of marble, ceramic, or resilient flooring may not be as effective on another.

EXHIBIT 6.1
Slip-Resistant Treatment Study 2000

ESIS RISK **C**ONTROL **S**ERVICES

Table of Contents
- Introduction
- Goals
- Protocol
- Other Specifications
- Surface Preparation
- Test Notes
- Packing and Shipping
- Treated Tiles Testing
- Notes on Maintenance
- Notes on Appearance
- Other Notes
- Comparison of Results–Ceramic
- Comparison of Results–Marble
- Credentials and Claims
- Federal Laws and Standards
- Consensus Standards
- Independent Laboratory Testing
- Conclusions and Summary
- Exhibit: Test Protocol
- Exhibit: Statistical Analysis

Note: References to laws, regulations, standards and guidelines are not intended to be legal opinions concerning the interpretations of those documents. They are the authors' opinion only. The information contained herein is not intended as a substitute for advice from a safety expert or legal counsel you may retain for your own purposes. It is not intended to supplant any legal duty you may have to provide a safe operation, product, workplace or premises. ESIS makes no representation that either using or avoiding any of the products tested will reduce the frequency or severity of accidents.

Introduction
This study represents a refinement of earlier research completed in 1998. Both studies, however, were conducted in response to our loss control clients' need to develop a more proactive or solutions-oriented approach to reducing their slip and fall risk.

There are currently hundreds of floor surface treatment products on the market that claim to provide slip resistance. The true efficacy of these products—apart from the manufacturers' claims—is, however, unknown.

Most products tested provided some improvement in slip resistance. Others offered a dramatic improvement. In some cases, resistance was actually reduced. Overall, effectiveness varied widely, particularly on wet test surfaces.

The results offer the first clear evidence from a scientific study that noteworthy differences among slip resistance floor treatment products do exist. They also suggest that manufacturers' product information may not be a reliable guideline in selecting the right products to assist in reducing slip and fall loss costs.

Slip and fall accidents account for a majority of general liability claims in real estate, financial, and retail operations. They also represent the second most frequent type of occupational injury. Their causes and control are therefore critical to successful loss control.

The study findings are not intended as opinions on the quality, merchantability, or fitness for the intended purpose of any product. Neither ESIS nor the authors of the study express any opinion on whether any products are defective in any way, and the study draws no conclusion, inference or implication that any product is in any way dangerous, defective or manufactured or designed improperly. No opinion is intended or offered with respect to the adequacy of any product warning.

Goals

The goal of this study was to compare the relative effectiveness of a variety of slip-resistance floor treatment products.

This study was conducted for the purpose of increasing the existing body of knowledge among safety professionals concerning a particular type of safety hazard. All findings are relative only in that all comparisons are between and among the products tested. No opinions are offered with respect to any product not tested.

Overview

Two types of floor surfaces were selected for testing. Test surfaces were (1) glazed ceramic tile and (2) marble in 12-in. × 12-in. squares, purchased at a retail facility.

A number of slip resistance floor treatment product manufacturers were contacted and invited to participate. The ten (10) vendors who agreed to be a part of this study are identified as Group One through Group Ten.

Flooring and Floor Maintenance

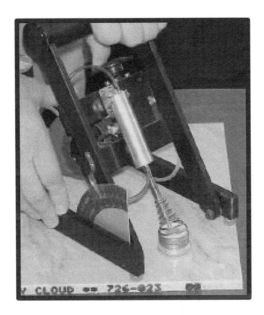

Each set of two surfaces was pretested as indicated below. One set of surfaces for each vendor was packed and shipped via UPS with instructions regarding application, documentation, and return transit.

Once the treated surfaces were returned to ESIS RCS from the vendors, they were retested under the same conditions. The results were recorded and compared. David Underwood, Ph.D., an analytical chemist and member of the American Society for Testing and Materials (ASTM) technical committee F-13, performed the statistical analysis. ASTM F-13, *Safety and Traction for Footwear,* develops standards and methodologies for slip resistance testing. Dr. Underwood compared the results obtained during pretesting with those obtained from the treated tiles.

Protocol

The test instrument for this study was a variable incidence tribometer (VIT), which was inspected and calibrated by the manufacturer prior to pretesting and again prior to the testing of treated tiles. The English XL has undergone a series of "round robin" workshops, conducted by the American Society for Testing and Materials (ASTM) over several years to demonstrate repeatability and reproducibility of test results.

The test foot material for this study was Neolite® test liner,* a generic and durable substance that is one of the most commonly used materials for slip resistance testing. Unlike leather, the properties of which are affected by moisture and wear, the characteristics of Neolite test liner do not change under normal conditions. Neolite test liner is recommended by the tribometer manufacturer and specified by ASTM

---

* Neolite® is a registered trademark with Goodyear Tire and Rubber Company.

D-5859, *Standard Test Method for Determining the Traction of Footwear on Painted Surfaces Using the Variable Incidence Tribometer.*

Testing was done in accordance with:

- *ASTM F-1679 Standard Test Method for Using a Variable Incidence Tribometer* (VIT), released in 1996, which recognizes the English XL VIT as a valid slip resistance field testing device for wet and dry surfaces.
- The most recent release (updated June 1999) of the instruction manual and supplement published by the manufacturer for the English XL Slip Resistance Tester.
- The test foot preparation protocol was done in accordance with the manufacturers' current specifications, using 180-grit silicon carbide sandpaper.

Other Specifications

Pretesting and treated testing were completed at a single location, in a temperature-controlled environment, and on the same level surface. Treated tile testing was performed on the same portions of the test quadrants identified and tested during pretesting. All testing was completed by the same operator, a member of ASTM F-13 and F-06 (*Resilient Floor Coverings*) who is ESIS-certified in slip resistance testing and has five (5) years of experience with the tribometer.

For statistical reliability, three sets of four readings (one for each quadrant) were taken.

Surface Preparation

Each surface was marked using an indelible marker on the underside of each tile, indicating:

- testing quadrant (A, B, C, D)
- group number (1–10)

Flooring and Floor Maintenance

The tiles were cleaned by running them through a dishwasher and were allowed to air dry.

Test Notes

- Tested each quadrant dry, then each quadrant wet
- Completed all dry testing for all surfaces first, then performed wet testing

	Dry Testing		Wet Testing
Relative humidity	39%–42%	Relative humidity	41%–43%
Temperature (F)	75°–77°	Temperature (F)	72°

Packing and Shipping

Each group (set of two pretested tiles) was packed and shipped to vendors in the following manner:

- Each tile was packed with three layers of bubble wrap, secured with cellophane packing tape.
- A layer of packing "peanuts" was placed at the bottom of the shipping carton.
- A layer of packing peanuts was placed between the tile packages and on top of the last tile to the top of the box.
- Shipping was done by UPS ground transportation.
- The tiles were returned by the same method shipped (UPS Ground Ship, prepaid).
- Return packaging used the same method and materials (e.g., bubble wrap and box) as the outgoing shipment.
- Condition of packing and tiles upon receipt was good: no visible damage was noted.

Treated Tiles Testing

	Dry Testing		Wet Testing
Date of test: 09/08/99		Date of test: 09/09/99	
Relative humidity	42%–47%	Relative humidity	41%–44%
Temperature (F)	72°–75°	Temperature (F)	72°–73°

Treated tile testing was completed using the same process and protocol as pretesting.

Notes on Maintenance

This study was completed under relatively ideal conditions. Treatments were applied to the test surfaces by the vendor. Testing of the treated tiles was completed fairly soon after application of the product. Since there was no wear due to use (e.g., walking on the surfaces), the treatment remained in pristine condition for testing.

Under real-life conditions, treated surfaces are continuously exposed to wear from pedestrian traffic, which can compromise the integrity and effectiveness of the product. As a result, proper maintenance is essential to sustaining effective slip resistance.

The extent and mechanics of maintaining treated surfaces varies considerably by the type of treatment. When selecting a given treatment, consideration should be given to the nature of the maintenance requirements. For effectiveness of a treatment over time, the manufacturer's instructions must be observed.

Notes on Appearance

A degree of dulling of the glossy surface occurred with most treatments, some more prominently than others. The amount of dulling depends upon a number of variables, including the type and quality of flooring material to which the treatment is applied. Appearance is another factor that must be considered when selecting the appropriate treatment for a given facility. To assure the suitability of the appearance, it is strongly recommended that treatments be applied to a sample of the surface to be treated, or in a small remote area prior to full-scale use.

Other Notes

There are a wide variety of treatments on the market, some of which have been developed to be effective only when used on specified types of flooring material. Care must be taken to assure that the treatment selected is suitable for the floor to which it is to be applied.

Another important factor to consider in selecting a floor treatment is the method of application. Some treatments involve the use of harsh chemicals and require expertise to apply properly, while others are nontoxic and are designed for use by the nonprofessional.

Comparison of Results–Ceramic

Ceramic–Dry

On dry ceramic, even the untreated tiles exceed the generally recognized 0.50 guideline. While all treatments increased slip resistance to some degree, some treatments (such as Groups 2, 6, and 7) provided substantial improvement, exceeding 0.9.

Flooring and Floor Maintenance

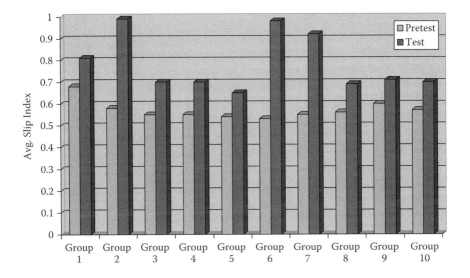

Ceramic–Wet

Wet ceramic untreated tiles were mostly between 0.1 and 0.2—very low slip resistance. All but one treatment (Group 8, which actually decreased) increased slip resistance, and Groups 1, 2, 6, and 9 provided improvement beyond 0.50. Additionally, Groups 2 and 6 did not slip, even at 1.0, the most horizontal position of the tribometer. This is a remarkable result, since both treatments managed to increase slip resistance from approximately 0.1 to more than 1.0, a ten-fold improvement.

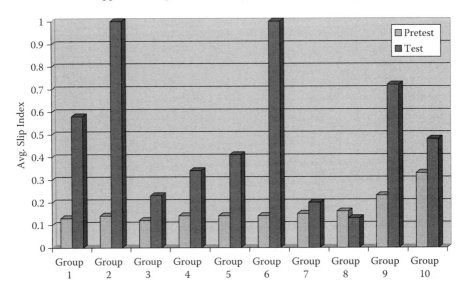

Comparison of Results–Marble

Marble–Dry

Like dry ceramic, untreated dry marble exceeds the generally recognized 0.50 threshold. While some treatments (such as Groups 1, 2, and 7) provided the most improvement (exceeding 0.8), others showed minimal improvement (Groups 4 and 5), and one demonstrated an actual decrease in slip resistance (Group 8).

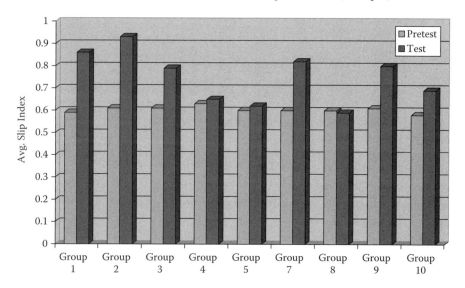

Marble–Wet

Wet marble untreated tiles pretested in a way similar to wet ceramic, mostly in the 0.1 to 0.2 range. It is clear that Group 1 and 2 treatments demonstrated dramatic improvements (beyond 0.8 and 0.9, respectively). Others, such as Group 5, 7, and 8, showed minimal increases. The Group 3 treatment showed a reduction in slip resistance from the untreated condition of the wet marble.

Note: Group 6 was omitted from marble results, since no testing was performed.

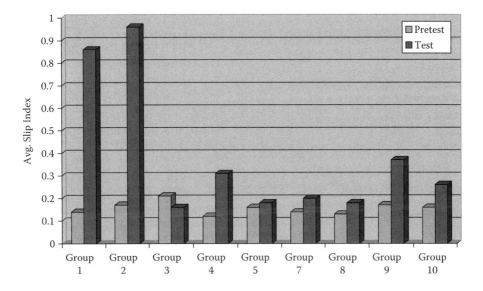

Credentials and Claims*

The product information available on products falls into one of two primary categories. Some vendors advertise that their products will meet or exceed applicable federal laws. Many state they meet industry consensus standards by engaging an independent testing firm or by testing the product in-house. In comparing the claims of vendors against the study results, there appeared to be low correlation between product claims and the efficacy of the product.

Federal Laws and Standards

The two most cited federal laws with regard to slip resistance are from the Occupational Safety and Health Administration (OSHA) and the Americans with Disabilities Act (ADA). Some vendors claim to meet or exceed the "standards" or "requirements" contained within these laws.

The OSHA "standard" for slip resistance is not a law, nor is it a standard. It is a proposed nonmandatory appendix item set by OSHA and was never adopted as a standard (it specifies slip resistance of 0.50 or higher for the workplace). While it is possible for an OSHA inspector to cite this guideline under the "General Duty Clause," we could find no evidence that this is done in practice. What makes this particularly difficult is that OSHA has specified no test protocol or device upon which to base a citation—effectively making it unenforceable.

The ADA "law," like OSHA, is also a guideline, since the specification appears in an appendix. And because no test protocol or device is specified, even this recommendation is difficult to apply. The ADA specifies slip resistance of at least 0.60

---

* The information for this section was gathered from additional and prior research, and is not limited to the vendors that participated in this study.

for level surfaces and 0.80 for ramps, where accessible by persons with disabilities. Subsequent to enactment of the ADA, it was determined that the study conducted to validate the specified level of slip resistance was faulty. The study used a laboratory force plate to measure traction demand for the handicapped.

The Architectural and Transportation Barriers Compliance Board (also known as the ATBCB or Access Board) of the U.S. Department of Justice was created to ensure federal agency compliance with the Architectural Barriers Act (ABA). The Access Board adopted the ADA recommendations and has stated these specifications as a guideline, not a requirement or a standard. Again, no test protocol (e.g., device or test method) is specified.

Consensus Standards

Meeting or exceeding a consensus standard is another often-cited feature in the advertising of slip resistance floor treatments. In some cases, these standards provide only a testing methodology, not a measure of safety; in others, the test method of a standard is modified or not followed properly. Finally, the standards themselves may be obsolete.

American Society of Testing and Materials (ASTM)

The ASTM is the most active organization in the development of standards for measuring slip resistance. The ASTM has promulgated standards for a number of tribometers, including the variable incidence tribometer (VIT or English XL) and the portable inclineable articulated strut slip tester (PIAST or Brungraber Mark II).

However, with the exception of Standard D-2047 (involving a specification of 0.5 for the James Machine—see UL below), the ASTM has never offered a slip resistance threshold of safety, making it impossible to "exceed" an ASTM slip resistance standard. Most ASTM standards are "test methods," or steps to follow in arriving at a measure of slip resistance. "Meeting" an ASTM test method standard only means that the proper steps were followed using the appropriate test device. It is not relevant to the results of the testing, just to the method of reaching those results.

Tests to demonstrate the effectiveness of a treatment are often done using a horizontal dynamometer pull-meter method, a device requiring the use of a 50-pound weight. It is known by experts for overestimating the slip resistance of wet surfaces due to a phenomenon known as "sticktion." This ASTM document, Standard Test Method for Determining the Static Coefficient of Friction of Ceramic Tile and Other Like Surfaces by the Horizontal Dynamometer Pull-Meter Method (C1028) has not been updated in many years. In addition, ensuring proper design and calibration of a self-constructed instrument brings into question the validity of the results.

One vendor cited ASTM D-56 in relation to the slip resistance qualities of the product. ASTM D-56 is titled Standard Test Method for Flash Point by Tag Closed Tester and is not related to slip resistance.

Flooring and Floor Maintenance

Ceramic Tile Institute (CTI)

Some vendors will market a slip resistance floor treatment on the strength of meeting or exceeding requirements as required by the Ceramic Tile Institute (CTI), an industry group representing the interests of the ceramic tile industry. CTI specifies use of ASTM C1028, which is a test method only and does not specify a level of safety. C1028 cannot be exceeded, since it specifies no "safe" or "unsafe" level of slip resistance.

Underwriters' Laboratories (UL)

Many vendors state that their products are classified or certified by Underwriters' Laboratories (UL) as to slip resistance.

While UL does not "certify" products as to slip resistance, it does "classify" products as such. UL will apply the treatment to a surface and test it to UL 410 using the James Machine. To qualify for classification and be permitted to use the UL Classification Marking, test results must show a slip resistance of greater than 0.50, and the manufacturer must pay a fee. Documentation of classification does not provide actual test results, but states only that it was greater than the minimum of 0.50.

The James Machine is a laboratory device designed to test under ideal conditions only for merchantability (the suitability of the product for sale) of new flooring products. This standard is inappropriate for application for field testing.

The James Machine is a complex device for which there are still no standard set-up, operating procedures, or precision and bias from any independent laboratory or consensus standards-making organization.

Complicating the situation, there have been several manufacturers of this device, and in each instance the apparatus was designed and built somewhat differently. Most important, the James Machine is not designed (or even listed by UL) for wet testing. Thus, testing done using this device can be considered questionable.

Independent Laboratory Testing*

The use of independent laboratories for testing products is a frequent component of marketing for slip resistance floor treatments. In most cases, these laboratories use the ASTM or UL standards. However, information demonstrating that the laboratory is qualified to perform such specialized testing is rarely available, and details on the conditions and results of testing are likewise unavailable.

Example:

> ... the increase in friction of wet surfaces was between 150% to 500%, based on independent testing with the James Machine.

Aside from the lack of details on the testing, the James Machine is not designed (or listed by UL) for wet testing, making these results irrelevant.

---

* The examples used in this section are not quotes from any specific product or vendor, but are intended to illustrate the types of claims a consumer may have to interpret.

Example:

... tested by one of America's leading independent testing laboratories ... results exceeded all national standards for dry floors, with a 300% increase on wet ceramic tile and 100% on wet marble.

Again, testing was done using the James Machine, inappropriate for all but dry laboratory testing of pristine surface materials. And while our tests also showed a 300% increase in ceramic tile when wet, our results showed 0.41, still below 0.50, the generally recognized safety level of slip resistance. For wet marble, our testing did not show an increase of 100%, but of 12.5%.

Example:

... thoroughly tested and approved by an independent testing laboratory.

The testing laboratory is a provider of insurance and financial services to automotive-related businesses. No details of test device, protocol, and results were provided.

Example:

Independent testing was done in 1993 in accordance with ASTM C1028.

Aside from the age of this testing, the testing laboratory specializes in providing environmental and geotechnical consulting services and does not appear to have expertise in slip resistance testing.

Conclusions and Summary

The measurement of slip resistance is an emerging field. New technology that was unavailable 10 years ago has pushed reliability far forward. But, it is essential that consensus organizations like the ASTM continue to oversee and refine standards and practices in slip resistance. At the same time, there needs to be a more sophisticated awareness about how current standards and technology might be applied appropriately, as well as their limitations.

With regard to slip resistance floor treatments, it is clear that more work needs to be done in evaluating the efficacy of these products. Hopefully, the results of this study will help demonstrate the wide range in effectiveness, as well as what is involved in interpreting marketing approaches used by manufacturers of these products.

Flooring and Floor Maintenance 231

EXHIBIT 6.2
Test Protocol

TEST PREPARATION

The English XL VIT was visually inspected before each testing session to ensure the security of fastenings, the alignment of the thrust cylinder, and the indicating pointer on the protractor.

Operating pressure of 25 PSI ±1.5 PSI was maintained for all tests.

Sanding was done according to the June 11, 1999 *Supplement to the XL Operations Manual.*

The Neolite test liner disk was sanded after each stroke that produced a slip. For wet testing, sanding was completed before each test session.

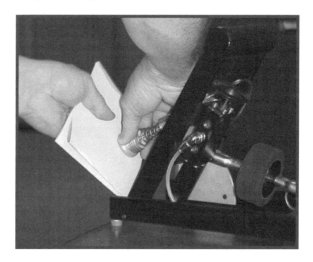

First, the XL was moved away from the tile (so that sanding dust would not fall onto the test zone).

The Neolite test liner disk was prepared by lightly sanding in a circulator motion for five cycles with 180-grit silicon carbide sandpaper with a hard backing. Fresh sandpaper was used when the paper became visibly worn. The test foot was then brushed off and returned to the testing position. In addition to sanding, the disk was rotated about ¼ turn after each slip. The combination of sanding and rotation of the disk avoids the potential for polishing of the disk, which could affect test results.

Wet Testing

Surfaces were first tested dry and then tested wet. Water was used for wet testing. Surfaces were wet in advance of actual testing to ensure that the surface material was adequately saturated. A thin, unbroken film of water was maintained on the surface.

Flooring and Floor Maintenance

Testing Process

The starting point of the mast angle was estimated conservatively, to minimize the potential of an immediate slip. A relatively low slip index (a more vertical mast, such as 0.2 slip index reading) was used, gradually working up to higher slip index (more horizontal mast).

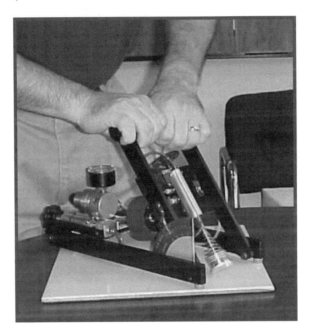

The hand wheel was turned about ¼ turn for each stroke. The actuating button was pressed for ½ second and released.

The process was repeated until the first full-stroke slip occurred. The results were then read from the slip index protractor (rounded to the nearest 0.01) and documented using the attached spreadsheet.

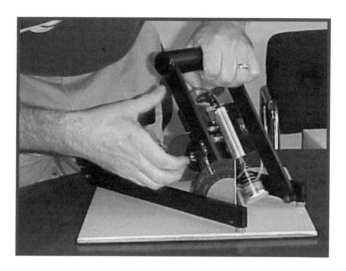

To ensure statistical reliability, three sets of four readings (one for each quadrant) were taken for each round of tests (e.g., readings on quadrants A, B, C, and D were taken three times for dry ceramic tile pretest).

EXHIBIT 6.3
Statistical Analysis

TEST RESULTS BY GROUP

The statistical analysis of test results was performed by David Underwood, Ph.D. A paired-comparison *t*-test was done, using the average from each set of four readings to obtain the probability of whether the numbers are the same. This probability ranges from 0 (no chance they are the same) to 1 (100% chance they are the same). Generally, if the probability is 0.05 or less, the numbers can be considered different. If a *p* value of 0.05 translates to $100 * (1 - 0.05) = 95\%$, there is 95% confidence that the numbers are different.

Where the confidence factor is in the 95% range, the results of the testing are considered to be statistically sound and reliable.

Sample	P Value	Confidence
Group One		
Ceramic dry	0.19	81%
Ceramic wet	<0.001	+99%
Marble dry	0.003	+99%
Marble wet	<0.001	+99%
Group Two		
Ceramic dry	<0.001	+99%
Ceramic wet	<0.001	+99%
Marble dry	<0.001	+99%
Marble wet	<0.001	+99%
Group Three		
Ceramic dry	0.02	98%
Ceramic wet	0.004	+99%
Marble dry	0.01	+99%
Marble wet	0.148	85%
Group Four		
Ceramic dry	0.001	+99%
Ceramic wet	0.002	+99%
Marble dry	0.10	90%
Marble wet	0.006	+99%
Group Five		
Ceramic dry	0.02	98%
Ceramic wet	0.002	+99%
Marble dry	0.34	66%
Marble wet	0.53	47%

Continued

Sample	P Value	Confidence
Group Six		
Ceramic dry	0.01	99%
Ceramic wet	<0.001	+99%
Marble dry	N/A	N/A
Marble wet	N/A	N/A
Group Seven		
Ceramic dry	<0.001	+99%
Ceramic wet	0.06	94%
Marble dry	<0.001	+99%
Marble wet	0.07	93%
Group Eight		
Ceramic dry	0.02	98%
Ceramic wet	0.13	87%
Marble dry	0.52	48%
Marble wet	0.03	97%
Group Nine		
Ceramic dry	0.03	97%
Ceramic wet	<0.001	+99%
Marble dry	0.02	98%
Marble wet	0.02	98%
Group Ten		
Ceramic dry	0.04	96%
Ceramic wet	0.02	98%
Marble dry	<0.001	+99%
Marble wet	0.06	94%

7 Overseas Standards

Rejoice not at thine enemy's fall—but don't rush to pick him up either.

Jewish proverb

7.1 INTRODUCTION

Organizations with overseas facilities may find that U.S. standards related to pedestrian safety are not well recognized. Even the research community is at odds when conducting studies on slip resistance and related issues. Whereas the United States has historically considered the static coefficient of friction (SCOF) as the most important measure related to pedestrian slip resistance, other countries have focused on the measurement of dynamic coefficient of friction (DCOF) as the most relevant measurement of slip resistance.

This chapter discusses the most commonly employed slip resistance test methods overseas, as well as some that are used infrequently. An overview of the overseas standards-making organizations involved in slip resistance and their work in the field is also detailed. Finally, profiles of industry and other independent overseas organizations involved in the study of slip resistance are presented.

7.2 SLIP AND FALL STATISTICS OVERSEAS

Perhaps not surprisingly, slip and fall accidents are one of the five leading causes of accidental death in other countries. According to the National Safety Council's International Accident Facts, falls are the number one (12 countries) or number two (20 countries) cause of death in 32 of the 37 countries that report such data. This is an average rank of 1.9, second only to motor vehicle accidents (with a rank of 1.3). Countries with the highest number of fall deaths are:

- Germany (10,052)—Rank 1
- Italy (9,624)—Rank 1
- France (9,564)—Rank 1
- Poland (4,805)—Rank 2
- Japan (4,690)—Rank 2
- Mexico (4,429)—Rank 2
- U.K. (4,369)—Rank 1

With the exception of Russia and Mexico, fall rates increase with age—mirroring experience in the United States.

Other statistics include:

- More than 30% of all nonfatal major injuries in U.K. workplaces are caused by a slip, trip, or fall on the same level. These accidents cost the U.K. economy up to an estimated £1.1 billion per year In the United Kingdom, slips and trips are responsible for, on average:
  - Over one-third of all reported major injuries
  - 20% of over three-day injuries to employees
  - 50% of all reported accidents to members of the public that occur in workplaces
  - More major injuries in manufacturing and in the service sectors than any other cause (HSE).
- In Germany, almost one in five work-related accidents is a result of a slip and fall, numbering almost a quarter million in 1997.
- Falls are a leading cause of occupational injuries in Taiwan. In 2002, almost 14% of such injuries were fall related, three-quarters of which were same-level falls (Li et al., 2006).
- In Australia, the average cost of a fall to an employer from 2004 through 2005 was approximately $18,900. Fatal falls have been estimated to account for 16% of all insurance claims and 26% of all costs. Slips and falls resulted in 400,000 annual hospitalizations throughout the 1990s, 7,000 deaths, and an estimated annual lifetime cost of $3.1 billion.
- In Japan, fatalities due to falls amounted to 6,702 during 2005, an increase of 791 deaths over 1995. However, over the same period, fatalities from accidents at work decreased by 900 deaths from 1995. In spite of the decrease in nonfalling accidents, the fatalities caused by falls have increased annually, have reached more than four times that of accidents at work, and are around 67% of traffic accidents. In Japan, 55% of fatal falls on the same level occur in the home, 17% on streets and roads, 11% in schools and public facilities, and 7% in apartment facilities (Nagata and In-Ju, 2007).

A fall is one of the external causes of unintentional injury. It is coded as E880–E888 in International Classification of Disease (ICD-9), and as W00–W19 in ICD-10. These codes include a wide range of falls, including falls on the same level, upper level, and other unspecified falls.

7.3 OVERSEAS STANDARD DEVELOPMENT

What about slip resistance testing methods in other countries? Could researchers elsewhere have developed and validated a method that we could adopt in the United States?

Generally, U.S. standards for tribometers are the result of full consensus, requiring a "balanced" committee. In the case of ASTM International, no more than 50% of the committee can be producers of related products, and committee representation

Overseas Standards

must be from a wide range of interests, including footwear, flooring, consultants, industries at risk, and the general public. This is known as balancing a committee, intended so that no single interest group is able to exert undue influence over the development (or lack of development) of a standard or its requirements. In addition, there is an appeal process for members with objections to compliance with ASTM's procedures in development of the standard. NFPA International has similar requirements. The full consensus approach is intended to arrive at viable standards that provide protection to the public while being reasonable enough to be implemented by industry. Unfortunately, even the full consensus approach can be manipulated. See the section entitled "Not Infallible" in Exhibit 7.2, Approaches to Development of Standards.

Few overseas organizations develop standards by the U.S. definition of "full consensus." They may be substantially funded, written, and published by groups with a strong financial interest in industry-friendly standards. Although these organizations may welcome the participation, there is generally no requirement to maintain a specified balance of interests. In many committees, end users are underrepresented. Without balance, there can be an understandably lower level of credibility and acceptance of resulting standards.

Although not seen as an issue overseas, safety professionals in the United States would expect such an approach to produce weak or biased standards. From the U.S. perspective, overseas standards developed primarily by vested interests should be approached with caution. None of the apparatuses discussed in this chapter (e.g., ramp test, pendulum, Tortus) has a comparable U.S. standard.

Another critical difference between ASTM standards and overseas standards has to do with validation of the test method. Whereas ASTM test methods require that interlaboratory studies be conducted and research reports be developed to document the degree of repeatability and reproducibility of the method, few overseas standards-making organizations include such requirements. In essence, ASTM requires proof that the method works as intended through a process of testing and study. In contrast, many overseas standards are issued and widely used without such validation.

Note: The following information on standards activities must be read as a snapshot in time. Standards are continually developed and updated. It is recommended that the reader consult the Web resources provided to determine the status of standards activities and obtain the current edition of standards cited herein.

7.4   RAMP TESTS

Ramp tests originated in Germany (see Figure 7.1) and have been adopted by the Australia/New Zealand cooperative. DIN 51097 and DIN 51130 require test persons to walk on various wet tiles. The subject walks forward or backward (as specified by the test method) while wearing a safety harness attached to an overhead gantry. Simultaneously, the table is tilted at a set speed by remote control. When the subject slips, a reading of the inclined angle is taken (Shipman).

FIGURE 7.1 A German ramp test in process. (Photograph courtesy of CSIRO.)

An interesting feature of these standards is the use of a classification system to provide guidance in selecting materials with the appropriate level of slip resistance for a given area, based primarily on the activities performed and the contaminants present or likely to be present.

7.4.1 Operational Issues

At first glance, this approach may make sense. Instead of using a device as a surrogate for walking, it uses actual people. If the concern is people slipping and falling, it would seem sound to go to the source.

Unfortunately, the ramp test approach presents a number of concerns. Experts agree that a person's awareness of a potentially slippery surface can have a strong influence on the way he or she traverses that area. Seeing ice or water, one will adjust one's gait accordingly and increase the chances of crossing the area without incident. It is when one is unaware of the hazard and expects the same level of traction that slips are most likely to occur. Regarding ramp tests:

- People selected to participate in a ramp test expect a slippery surface. No amount of preparation or instruction will change that. This makes it likely that they will alter their gait in anticipation, and so they will perform much better on a ramp test than when encountering an unexpectedly slippery surface in real life. K. Schuster from BIA (see Section 7.11.6) concluded that tests with human subjects under such conditions are irrelevant because they essentially eliminate the element of surprise.

- As is known by many researchers in biomechanics, people walk differently on an incline than on a level surface. Slipping at a certain point on an incline does not correlate to slipping on the same surface on a level plane. Because the biomechanical factors of human ambulation are so vastly different in such situations, this test method should not purport to be valid for a condition it does not measure.
- The length, speed, and biases in each person's gait vary, and those factors also change with age. Generally, ramp tests have involved only younger, able-bodied individuals, hardly representative of the population at large (and less so as the population ages). Although tribometers provide a consistent means of measurement by limiting variables to the surface being tested, it is difficult to argue that ramp tests do so.
- The test method specifies the use of as few as two test subjects, a statistically inadequate sample hardly sufficient to provide a basis for validating results. The two subjects selected could easily be anomalous, thus delivering measurements that bear no resemblance to actual conditions. It is difficult to imagine how a qualified researcher in any field would regard results derived from two subjects to be valid.
- Ramp tests refer to "standard test shoes." This is yet another variable that must be accounted for. In fact, the German DIN committee has been faced with the inability to obtain the specified type and quality of Bottrop footwear needed for the test.
- Another barrier to the use of this method as a reasonable test standard is the cost. The construction of ramp facilities and the use of test subjects plus operators make it cost prohibitive to most. The calibration board tiles alone cost nearly $7,000. In addition, at last report, the reserve stock of these tiles is almost depleted.
- In some instances, pretesting of tiles with an alternate method (e.g., pendulum) indicates low slip resistance. During subsequent ramp tests, these suspect tiles are deliberately placed at the bottom of the ramp to ensure that they are not being stood upon at the beginning of each walk down the ramp. This can result in unrealistically favorable test results.

The test method requires standardized shoes and calibration boards. The specified Bottrop footwear has not been produced for several years and has not been available for purchase since 1998. A study has found substitute footwear (Lupos Picasso S1 boots) that provides similar results. The results using the Lupos shoes were about 0.3 degrees higher than those for the Bottrop boots (Bowman, R., 2004).

According to the U.K. Health and Safety Executive (HSE):

- In the DIN 51130 method (which specifies cleated safety boots and motor oil contamination), floors that perform well in the test do not necessarily perform well with water contamination.
- In the DIN 51097 method (which specifies using barefoot operators with soapy water as the contaminant), floors that perform well in the test do

TABLE 7.1
DIN 51130 R-Value Slipperiness Classifications

Classification	R9	R10	R11	R12	R13
Slip angle (°)	6–10	10–19	19–27	27–35	>35

TABLE 7.2
DIN 51097 Slipperiness Classifications

Classification	Typical Applications	Critical Angle
Class A	Barefoot, but mainly dry aisles and walkways, dry changing areas	≥12°
Class B	Shower rooms, pool surrounds, wet changing areas, disinfectant spray areas (plus all areas covered by Class A)	≥18°
Class C	Areas constantly under water, e.g., steps into pools, foot baths, inclined pool surrounds, jacuzzis (plus all areas covered by Class A and B)	≥24°

Note: DIN 51097 is used to assess the slip resistance of floor tiles under wet barefoot conditions. It classifies tiles for areas where people walk barefoot.
Source: Porcelanosa Group Limited, 2005.

not necessarily perform well with clean water contamination (Health and Safety Executive, Slips and trips—Relevant laws and standards).

Even between the types of ramp tests there is disagreement. The U.K. Health and Safety Executive (HSE) has reservations about the DIN test methods, because neither uses contaminants representative of those commonly found in the workplaces and due to the way in which the results are interpreted and applied.

The DIN classification schemes outlined herein (Table 7.1 and Table 7.2) have led to some confusion, resulting in occasional installation of improper floor surfaces.

A common problem is the misconception that the R scale runs from R1 to R13, where R1 is most slippery and R13 least slippery. In some cases, R9 floors have been specified as anti-slip surfaces. In reality, the R scale runs from R9 to R13, where R9 is most slippery and R13 least slippery. Floor surfaces classified by the DIN 51130 standard as R9 (or in some instances R10) are likely to be unacceptably slippery when used in wet or greasy conditions. Further problems may arise from the wide range of COF within a given classification. For example, R10 covers a COF range of 0.18 to 0.34, a wide range.

In summary, ramp tests involve many biases and uncontrolled variables, bringing into question whether they can be considered a viable method for evaluating the traction of level walkway surfaces. This being the case, reasonable effort should be made to minimize those variables. Part of that effort should include well-designed and conducted ruggedness testing and an interlaboratory study. Because of the numerous and complex variables involved and the ease with which results could be manipulated, such studies are essential to the improvement and validation of this test method.

7.5 PENDULUM TESTERS

The first pendulum-class tester, the Sigler, was developed by Percy Sigler at the U.S. National Bureau of Standards (NBS) in the 1940s and dubbed the "NBS Standard Dynamic COF Tester." The Sigler has long since fallen out of U.S. standards for pedestrian slip resistance (there was, at one time, a federal and an NBS standard for this device) due to the inability to correlate test results with human perception of slipperiness. In the 1950s, ASTM Committee D21 on Polishes studied the Sigler and the James Machine as potential test methods for slip resistance. However, only the James Machine standard (in the form of test method D2047) was ever developed and published.

The U.K. Pendulum Test was developed to provide highway engineers with a method to assess the skid resistance of wet road surfaces. It is preferred by Australians and the U.K. Slip Resistance Group to measure pedestrian slip resistance, and is also often preferred by U.K. scientists/consultants involved in the tiling industry and the London Underground. Over the last decade, two opposing camps have been firmly entrenched, one siding with the Tortus and the other favoring the pendulum. The Health and Safety Executive (HSE) refuses to accept that the Tortus is accurate under wet conditions and relies on the pendulum test instead (Porcelanosa Group Limited, 2005).

7.5.1 OPERATION

The pendulum is based on the Izod principle. A pendulum rotates on a spindle attached to a vertical pillar. The end of the pendulum arm holds a foot fitted with a material (usually rubber) ranging from 1.25 to 3 in. The pendulum is released from the horizontal position so that it strikes the test foot under a constant velocity. The distance the test foot travels after striking the surface determines the friction of the floor surface, which is read from the scale on the apparatus.

The sample is tested in three directions. The first three readings in each set are ignored. The median is calculated using the five readings remaining from each set. The pendulum data are routinely supplemented by Rz microroughness measurements (see Chapter 3, Principles of Slip Resistance). The pendulum test is the only portable slip resistance measurement method used by U.K. Health and Safety Laboratory (HSL) on behalf of the HSE (see Figure 7.2 and Table 7.3).

The U.K. Slip Resistance Group developed the slip potential classification shown in Table 7.4, based on pendulum test values (PTV) in 2005, which is endorsed by the Health and Safety Executive (HSE).

The New Zealand/Australia standards use still another set of criteria, known as the Mean British Pendulum Number (MBPN) from wet pendulum tests (Table 7.5).

7.5.2 ASTM E-303

The pendulum provides readings between 0 and 150, often referred to as "pendulum numbers." Because pendulum numbers are unique and thus difficult to correlate with other measurements, guidelines by the Greater London Council can be used to interpret test results (see Table 7.3). A subsequent variation of the Sigler, the

FIGURE 7.2 A Stanley pendulum tester, quite similar to the British Portable Skid Tester (BPST). Photograph courtesy of CSIRO.

TABLE 7.3
Greater London Council Pendulum Thresholds

Original GLC Guideline (TRRL Test Foot)		Guideline as Modified by James (4S Test Foot)	
Pendulum Reading	Interpretation	Pendulum Reading	Interpretation
75 and above	Excellent	65 and above	Very good
40 to 74	Satisfactory	35 to 64	Good
20 to 39	Marginal	25 to 34	Marginal
19 and below	Dangerous	24 and below	Unsatisfactory

TABLE 7.4
UKSRG Thresholds

Interpretation	PTV
High slip potential	0–24
Moderate slip potential	25–35
Low slip potential	36+

TABLE 7.5
NZ/A Thresholds

Classification	MBPN
V Class	>54
W Class	45–54
X Class	35–44
Y Class	25–34
Z Class	<25

TRRL Tester, also known in the United States as the British Portable Skid Tester (BPST) does fall under an ASTM standard, but not for pedestrian slip resistance. ASTM E-303, Method for Measuring Surface Frictional Properties using the British Pendulum Tester, is approved for the evaluation of the skid resistance of roadways for vehicular use. The term "skid-resistance" is defined to correlate the performance of a vehicle with patterned tires braking with locked wheels on a wet road at 50 km/h (30 mph). To reinforce this point, the standard was developed in ASTM Technical Committee E-17 on Vehicle-Pavement Systems, under the jurisdiction of subcommittee E17.23 on Surface Characteristics Related to Tire Pavement Slip Resistance.

7.5.3 Issues

The BPST was developed by a British government organization, Road Research Laboratory (or Transport Research Laboratory). For more information on the BPST, see http://home2.btconnect.com/Munro-group/skid/index.html. In the United Kingdom, the BPST is now used for evaluating pedestrian slip resistance, despite the fact that at eight miles per hour (about 200 times faster than the movement of the Tortus test foot), the slider speed is much greater than that of a person's heel striking the floor during walking.

It is claimed that the pendulum simulates movement of the foot at heel strike, but in reality it does not reflect the type of contact or the forces involved in the interaction between foot and surface at and immediately after heel strike. The foot does not swing forward from the knee as a pendulum does, and the sliding pendulum pad does not emulate the foot's movement or direction of forces following heel strike (Adams, 1997).

Based on substantial research by SATRA, it was determined that irrecoverable slips occur while the foot is moving in the forward direction by only two means:

- Heel contact only, in which a slip is virtually irrecoverable once initiated (influenced by DCOF)
- Heel contact, quickly followed by sole contact as the slip starts (influenced by SCOF). SATRA research suggests that SCOF and DCOF are almost equal at this stage (Greater London Council, 1985).

According to the Greater London Council, the type of slip measured by the pendulum corresponds with the landing phase of ambulation (Greater London Council,

1985). This means that the pendulum provides a reading that relates to a phase of walking during which slipping is rare and brings into question the usefulness of such measurements.

Practical problems with this device for measuring pedestrian slip resistance include the dynamics and operation of pendulum devices in general. Of particular concern is the excessive velocity at which the machine operates, bearing no relation to that of human ambulation. Research conducted in the 1970s by ASTM F15.03 (Consumer Safety) determined that the pendulum devices showed significant variation across the test surface, making a reasonable correlation of these results to a single slip resistance value impractical. The same conclusion was reached in the Bucknell ASTM workshop in 1991, and in a separate research project by the NBS in the late 1970s (in which NBS concluded that it would be difficult, if not impossible, to relate friction directly to the energy loss).

Although it is known that slip resistance increases with surface roughness, a study by Rubber and Plastics Research Association (RAPRA) (see Section 7.11.1) of the pendulum concluded that there is a poor correlation between wet friction as measured by this instrument and surface roughness. Likewise, pendulum instruments are unable to accurately meter surfaces that slope by more than five degrees, or surfaces with irregular profiles (e.g., nonsmooth surfaces).

Still another issue is usability. Because the device is complex and difficult to set up and operate, results are highly subject to operator influence.

The sensitivity of the pendulum to the tension in its friction discs and the condition of the slider, and the ease with which its adjustment can be disturbed in transport, during assembly, and in set-up make its reliability questionable (Adams, 1997).

A full interlaboratory study by Transit New Zealand, which examined repeatability and reproducibility, indicated that operator practice can have a significant effect on results. The range of results for repeatability was found to be moderate to large (ranging from 4 to 25 pendulum numbers), and the range for reproducibility was always large (at 20 to 34 pendulum numbers).

In a study evaluating the validity, repeatability, precision, and consistency of portable tribometers, the precision and consistency of the pendulum (Portable Skid Resistance Tester model) was found to be reduced at higher COF levels, and researchers stated that, "the validity of the PSRT for pedestrian slip resistance measurement must be improved" (Grönqvist et al., 1999).

This result was corroborated by an interlaboratory study coordinated by Commonwealth Scientific and Industrial Research Organisation (CSIRO) involving 27 laboratories and six sets of tiles. The largest result was typically 30% above the mean, and the smallest was 15% below. Thus if a tile had a mean of 40 BPN, the results could range from 34 to 52 units (Bowman, Understanding the new slip resistance standards and its implications).

In terms of proper adjustment, possible faults in apparatus setup include the following:

Arm issues
- Arm may not release freely from the catch block.

- Arm may not swing parallel to the main frame.
- Foot may not be square with the main frame.

Pointer issues
- Pointer may not be parallel to the arm.
- Pointer may not reach zero.

If results are lower than expected
- Lifting handle adjustment screw may be over tightened.
- Bearings in the foot may be contaminated and need cleaning.
- The pressure disk abutting the felt washer on the pointer may not be fixed.

There is a high dependency on the degree of slope as well as the direction of travel (up or down). Measuring up a slope of 15% yields results 30% to 50% higher than level surface measurements. Measuring down a similar slope yields results 20% to 30% lower than those on level surfaces (Dravitzki and Potter, 1997).

The calculation and interpretation of readings is intricate, requiring the insertion of three values into an algebraic formula. In fact, in the NBS Preliminary Study of the Slipperiness of Flooring (NBSIR 74-613, 1974), the results of the BPST had to be expressed in British Portable Skid Tester Numbers, because they are not readily convertible to coefficient of friction numbers. The external calibration requirements for this complex instrument allow up to two years between calibrations.

Understandably, the pendulum tester is not suitable for assessing the slip resistance of resilient flooring materials. Because it relies upon a loss of energy to obtain a measurement of slip resistance, and a resilient surface tends to absorb energy, unreliable results can be expected. The tendency of resilient materials to deform under the force of the pendulum may also contribute to lower-than-expected readings obtained from these materials. Like the Tortus, the pendulum cannot be used to obtain valid results on profiled surfaces or those with a gradient of more than five degrees.

The pendulum tester (particularly the BPST) was selected for European (and therefore international) standardization purely by a process of elimination. A field test method for wet measurement was needed, and preferred overseas technologies were not viable. The ramp test is a laboratory apparatus, and the Tortus (see Figure 7.3 and Figure 7.4) was acknowledged to be unsuitable for wet testing, leaving only the pendulum.

7.6 DIGITIZED DRAGSLEDS

Originally developed in 1980 by British Ceramic Research Association (Ceram) as a bench test for ceramic floor tiles, the patent for the original Tortus has expired, and there are several variations commercially available. The Tortus II and III (British), Sellmaier (German), FFT (Floor Friction Tester), Gabbrielli SM (Italian), UWT BOT 3000 (U.S.), and Floor Slide Control (FSC) 2000 are variations of the same technology.

The Tortus and its successors are four-wheeled, self-propelled, electronic devices based on dragsled principles. Test pad material varies (the Tortus II uses 4S rubber, while the Sellmaier uses several different materials). As the test pad is dragged

FIGURE 7.3 A subsequent incarnation of the original, the Tortus II, is powered by an internal battery.

FIGURE 7.4 The underside of the Tortus II, showing the wheels that propel the device and the test foot that is pulled along the floor.

TABLE 7.6
Tortus Round Robin (Dry)

Tile	High Reading	Low Reading	Average	SD
A (Smooth)	0.86	0.44	0.66	0.11
B (Textured)	0.94	0.70	0.87	0.06
C (Textured)	0.78	0.63	0.71	0.05

across the floor, it records frictional forces and displays and prints the values. Mean values are estimated from this data. The drag on the 4.5-mm circular disc is measured through leaf springs. The original Tortus required an electrical connection, but subsequent versions are battery powered. The contact area and load was intended to reproduce the area and pressure of a heel striking the ground. However, the movement of the Tortus test foot along the floor is quite unlike human ambulation in that it is dragged in continuous contact with the walkway surface.

7.6.1 Issues

Also known as "sled tests," HSE laboratory-based assessments show that some smooth flooring appears to be less slippery in wet conditions than when dry, clearly a counterintuitive result. HSE found that such tests may give credible results in dry conditions.

Force plate data from the 1991 Bucknell workshop conducted by ASTM F-13 (in which several other tribometers were also evaluated) demonstrate the erratic and unstable output of this class of tester. The Bucknell workshop also produced results similar to the variable results of Horizontal Pull Slip Meter, another dragsled device, when performing wet testing. This was revealed by forceplate analysis because the instrument has a built-in filter that averages needle indications to make them appear less erratic. This averaging of greatly fluctuating readings leads to inconsistent and questionable results.

A British study likewise reported that this instrument tends to overestimate slip resistance due to the high traction properties of the standard slider and its inability to account for hydroplaning on wet surfaces. In addition, an Australian study found high variability between the test results of different operators (Bowman, 1992). Like the pendulum, the Tortus is also incapable of accurately measuring nonsmooth surfaces or those with slopes exceeding five degrees.

Even under dry conditions, the precision of the Tortus is questionable. A paper presented in 1997 by a distributor of the Tortus provides the results of a dry test round robin involving 12 machines, one-third of which were completed at the same laboratory (Martin and Dimopoulos, 1997). Even so, the results demonstrated poor precision (see Table 7.6).

The paper goes on to state, "[i]t is difficult to explain these fluctuations other than to attribute them to tile preparation . . ." (see Figure 7.3). This device has several disadvantages. Major issues include:

- As a dragsled class of tribometer, the instrument in no way emulates human ambulation, so its readings are not meaningfully related to the human perception of walking.
- Because the instrument is considered to be simulating the action of a pedestrian moving cautiously across a floor, this becomes a test of greatly limited value.
- In a study of the Tortus at the ASTM F-13 Bucknell workshop in 1991, readings from this instrument were the result of a series of stick-slip events caused by the foot sticking then jumping and sticking again continuously. The actual readout is some average between the force buildup before the foot jumps (high slip resistance) and the drift while it moves to a location where it remains stationary (low slip resistance).
- Significant anomalies in readings occur in that readings often do not match subjective observation. In one example, the Tortus delivers very high readings on very smooth surfaces. Surfaces that are patterned with raised pyramid, stud, or rib designs, and those with large grit deliver unexpectedly low slip resistance readings. It is theorized that the reason for some of these anomalies is that the small size (and configuration) of the test foot at such a slow pace allows it to travel "over and down the profile without measuring its true effect in providing a slip-resistant surface" (Dravitzki and Potter, 1997).
- The Tortus is not reliable for wet testing, due in part to the lack of adequate wheel traction and the problem of sticktion. Even the new AS/NZ 4586 standard (Slip Resistance Classification of New Pedestrian Surface Materials) does not specify this device for wet testing. This has been shown as an adhesion (or sticktion) problem developing between the test foot and the walkway surface due to the time delay between the application of horizontal and vertical forces (Bowman, 1992). In addition, studies have demonstrated poor correlation with subjective assessment of slip resistance. In these studies, the Tortus readings on virtually all wet surfaces were in the dangerous range (Harris and Shaw, 1988; Proctor and Coleman, 1988; Strandberg, 1985). A study by RAPRA (see Section 7.11.1) resulted in the conclusion that "the Tortus instrument is not at all reliable in wet conditions."
- The appropriateness of the test pad is questionable because, unlike heels, they are convex or cylindrical. It has been noted that the test foot contacts the floor not flatly, but at one edge. The walkway surface would need to be perfectly plane for flat contact to be made and maintained with the low pressure applied.

Heavily profiled industrial tiles, designed as slip resistant have relatively low wet Tortus results, which effectively demonstrates the inappropriate nature of the Tortus test method (Bowman, 2005).

It is widely accepted that the Tortus significantly inflates the dynamic COF for most surfaces. This overestimation of the available friction may be a function of the exceptionally slow slider movement of the device in combination with its slider's tiny contact area (Adams, 1997).

The Australian standards committee BD/44/3 found that the Tortus simulates the action of pedestrians moving slowly or cautiously across a floor, virtually eliminating the tendency to aquaplane under wet conditions. A British study of *in situ* methods of measurement for liquid-contaminated floors reported that the Tortus does not properly assess the effect of aquaplaning on smooth floors, overestimating the level of traction (Bowman, "Legal and Practical Aspects").

At the 2003 ASTM F13 USC Slipmeter Workshop, the Sellmaier version of this device delivered excessively high readings on dry and wet surfaces. It was also noted that the further the machine travels, the lower the average of the dynamic coefficient of friction.

Although the original Tortus is no longer manufactured, information about the Tortus III, based on the original Tortus can be found at http://www.tortus.co.uk. Severn Science, the manufacturer of the Tortus III, is now part of Wessex Engineering. Although they offer the Tortus III on its Web site, Wessex has no plans to continue manufacturing the instrument.

In 2004, Vario Systems obtained the rights to a version of the Tortus and formed Universal Walkway Testing LP (later renamed Regan Scientific) to manufacture and distribute a U.S. version called the BOT-3000.

7.7 OTHER DRAGSLEDS

As in the United States, a variety of other dragsled-class instruments are used overseas, few of which have gained prominence. Among them are:

- The Schuster Machine (Germany) is similar to the ASTM C1028 (see Chapter 4, U.S. Tribometers), in that it is a manually pulled sledge, although the dynamometer is built in.
- The Hoechst Device is more like an HPS (ASTM F609), a mechanically driven instrument with three small circular test feet (chromium plated). It is also of German manufacture.
- The Gabbrielli (GABTEC) Slipperiness Static Pull-Meter SS (http://www.gab-tec.com) is intended to test wet and dry static coefficient of friction in accordance with ASTM C-1028 and a now withdrawn ISO proposed draft standard. It differs in several ways from the apparatus specified by ASTM in that it uses a separate electronic meter and can connect to a personal computer or printer. It also uses a 55-lb (25-kg) weight instead of the ASTM-specified 50-lb (23-kg) load.
- In a study evaluating the validity, repeatability, precision, and consistency of portable tribometers, the Gabbrielli was found to be poor. Performance was studied against a force platform and a bio-mechanically validated slip simulator as reference equipment. The researchers added that, "[t]he validity of this instrument can be seriously questioned" (Grönqvist et al., 1999).
- The CEBTP Skidmeter consists of a block of tire rubber pulled along two cylindrical rails by an electrically braked motor and a reduction gearbox. Developed in France by the Centre Experimental de Recherches et d'Etudes

du Batiment et des TRAVAUX Publics (CEBTP), this presumably portable device consists of a substantial amount of equipment to operate.

As with any tribometer of this class, these dragsleds are unsuitable for wet testing due to "stick-slip" (see Chapter 3, Principles of Slip Resistance, and Chapter 4 U.S. Tribometers). Even under dry conditions they do not relate to the mechanics of human ambulation. Although Germany has a DIN standard for a dragsled meter, none is being considered for CEN or ISO standards at this time.

7.8 ROLLER-COASTER TESTS

The "roller-coaster test" is relatively new on the market. For lack of a better classification, it could be considered a dynamic coefficient of friction dragsled. The only commercial manufacturer is SlipAlert LLP. Like the pendulum, friction is measured by indirect means, determining the loss of friction from an initial specific velocity.

SlipAlert is marketed as a portable device. It is a gravity-based trolley which rolls down a prescribed ramp. Once in contact with the floor, SlipAlert rolls on two front wheels and a rubber slider in the rear until it skids to a stop.

7.8.1 SLIPALERT OPERATION

SlipAlert consists of the device and an aluminum ramp. The ramp is set up on the floor to be measured. SlipAlert is switched on and placed on the ramp, pulled manually to the top of the ramp, and released. SlipAlert runs down the ramp on the four wheels. When it meets the floor, the front wheels stay in contact with the floor, and the rear wheels lift up to allow the slider pad underneath to contact the floor. With an optional inclinometer, SlipAlert can be used on a slope.

It has a built-in digital display which produces a proprietary output (e.g., SlipAlert Number or SAN). This number is converted to coefficient of friction using a chart attached to the ramp. An approximate relationship between the SlipAlert SAN and the British Pendulum BPN was developed, which is expressed as the British Pendulum Number Equivalent (BPNE).

The device is used with sliders (or test feet) of three materials: 4S (for maximum correlation to the pendulum), "Durable SlipAlert Slider" (an unspecified material), and TRL (to simulate soft heels and bare feet).

The test protocol uses the median of the last five of eight runs as per the UKSRG guidelines for the pendulum test.

There has been little peer-reviewed research of this test method and its results. A study by the HSE Health & Safety Laboratory (Hallas and Shaw, 2006) evaluated SlipAlert and a similar custom-built roller-coaster tester (the Kirchberg Rolling Slider) to results obtained by the British Pendulum, the method of choice by the HSL/HSE and the U.K. Slip Resistance Group.

The Kirchberg Rolling Slider was constructed by HSL according to the design described by Kirchberg et al. (1997 in HSE 2006) and has three rubber sliders arranged like those on a GMG100 dragsled. While SlipAlert has a single-height ramp, the Kirchberg ramp has three different heights (and therefore speeds). The distance traveled by the trolley after initial contact with the floor is used to determine the level of dynamic friction available. This is determined by tape measure from the point of contact of the rear of the trolley with the floor to the rear of the trolley when stopped.

The study pointed to several operational issues and interesting conclusions on the results.

7.8.2 OPERATIONAL ISSUES

- The trolley tends to veer off to one side where there are imperfections in the flooring, which can produce misleading results.
- In heavily contaminated conditions, the trolley wheels become contaminated, making operation more difficult. It has been noted that the manufacturer has since introduced a new aluminum ramp to reduce this effect.
- Some SlipAlert operators compare the digital readout value without converting it to COF. Since conversion is an inverse relationship, the less tractive floor may read as more tractive. The same operators were unaware of the difference between 3S and 4S rubber, which impacts results.
- Some bouncing was observed on softer flooring materials (e.g., vinyl and safety vinyl). TRRL rubber slides accentuated this tendency, impeding proper operation.
- On slippery floors, the device can travel up to 2 meters. In many cases, there is insufficient floor space to use the tests correctly.
- When testing the effect of spills with the roller-coaster tests, the entire length of the path traveled by the slider must be contaminated to obtain a representative reading.
- When traveling long distances, the instrument can turn, altering results. The effect on the distance counter readings can be significant, with readings of 185 where the spinning is observed, and readings of 210 in the same conditions when no turning occurs.
- For accident investigation, several limitations would inhibit its use. It is not possible to measure a small spill.
- Floors with variability within a small area are difficult to accurately assess (e.g., floors of a resin matrix and a fine aggregate or where a variety of tile is used). On such surfaces, SlipAlert only provides a mean COF. If such variability is not taken into account, faulty conclusions could be reached regarding the slip potential of an area.

7.8.3 RESULTS

- A prior study (Hallas et al., 2005) noted that SlipAlert "tends to give failsafe readings, underestimating the friction available." In the HSL study, SlipAlert has overestimated the available friction on several floors.
- The correlation between the pendulum and the SlipAlert above a COF of approximately 0.3 is not as close as on more slippery surfaces.

- When using the 3S slider, SlipAlert misclassifies floors more often than the pendulum, including overestimating available friction on terrazzo.
- Under dry conditions, the correlation between SlipAlert and the pendulum is "not so good." Where a small amount of dust is present on a smooth floor, SlipAlert exaggerates the effect of the dust.

Despite these numerous shortcomings, HSE reports that it correlates well with the pendulum test when using a specified test foot material.

SlipAlert LLP was founded in 2003 and was designed by Malcolm Bailey, the secretary of the U.K. Slip Resistance Group. The apparatus sells in the United Kingdom for approximately 5,600 U.S. dollars, plus the cost of sliders and the inclinometer. It is patented under European Patent EP1634056.

7.9 PORTABLE FRICTION TESTER

Developed by the Swedish National Road and Transport Research Institute, the Portable Friction Tester (PFT, shown in Figure 7.5) was originally designed to measure the friction on road markings (see http://www.tft.lth.se/kfbkonf/5bergstromnew.

FIGURE 7.5 The Swedish portable friction tester (PFT or FIDO).

PDF). Its use was later expanded to include the friction of bicycle paths, and finally expanded further still as a means to measure pedestrian slip resistance. It has been referred to as a braked-wheel/skiddometer instrument in which there is an axle torque from braked rolling wheel.

Also known as FIDO, this instrument was found to be highly operator dependent, particularly with regard to maintaining a constant horizontal velocity. It is clear that the friction model has little to do with human locomotion, yet it is still used by some overseas for pedestrian safety studies and floor slip resistance evaluation. The Finnish Institute of Occupational Health (FIOH) based a prototype tribometer on the same principles as FIDO in which six test pads are mounted on a pneumatic wheel driven by a direct current motor. There is no U.S. standard, nor is one under consideration. In fact, there is no known overseas standard either.

7.10 INTERNATIONAL STANDARDS

The standards that follow primarily deal with measurement and assessment of slip resistance of walkway surfaces, or are referenced by such standards. Footwear related slip resistance standards are not the focus of this section (see Chapter 8, Footwear).

Three primary sources are available to access international standards, regulations, and related documents. Each source can be searched by document number, title, or key words:

- **Global Engineering Documents (GED)** at http://global.Ihs.com—Established in 1959, GED serves as a comprehensive source of technical industry standards and government and military standards. Most documents are available for download in PDF format.
- **NSSN: A National Resource for Global Standards** at http://www.nssn.org—This is a comprehensive data network on developing and approved national, foreign, regional, and international standards and regulatory documents. NSSN standards are products of the American National Standards Institute (ANSI). Part of NSSN is the STAR Service (Standards Tracking and Alerting Service), which allows users to establish profiles and track development standards.
- **ANSI Electronic Standards Store** at http://webstore.ansi.org/default.aspx. Here, U.S. and international standards can be obtained. Standards can be searched from over 80 standards publishers by keyword or document number and most can be downloaded in PDF format.

7.10.1 EUROPEAN STANDARDS

In Europe, the ramp and pendulum test methods for measuring slip resistance have a strong following; however, the European Union has yet to reach consensus. Some countries prefer a Tortus-style instrument, others are promoting standards for dragsled devices, and still others split the difference and recognize different devices under different conditions. Consensus on this issue in the EU does not appear to be achievable in the near future (see http://www.cen.eu/cenorm/homepage.htm).

7.10.1.1 CEN

The Comite Europeen de Normalisation (European Committee for Standardization) (CEN) was founded in 1961 by the national standards bodies in the European Economic Community and European Free Trade Association countries. Its mission is to promote voluntary technical harmonization in Europe in conjunction with worldwide bodies and its partners in Europe. Many see this as an initiative to protect European industry from foreign incursions. The national members are the "only effective" members according to Belgian law, under which CEN is registered as a nonprofit, international scientific and technical organization. Members consist primarily of the national standards bodies of the EU and European Fair Trade Association (EFTA, http://secretariat.efta.int) countries.

European standards related to slip resistance are developed by a variety of different committees of CEN. Following is a list of committees and slip resistance standards in process.

7.10.1.2 CEN Standards Process

The members develop and vote for the ratification of European Standards. They must implement such standards as national standards, withdrawing any conflicting national standards on the same subject. In time, all European countries are expected to have CEN standards in place.

Through the Vienna Agreement, CEN and ISO have agreed to develop standards jointly. Once developed, CEN standards will almost certainly also become ISO standards. For more information on the relationship of CEN with ISO, see Section 7.10.7.1.

ANSI is the U.S. representative to CEN. As a nonmember nation, ANSI does not have a vote. Having a vote would obligate the United States to implement CEN standards and withdraw comparable U.S. standards, as CEN member nations are required to do.

The process begins with the establishment of a technical committee and project slate of standards. Draft standards are developed or adapted from one of the member countries. Next, each country's designated voting entity within CEN (for example, DIN represents Germany) ballots the draft in its own country. The technical committee then meets to finalize the document.

The designation "EN" signifies that the document is an approved standard that has been implemented by member countries. The designation "ENV" is a European pre-standard, which is a prospective or provisional standard. ENVs are used when accelerated standards development is needed due to high rates of innovation (e.g., computer technology) or when there is an immediate need for the information. The designation "prEN" signifies a draft document that has not yet been approved or implemented.

7.10.1.3 CEN Slip Resistance Standards and Drafts

In many instances, CEN slip resistance standards are developed for different materials (e.g., a ceramics method, a natural stone method), long before unification into a single set of standards. BSI (British Standards Institute) estimates that it will take until around 2011 before these standards are harmonized. To search for the status of a given standard, or to obtain listing of which members have developed a corresponding country standard, visit http://www.cen.eu/esearch.

7.10.1.3.1 CEN/TC 339 Slip Resistance of Pedestrian Surfaces—Methods

As of this writing, this committee has no published standards. The secretariat is Instituto Português da Qualidade (IPQ).

> prEN 15673-1 Determination of slip resistance of pedestrian surfaces—Method of evaluation—Part 1: Reference method per EU Directive: 2001/16/EC
> Status: Under Approval 2009-01

This document describes a reference method incorporating three procedures for the determination in the laboratory of the slip resistance of floorings in the three most commonly encountered situations in which pedestrians walk (normal flooring, barefoot, and industrial situations). It specifies a laboratory reference method based on the subject-based inclined ramp method. This standard has already been adopted by Germany (2007).

7.10.1.3.2 CEN/TC 134 Resilient Textile and Laminate Floor Coverings

The secretariat is British Standards Institute (BSI).

> EN 13893:2002—Resilient, laminate and textile floor coverings—Measurement of dynamic coefficient of friction on dry floor surfaces per EU directive: 89/106/EEC

This document outlines the protocol for obtaining DCOF readings, but does not specify a test instrument. Rather it is designed to be a framework for any DCOF instrument (e.g., pendulum), specifying a leather test pad prepared with 320-grit sandpaper. It is important to note that the scope of this standard is limited to dry testing only.

Approved in 2002, this standard is primarily a product of the British Standards Institute (BSI), which also acts as the secretariat for this technical committee. To date, 33 EU members have developed a corresponding country standard.

> EN 13845:2005 Resilient floor coverings—Polyvinyl chloride floor coverings with particle based enhanced slip resistance—Specification per EU Directive: 89/106/EEC

This method is a laboratory-based ramp test specifically for resilient floor coverings with enhanced slip resistance. The test uses "standard" footwear and soapy water. According to the HSE, floors that perform well on this test do not necessarily perform well under clean water conditions.

To date, 31 EU members have developed a corresponding country standard.

EN 13553:2002 Resilient floor coverings—Polyvinyl chloride floor coverings for use in special wet areas—Specification per EU Directive: 89/106/EEC

To date, 13 EU members have developed a corresponding country standard.

7.10.1.3.3 CEN/TC 246 Natural Stones
The secretariat is Ente Nazionale Italiano di Unificazione (UNI).

> EN 14231:2003 Natural stone test methods—Determination of the slip resistance by means of the pendulum tester per EU Directive: 89/106/EEC

To date, 13 EU members have developed a corresponding country standard.

7.10.1.3.4 CEN/TC 67 Ceramic Tiles
The secretariat is UNI.

> EN 1308:2007 Adhesives for tiles—Determination of slip per EU Directive: 89/106/EEC

Based on an earlier draft of this document, this standard specifies the use of any of several test instruments, including the dynamic slider (e.g., Tortus), static slider (e.g., C1028), inclined platform (e.g., ramp test), and pendulum (e.g., BPST). Except for the inclined platform, all are permitted for field testing in dry or wet conditions. It specifies a 4S rubber test foot prepared with 400-grit sandpaper. The inclined platform is essentially the DIN oil-coated ramp method, a laboratory test that must be done with 10W-30 engine oil. To date, only Britain (December 2007) has developed a corresponding country standard.

To date, 18 EU members have developed a corresponding country standard.

7.10.1.3.5 CEN/TC 217 Surfaces for Sports Areas
The secretariat is British Standards Institute (BSI).

> EN 14837:2006 Surfaces for sports areas—Determination of slip resistance

To date, 27 EU members have developed a corresponding country standard.

> prEN 14903 Surfaces for sports areas—Determination of rotational friction
> Status: Under Approval 2002-02

This standard has already been adopted by Germany (February 2006).

7.10.2 German Standards

According to German regulations (see http://www.en.din.de), walkways must be slip resistant to comply with the Workplace Order and the Accident Prevention Regulation "General Regulations" (VBG 1). Standards in Germany are promulgated by the Deutsches Institut fur Normung e.V. (German Institute for Standardization). Also known as DIN, this registered association was founded in 1917. Its head office is in Berlin. Since 1975, it has been recognized by the German government as the national standards body and represents German interests at the international and European levels.

External experts, numbering 26,000, carry out standards work by serving as voluntary delegates in more than 4,000 committees. Draft standards are published for public comment, and all comments are reviewed before final publication of the standard. Published standards are reviewed for continuing relevance at least every five years.

Standards related to slip resistance are the responsibility of Committee NA 062-08-82 AA—Prüfung der rutschhemmenden Eigenschaft von Bodenbelägen, which corresponds to the CEN/TC 339 Committee (see Section 7.10.1.3.1). NA 062-08-82 AA falls under the DIN Normenausschuss Materialprufung (Materials Testing Standards Committee).

7.10.2.1 DIN 18032 P2 DIN V 18032-2 Sport Halls—Halls for Gymnastics, Games and Multi-Purpose Use—Part 2: Floors for Sporting Activities; Requirements, Testing

DIN 18032 P2 was updated April 2001 and is available in German only.

This standard specifies the use of the Stuttgart-Tester (or SST) for the measurement of friction. Originally developed in the 1960s, this semiportable instrument consists of a vertical shaft supported by a frame. The shaft moves downward when turned clockwise and upward when turned counterclockwise. A test foot is mounted on a pivot to the base of the shaft. A constant torque is applied by a steel wire wound over a winding drum and down onto the shaft. The wire runs over a guide pulley and is tensioned by a freely suspended 5-kg weight, which drives the shaft. The test foot is equipped with a strain gauge or piezo-electric device for measuring torque. The test foot surface consists of three skids that are 20 mm wide and 45 mm long, of a 50 mm diameter cylinder, and is surfaced with leather. To operate, the shaft is raised (causing the steel wire to wind onto the drum) and is released so that the weights drive the shaft downward. The test foot contacts the surface, and the rotation of the shaft is braked by the friction, measured as torque. There is no known manufacturer of the instrument.

7.10.2.2 DIN 51097 Testing of Floor Coverings; Determination of the Anti-Slip Properties; Wet-Loaded Barefoot Areas; Walking Method; *Ramp* Test

DIN 51097 was published in November 1992.

This standard is intended to provide a means to measure floor-covering materials intended for use in areas which are normally wet and walked upon barefoot (e.g., hospitals, changing rooms, washrooms, showers, swimming pools). It calls for a minimum of two subjects to walk on a gradually increasing inclined plane barefoot on a wet ramp. The results are rated in Table 7.7.

7.10.2.3 DIN 51130 Testing of Floor Coverings; Determination of the Anti-Slip Properties; Workrooms and Fields of Activities with Raised Slip Danger; Walking Method; Ramp Test

DIN 51130, with a publication date of November 1992 and updated in June 2004, is available in German only.

TABLE 7.7
DIN 51097 Valuation Groups

Valuation Group	Angle of Inclination
A	>12 degrees
B	>18 degrees
C	>24 degrees

TABLE 7.8
DIN 51130 Valuation Groups

Valuation Group	Angle of Inclination	Rating
R9	>3–10 degrees	Low static friction
R10	>10–19 degrees	Normal static friction
R11	>19–27 degrees	Increased static friction
R12	>27–35 degrees	High static friction
R13	>35 degrees	Very high static friction

This standard is intended to provide a means to measure floor-covering materials intended for use in workrooms and public areas. It calls for a minimum of two subjects to walk on a gradually increasing inclined plane with specific footwear on an oil-coated ramp. The results are rated in Table 7.8.

7.10.2.4 Draft Standard DIN 51131 Testing of Floor Coverings—Determination of the Anti-Slip Properties—Measurement of Sliding Friction Coefficient

DIN 51131 is a draft standard with a publication date of June 2006 and is available in German only. This test method specifies the use of a dragsled class of tribometer.

7.10.3 BRITISH STANDARDS

British standards are the responsibility of BSI (http://www.bsi-global.com/en/Standards-and-Publications/About-BSI-British-Standards), the British Standards Institute. Formed in 1901 under a royal Charter, BSI was created to help British industry compete in European and international trade markets. The BSI Group began life as a committee of engineers determined to standardize the number and type of steel sections. Currently, BSI operates in 112 countries and represents the United Kingdom for European and international standardization.

BSI British Standards has 27,000 current standards with approximately 6,000 standards in development at any one time. Six thousand people from 1,800 organizations are involved in helping BSI to make approximately 1,700 standards each year.

British standard BS 0 specifies how BS committees are to set up and operate. British standards are subject to public comment before being published, although drafts must be purchased for review.

In terms of regulation, the U.K. Health and Safety at Work etc Act 1974 (HSW Act) requires employers to ensure the health and safety of employees. The Management of Health and Safety at Work Regulations 1999 build on the HSW Act and include duties on employers to assess slip and trip risks to employees and others affected by their work activity and act to control these risks.

The Workplace (Health, Safety and Welfare) Regulations 1992 require floors to be suitable for the purpose for which they are used and free from obstructions and slip hazards. The Provision and Use of Work Equipment Regulations 1998 require work equipment (e.g., scrubber-drier, mop) to be well maintained and suitable, and the provision of training in its use (HSE information sheet).

7.10.3.1 Committee B/556

Although other committees also develop pedestrian slip-related standards, Committee B/556 is the primary committee for pedestrian slip measurement and is represented by manufacturers, building/materials research organizations, testing laboratories, the Health & Safety Executive (HSE), architects/consulting engineers, and certifying organizations; Committee B/556 is said to be well balanced. Committee B/556 coordinates pedestrian slip resistance for the following set of standards:

BS 7976-1:2002 Pendulum testers. Specification
BS 7976-2:2002 Pendulum testers. Method of operation
BS 7976-3:2002 Pendulum testers. Method of calibration

These standards describe the specification, operation, and calibration of the pendulum test, used to assess floor surface slipperiness under dry and contaminated conditions. The results are reported as pendulum test value or slip resistance value and are approximately 100 times the coefficient of friction. The set was published in August 2002.

7.10.3.2 Committee B/208

Committee B/208 oversees a wide range of stair- and walkway-related standards, including those for stairs, ladders, and walkways; industrial-type flooring and stair treads; fixed ladders; roof access including walkways, treads, and steps; machinery access including stairways; and stepladders, guardrails, walkways, and working platforms.

> BS 5395, Part 1 Stairs, Ladders & Walkways. Code of Practice for the Design, Construction and Maintenance of Straight Stairs and Winders

Published in June 2000, this standard refers to the British Portable Skid Tester (BPST), a pendulum class tribometer apparatus used to measure the slip resistance of walkway surfaces. Guideline thresholds are listed in Table 7.9.

7.10.3.3 Committee B/545

Committee B/545 oversees natural stone; it is responsible for BS EN 14231, (CEN standard EN 14231).

TABLE 7.9
BS 5395 BPST Slip Resistance Guidelines

>0.75	Good (suitable for "high risk" areas)
0.40–0.75	Adequate (suitable for "normal use")
0.20–0.39	Poor, may be unsafe
<0.20	Very poor, unsafe

7.10.3.4 Committee B/539

Committee B/539 oversees ceramic tiles and other rigid tiling. This committee ballots CEN standards relating to ceramic tiles.

7.10.3.5 Committee PRI/60

This committee ballots CEN standards relating to resilient floor coverings.

7.10.4 Swedish Standards

The Swedish Standards Institute (SIS) is an independent, nonprofit association. The Swedish Standards Board (NSS) approves Swedish standards, which are prefixed SS. SIS now develops few standards, instead adopting standards and working with CEN and ISO representing Sweden (see http://www.sis.se/defaultmain.aspx?tabid=741).

7.10.4.1 SS-EN 1893 Resilient, Laminate and Textile Floor Coverings—Measurement of Dynamic Coefficient of Friction on Dry Floor Surfaces

Published in June 2003, this European standard specifies the method for the measurement of dynamic coefficient of friction on surfaces of resilient, laminate, and textile floor coverings, usually walked on with shoes. Measurements are made in a laboratory on ex-factory dry floor covering surfaces only. The method is not suitable for testing on wet or contaminated surfaces.

The standard can be obtained here: http://www.sis.se/DesktopDefault.aspx?tabName=%40DocType_1&Doc_ID=34178.

7.10.5 Australia/New Zealand Standards

In Australia, most standards are published by Standards Australia (SA), which is the trading name of Standards Australia International Limited, a company limited by guarantee (see http://www.standards.org.au). It is an independent, nongovernmental organization; however, through a Memorandum of Understanding, SA is recognized by the Commonwealth Government as the peak nongovernmental standards body in Australia and represents Australia in ISO.

Standards Australia, originally called the Australian Commonwealth Engineering Standards Association, was founded in 1922, became the Standards Association of Australia in 1929, and was incorporated under a Royal Charter in 1951. In 1988, the name was changed to Standards Australia, and in 1999 it became Standards Australia International Limited.

SA develops and maintains around 7,000 Australian Standards and related publications prepared by over 1,500 committees involving more than 8,000 committee members.

Because they are issued by a nongovernmental organization, Australian standards have no independent authority. Many of them are called up in federal or state legislation, however, and then become mandatory.

Standards Australia derives 97% of its revenues from normal commercial activities and dividends from QAS Pty Ltd, a certification subsidiary. The remaining 3% represents fees for service from the commonwealth government for activities in the national interest (primarily international standardizing activities).

As part of the Closer Economic Relations agreement, Standards Australia maintains strong links with Standards New Zealand, with whom there is a formal agreement for preparing and publishing joint standards where appropriate. Standards Australia has a policy of adopting international standards wherever possible, in line with Australia's obligations under the World Trade Organization's Code of Practice, which requires the elimination of technical standards as barriers to international trade. Approximately one-third of current Australian standards are fully or substantially aligned with international standards.

Australian Standards (and others) are sold and distributed worldwide by SAI Global Limited (http://www.saiglobal.com/shop), which also sells ISO standards on behalf of Standards Australia. SAI Global has exclusive license over the distribution and sale of Australia standards.

7.10.5.1 The Standards Process

A request for initiating activity on a standard is initiated by industry or government, and the public support is determined before work begins. Actual committee members are selected by the representative association and not by Standards Australia. Draft standards are available to the public along with a two-month public comment period. Publication requires the support of 67% of the committee or 80% of those voting (whichever is less), with no major dissenting interest.

7.10.5.2 Australian Standards

Committee BD/94 Slip Resistance of Flooring Surfaces is a joint committee of Standards Australia and Standards New Zealand.

Australian standards specify the use of three instruments: pendulum, Tortus, and ramp test. CSIRO (see Section 7.11.4), the leading nongovernmental testing organization in Australia, is the sole owner of the complex and costly ramp facilities required for testing.

7.10.5.2.1 AS/NZS 4586:2004

First published as part of Joint Standard AS/NZS 3661.1:1993, this is the second edition. AS/NZS 4586:2004 is the Slip Resistance Classification of New Pedestrian Surface Materials standard, prepared by a joint standards committee of Australia and New Zealand, Committee BD/94, Slip Resistance of Floor Surfaces.

The standard provides means of classifying walkway surface materials (excluding carpet) wet or dry according to frictional characteristics using a range of five different laboratory test methods.

- Wet pendulum test method
- Dry floor friction test method
- Wet/barefoot ramp test method
- Oil-wet ramp test method
- Displacement volume test method

It is also intended for evaluating surface applications and treatments including sealers, polishes, and etching, which may modify surface characteristics. It calls for the pendulum (e.g., BPST) using TRRL or 4S rubber for wet testing, specifying a minimum of 0.4. Floors measuring less than 0.4 receive a ZG rating, meaning the floors are not slip resistant and the risk of slipping when wet is very high.

The dry floor friction test method called for is the floor friction tester (e.g., Tortus) using 4S rubber for dry testing. Floors tested by the dry method and that have not been tested wet with the pendulum receive a Z rating, meaning the "contribution of the floor surface to the risk of slipping when wet is very high."

The standard also includes the wet/barefoot ramp test (e.g., DIN barefoot ramp test) and oil-wet ramp test (e.g., DIN oil-wet ramp test), as well as a classification system for the results, based on the test method employed. The inclining ramp test methods are used to measure the slip resistance of gratings, heavily profiled surfaces, and resilient surfaces.

The Building Code of Australia (BCA) has called up this standard, making it a legally binding document.

7.10.5.2.2 AS/NZS 4663:2004

AS/NZS 4663:2004 is the Slip Resistance Measurement of Existing Pedestrian Surfaces standard. This standard provides means of measuring the frictional characteristics of existing pedestrian surfaces in wet and dry conditions. It is also intended for evaluating surface applications and treatments including sealers, polishes, and etching that modify walkway surface characteristics.

Similar in many ways to AS/NZS 4586, it specifies a pendulum instrument for wet testing (using 4S or TRRL rubber, prepared with 400-grit sandpaper). Clay and concrete pavers are tested using TRRL rubber, whereas Four S rubber is used for other pedestrian surfaces. The Tortus device is specified for dry testing (using 4S rubber prepared with 400-grit sandpaper). Ramp tests are excluded because the focus of the standard is evaluation of existing walkway surfaces, including surface applications and treatments.

The standard indicates that it may be unsuitable for measuring some walkway surfaces such as highly profiled surfaces.

7.10.5.2.3 AS/NZS 3661.2:1994

AS/NZS 3661.2:1994 is the *Slip Resistance of Pedestrian Surfaces—Guide to the Reduction of Slip Hazards* standard. Originally published in New Zealand as NZS

5841.1:1988, the purpose of this document is to provide guidance on selection, installation, and maintenance of walkway surfaces in residential, commercial, and public areas. It is not intended for use in industrial settings.

This standard, with the exception of Sections 5 and 6, comes under the jurisdiction of the New Zealand Building Act and is intended as a solution to the corresponding provisions of the New Zealand Building Code.

7.10.5.2.4 HB 197:1999

HB 197:1999 is the *Introductory Guide to the Slip Resistance of Pedestrian Surface Materials.* This handbook provides guidelines for the selection of walkways surfaces as classified by AS/NZS 4586:1999. In part, it recommends minimum floor surface classifications for a variety of locations and includes a commentary on test methods set out in AS/NZS 4586, as well as information on the consideration of ramped surfaces. It is published in concert with CSIRO (see Section 7.11.4). Primary topics include:

- Use of AS/NZS 4586 Classifications in Selecting Pedestrian Surface Materials
- Which wet slip test should be used as the basis for specifications
- Ramp and pendulum classifications
- Requirements for ramps and other sloped surfaces
- Selection of pedestrian surface materials according to ramp tests

7.10.5.2.5 AS/NZS 1141.42:1999

AS/NZS 1141.42:1999 is the standard for Methods for Sampling and Testing Aggregates—Pendulum Friction Test. This standard establishes a method for determining the friction value of a surface using the pendulum friction tester on test specimens of materials intended for use as pavement surfacing (e.g., skid resistance for roadways) material. It is included herein as it is referenced by AS/NZS 4663:2002 and AS/NZS 4586.

7.10.5.2.6 AS 1657-1992

First published as AS CA10-1938, *AS 1657 Fixed Platforms, Walkways, Stairways and Ladders—Design, Construction and Installation* is now in its third edition. This standard establishes requirements for the design, construction, and installation of fixed platforms, walkways, stairways, and ladders intended to provide means of safe access at areas normally used by operating, inspection, maintenance, and servicing personnel. Included in the appendices are methods for testing guardrails and posts and typical component dimensions and spacing for guardrails.

7.10.6 Italian Standards

7.10.6.1 DM 14 Guigno 1989 n. 236

As with most European countries, Italy measures pedestrian slip resistance as the dynamic coefficient of friction (DCOF). This is a law specifying that measurement of the DCOF should be performed using the Tortus. It specifies 0.4 as a threshold of safety. Some believe that development of an international slip resistance standard for ceramic tile has been impeded due in part to the Italian need to use wet Tortus test results to demonstrate compliance with a national provision for disabled access in some public areas (Bowman, 2005).

UNI, the Italian Organization for Standardization (http://www.uni.com/uni/controller/en/) is a private association founded in 1921 and appointed by the Italian government and the EU to develop, approve, and publish technical standards. UNI also published UNI EN 13893:2005 Resilient, laminate and textile floor coverings—Measurement of dynamic coefficient of friction on dry floor surfaces.

7.10.7 INTERNATIONAL ORGANIZATION FOR STANDARDIZATION (ISO)

The International Organization for Standardization (ISO) (http://www.iso.org/iso/standards_development.htm) is a worldwide federation of national standards bodies from 157 countries, one from each country. ISO is a nongovernmental organization established in 1947. ISO's work results in international agreements that are published as international standards. To date, ISO's work has resulted in some 17,000 international standards.

The major responsibility for administrating a standards committee is accepted by one of the national standards bodies that make up the ISO membership. A member body of ISO is the national body in a country that is "most representative of standardization in its country." Only one such body for each country is accepted for membership of ISO. Member bodies are entitled to participate and exercise full voting rights on any technical committee and policy committee of ISO. Each member body interested in a subject has the right to be represented on a committee. Governmental and nongovernmental international organizations, in liaison with ISO, also take part in the work.

ANSI is a founding member of and the United States representative of ISO. ANSI participates in 78% of all ISO technical committees and is responsible for appointing U.S. Technical Advisory Groups (TAGs), whose primary purpose is to develop and communicate U.S. positions on activities and ballots.

7.10.7.1 ISO Standards Process

To obtain final approval of an international standard draft requires approval by two-thirds of the ISO members that have participated actively in the standards development process and approval by 75% of all members that vote. Whether a member country votes for or against passage of an ISO standard, there is no requirement to actually implement any ISO standard as a national standard.

Through the Vienna Agreement, ISO and CEN have agreed to share work in progress and have provisions for one or the other development standards for both. About 25% of ISO standards originate with CEN, and about 40% of CEN standards originate with ISO. This historic agreement was made between ISO and CEN for several reasons, the most salient of which were:

1. The defection of western European national standards bodies and associated manpower from ISO in favor of CEN, thus resulting in a shortage of technical expertise for ISO

2. ISO obtaining buy-in and participation from the major world market segment of the EU and avoiding duplication of work effort

It is important to understand that 80% of all ISO technical committees have 50% or more voting members from CEN and CEN-affiliated countries. ASTM has noted that ISO committees with CEN secretariats reduce ISO efforts until CEN work projects are completed. This suggests an intention to produce a completed CEN standard that can be moved through ISO with the procedural advantages offered by the Vienna Agreement.

7.10.7.2 ISO Concerns

Biases built into the ISO structure raise several concerns:

- Through the Vienna agreement, CEN standards are permitted to go to final ballot (e.g., fast-track procedure), thus limiting the extent of changes that can be made in the document. This gives CEN members a strong advantage over other ISO members, including the United States and Australia.
- Each country member of CEN is also a member of ISO, and CEN requires its members to vote with CEN. Due to this substantial and solid voting block of 18, there is an inherent advantage for CEN standards to pass. There is also an inherent disadvantage to ISO members such as the United States and Australia, should CEN direct members vote against their positions. Given these circumstances, it would appear more appropriate for CEN to have a single voting interest in ISO.
- The process tends to dilute the work of more advanced country members because standards must be suitable for less advanced country members. This can result in standards of less significant technical merit.
- Because the purpose of ISO is to promote trade, it is difficult to balance the political implications of a given standard in each country with the technical soundness and appropriateness of the standard.

7.10.7.3 ISO Flooring Committees

7.10.7.3.1 *TC 219 Floor Coverings*

(http://www.iso.org/iso/iso_technical_committee.html?commid=54988)
Secretariat: BSI
Working Groups (e.g., subcommittees):
- TC 219/WG 1 Textile floor coverings (NBN)
- TC 219/WG 2 Resilient floor coverings (ANSI)
- TC 219/WG 3 Laminate floor coverings (SIS)

There are 22 countries participating on this committee. Of the current 56 published standards, none address slip resistance or coefficient of friction, nor are any shown as in process.

7.10.7.3.2 TC 189 Ceramic Tile

(http://www.iso.org/iso/standards_development/technical_committees/list_of_iso_technical_committees/iso_technical_committee.htm?commid=54320)
Secretariat: ANSI
Working Groups (e.g., subcommittees):
- TC 189/WG 1 Test methods (UNI)
- TC 189/WG 2 Product specifications (ANSI)
- TC 189/WG 3 Products for installation (ANSI)

There are 18 countries participating on this committee. Of the current 28 published standards, none address slip resistance or coefficient of friction, nor are any shown as in process.

7.11 OVERSEAS ORGANIZATIONS INVOLVED IN SLIP RESISTANCE

7.11.1 The U.K. Slip Resistance Group

Rapra Technology, Ltd., established the U.K. Slip Resistance Group (UKSRG) in 1986. Rapra Technology was formerly the Rubber and Plastics Research Association (RAPRA), a trade group of the rubber manufacturing industry, established in 1919. Some of its research in the field of pedestrian slip resistance is conducted for and funded by the Health and Safety Executive (HSE), the British equivalent of OSHA; however, this work is often not available to the public.

Ostensibly, the initial goal of this group was to develop a single standard rubber to be used for slip meters used in the United Kingdom. From that first objective, the group has continued to evolve, with their mission including other aspects of slip resistance measurement. In January 2002, the members of UKSRG decided to become a fully independent organization.

There are currently 35 members, primarily related to flooring and associated industries. Because it was conceived and controlled by a trade group organization, the UKSRG tends to promote materials manufactured by the trade group.

Membership in the UKSRG is by invitation only, and the members' deliberations are not made public. Application for membership requires two sponsors from the existing membership. Members are drawn from flooring manufacturers, health and safety organizations, engineering consultants specializing in slipping accidents, testing and research organizations, and slip test equipment manufacturers.

The UKSRG maintains an extensive library of technical papers relating to slip resistance issues available for a nominal fee to members only. One publication available for purchase is *The Assessment of Floor Slip Resistance—The U.K. Slip Resistance Group Guidelines,* January 2006. It provides information on the rationale for selecting the pendulum as the tribometer of choice, and how to calibrate, use, and maintain the apparatus. It also discusses the use of roughness meters, and guidelines on assessing test results.

The group is not reticent about endorsing commercial products, including a patented device known as SlipAlert (see Section 7.8). It should be noted that the

Overseas Standards   269

managing director of SlipAlert LLP is secretary of the UKSRG, and another member is U.S. distributor for this product. The UKSRG also endorses the use of a specific roughness meter, the Taylor Hobson Surtronic Duo. They also maintain a link to both product manufacturers (as well as both pendulum manufacturers) on their Web site.

7.11.2 HEALTH AND SAFETY EXECUTIVE (HSE)

The Health and Safety Commission is responsible for health and safety regulation in Great Britain. The Health and Safety Executive (HSE, http://www.hse.gov.uk) and local government are the enforcing authorities who work in support of the commission.

HSE's job is to help the Health and Safety Commission ensure that risks to people's health and safety from work activities are properly controlled.

The HSE is active in research and the development of standards related to slips, trips, and falls. They frequently partner with the UKSRG on slip resistance test methods, roughness guidelines, and other related topics. HSE's Web page titled "Preventing slips and trips in the workplace" (http://www.hse.gov.uk/slips) provides a wealth of information, including the following:

- Statistics
- Research papers focusing on:
  - Floor/working surfaces, surface coatings and coverings
  - Unevenness and obstructions
  - The efficacy of cleaning regimes
  - The effect of contaminants
  - The performance of protective systems and aids, including footwear
  - Human factors issues underlying slips and trips
  - Reliable and reproducible slip resistance measurement techniques for flooring and footwear
- Causes of slips and trips, including case studies
- Resources to help prevent slips and trips with a focus on management, employees, architects, and footwear and flooring manufacturers
- Slip assessment tool (SAT, http://www.hsesat.info)—HSE and HSL produced software to assess the slip potential of level pedestrian walkway surfaces. The SAT prompts the user to collect surface microroughness data from the test area. The SAT supplements the surface microroughness data (R_z) with other relevant information from the pedestrian slip potential model. This includes the causes of floor surface contamination, the floor cleaning regimen, the footwear types worn, and related human and environmental factors. Upon completion, a slip risk classification is provided, indicating the potential for a slip.

7.11.3 SATRA FOOTWEAR TECHNOLOGY CENTRE

SATRA Technology Centre (http://www.satra.co.uk), an independent research and technology organization for footwear and other consumer product industries, is similar in some ways to Underwriters Laboratories. Formed in 1919 to serve the footwear industry (and formerly the Shoe and Allied Trades Research Association), SATRA currently employs 180 scientists, technologists, and support staff and serves more than 1,600 member companies in 70 countries. Its main objective is to increase the profitability of its members by offering them exclusive access to research, products, and services.

SATRA has played a major role in developing national and international standards for safety footwear, assessing products against the standards, and providing technical support and assistance to manufacturers (see Chapter 8, Figures 8.3 and 8.4). With the introduction of European CE marking, SATRA has been involved in developing European standards for safety and protective footwear through CEN Technical Committee TC161. SATRA is the leading notified body for CE marking of safety footwear.

Since 1974, SATRA has conducted much research in the field of slip resistance. The SATRA Whole-Shoe Tester™ is used by much of the footwear industry (see Chapter 8, Figures 8.3 and 8.4). An intricate laboratory instrument first developed in 1977, SATRA offers testing services (currently operating four of these testers) and also markets the device for sale. SATRA developed test method SATRA TM 144 for measuring dynamic coefficient of friction, which specifies a threshold of 0.4 under wet or dry conditions for safety footwear. This test method has since become the basis for the ISO footwear standard EN ISO 13287 (see Chapter 8, Footwear).

In May 2007, SATRA built a floor coverings laboratory to assess products for compliance with national and international regulations, including slip resistance. The laboratory test machines include an inclined ramp and a TRL pendulum.

The laboratory is also the base for SATRA's slip team, which performs on-site testing using a portable slip tester machine (see below), which gives a direct reading of the coefficient of dynamic friction with various surface contaminants.

In 2002, SATRA's Floor Coverings Test Division announced the development of a portable floor tester. In June 2006, it was announced that SATRA had taken delivery of the new instrument. Although few details are available, it is based on principles similar to the laboratory machine and uses smooth or patterned test feet of 4S rubber. The device is also used on ramps and profiled floors as well as dry, wet, and contaminated surfaces. SATRA indicates the method is a "... unique test for measuring the slip resistance properties of floor surfaces in situ." No information has been obtained regarding validation of the instrument, or its correlation to existing test methods.

7.11.4 COMMONWEALTH SCIENTIFIC AND INDUSTRIAL RESEARCH ORGANISATION (CSIRO)

Australia's Commonwealth Scientific and Industrial Research Organisation (CSIRO, http://www.dbce.csiro.au) is an independent statutory authority created and operated under the provisions of the Science and Industry Research Act 1949. CSIRO is an agency in the Innovation, Industry, Science, and Resources government portfolio. It provides independent expert advice to the government and, thus, is influential in forming policy relating to science and technology.

Overseas Standards

For the most recent 4-year period, the federal government is budgeted to provide over A$2.8 billion for CSIRO. CSIRO Australia has a staff of more than 6,000. Globally, CSIRO is active in more than 70 countries with over 700 current or recently completed projects.

CSIRO typically has over 3,000 active research contracts each year serving small, medium, and large businesses in Australia and overseas, as well as public sector agencies, national and state governments, and other research organizations. More than 150 spin-off companies are based on CSIRO-generated intellectual property and expertise. CSIRO enters into commercial arrangements private and public organizations for such transactions as contract research, commercial licensing, and consulting and technical services. The Materials Science and Engineering Division, under which slip resistance issues fall, is one of 18 CSIRO divisions.

CSIRO employs a variety of slip resistance testing equipment, including the German ramp tester, British pendulum (Stanley and Wessex), floor friction testers (Tortus and Gabbrielli), ASTM C1028 horizontal dynamometer pull meter, Whiteley HPS, and the SATRA STM 603. CSIRO is a major contributor to the development of Australian standards (see Section 7.10.5).

The Australian Tile Council, as the "official" laboratory for determining product quality and resolving industry disputes, recognizes CSIRO.

7.11.5 INRS NATIONAL RESEARCH AND SAFETY INSTITUTE

The INRS is a nonprofit organization (http://en.inrs.fr) created in 1947 under the auspices of the CNAMTS. It was originally named Institut National de Sécurité (INS, National Safety Institute) and took on its current name, Institut National de Recherche et de Sécurité (INRS, National Research and Safety Institute), in 1968. With a staff of about 650 working in 20 units and divisions, the INRS operates on behalf of the employees and companies organized under the general Social Security scheme.

Its budget of about 400 million French francs comes almost entirely from the National Occupational Accident and Disease Prevention Fund. Its activities are programmed in accordance with directives from the national salaried workers' health insurance fund and policies defined by the Ministry of Employment and Solidarity. A joint board of directors, representing employers and employee trade unions, manages it.

Regarding slips and falls, INRS has a standing committee known as A.4 Hazards Associated with Falling (Persons or Objects). Among other research, this committee has done work in the following areas, mostly in French:

- Analysis of 600 *in situ* slip resistance measurements
- The slippage of footwear and surfaces—what measurement techniques to use
- Analysis of measurements of slip resistance of soiled surfaces on-site
- Comparison of seven methods for the evaluation of the slip resistance of floor coverings—contributions to the development of standards
- The prevention of slipping accidents—a review and discussion of work related to the methodology of measuring slip resistance

7.11.6 BERUFSGENOSSENSCHAFTLICHES INSTITUT FUR ARBEITSSICHERHEIT (BIA)

BIA is the primary research and testing organization in Germany for statutory accident insurance and prevention (http://www.hvbg.de/d/bia/starte.htm). Its work also involves certification of products and quality management systems. The BIA database of approximately 1,100 technical publications includes several relating to slip resistance, and all are available at no charge via postal mail or download from the BIA Web site. Although the database is searchable in English, French, and Spanish, many publications are currently available only in German. Slip resistance research papers include work related to walkway surfaces and footwear, some using proprietary apparatus that is not well known outside of Germany, such as the GMG 100 and the Schuster Machine (both dragsled class instruments), as well as ramp tests.

7.11.7 FINNISH INSTITUTE OF OCCUPATIONAL HEALTH (FIOH)

FIOH is a research and advisory institute for occupational health and safety (http://www.ttl.fi/Internet/English/default.htm). With a total of 850 employees, 10 health and safety disciplines are covered, providing services to all of Finland. Among the research projects undertaken by FIOH are the evaluation of the slip resistance of footwear and floor surfaces, slips on ice, and methods of reducing slipping accidents. These projects are funded by various organizations, and some are done in collaboration with other organizations such as SATRA and INRS. FIOH also offers testing and certification services for slip resistance of footwear and walkway surfaces.

7.11.8 INTERNATIONAL ASSOCIATION OF ATHLETICS FEDERATIONS (IAAF)

The International Association of Athletics Federations was founded in 1912 by 17 national athletic federations who saw the need for a governing authority for an athletic program, standardized technical equipment, and world records (http://www2.iaaf.org/TheSport/Technical/Tracks/Appendix4.html). The number of affiliated federations grew from 17 in 1912 to 210 in 1999. The IAAF is headquartered in Monaco and staffed by over 40 full-time, multinational professional staff. The IAAF has used corporate sponsorship as a means to better promote and develop the sport worldwide. In 2001, the IAAF Congress voted unanimously for the organization's name to be changed to the International Association of Athletics Federations.

The IAAF Performance Specifications for Synthetic Surfaced Athletics Tracks, under Section 1.6 of Specifications, specifies the use of a pendulum tester (Skid Resistance Tester). It sets a threshold for synthetic surface friction of no less than 0.5 wet, corresponding to a 47 reading on the pendulum. Appendix 4 of this Specification deals with the two acceptable methods of measuring friction.

- Method A provides instructions for measuring friction using the pendulum.
- Method B specifies the use of the Stuttgart Sliding Test apparatus (or SST), an obscure and arcane device of large size and little note (see Section 7.10.2.1).

EXHIBIT 7.1
National Standards Bodies

Although it is not comprehensive, the following is a listing of many national standards bodies. Not all have activity related to slip resistance or fall prevention. Note that Web sites and memberships are subject to change.

Afghanistan
　　Afghanistan National Standardization Authority
　　No Web site
　　Membership: ISO (Correspondent Member)

Albania
　　General Directorate of Standardization (DPS)
　　http://www.dps.gov.al/index_eng.html
　　Membership: ISO (Correspondent Member)

Algeria
　　Institut Algerien de Normalization (IANOR)
　　http://www.ianor.org (No English)
　　Membership: ISO (Member Body)

Angola
　　Instituto Angolano de Normalização e Qualidade (IANORQ)
　　No Web site
　　Membership: ISO (Correspondent Member)

Antigua and Barbuda
　　Antigua and Barbuda Bureau of Standards (ABBS)
　　http://www.ab.gov.ag/gov_v1/bureau/aboutus.htm
　　Membership: ISO (Subscriber Member)

Argentina
　　Argentine Institute of Standards (IRAM)
　　http://www.iram.com.ar (No English)
　　Membership: COPANT (Active Member), ISO (Member Body)

Armenia
　　Department for Standardization, Metrology and Certification (SARM)
　　http://www.sarm.am/?LanguageID=1
　　Membership: ISO (Member Body)

Australia
　　Standards Australia (SA)
　　http://www.standards.org.au
　　Membership: ISO (Member Body)

Austria
Osterreichisches Normungsinstitut (ON)
http://www.on-norm.at/publish/index.php?id=home&L=1
Membership: CEN, ISO (Member Body)

Azerbaijan
State Agency on Standardization, Metrology and Patents of Azerbaijan Republic (AZSTAND)
http://www.azstand.gov.az/index.php?lang=en
Membership: ISO (Member Body)

Bahrain
Bahrain Standards & Metrology Directorate (BSMD)
http://www.moic.gov.bh/moic/en
Membership: ISO (Member Body)

Bangladesh
Bangladesh Standards and Testing Institution (BSTI)
http://www.bsti.gov.bd
Membership: ISO (Member Body)

Barbados
Barbados National Standards Institute (BNSI)
No Web site
Membership: COPANT (Active Member), ISO (Member Body)

Belarus
State Committee for Standardization, Metrology and Certification of Belarus (BELST)
http://www.gosstandart.gov.by (No English)
Membership: ISO (Member Body)

Belgium
Institut Belge de Normalisation/Belgisch Instituut voor Normalisatie (IBN/BIN)
http://www.nbn.be/EN/home_en.html
Membership: CEN, ISO (Member Body

Benin
Centre Béninois de Normalisation et de Gestion de la Qualité (CEBENOR)
No Web site
Membership: ISO (Correspondent Member)

Bhutan
Standards and Quality Control Authority (SQCA)
http://www.sqca.gov.bt
Membership: ISO (Correspondent Member)

Bolivia
Bolivia Institute of Quality and Standards (IBNORCA)
http://www.ibnorca.org (No English)
Membership: COPANT (Active Member), ISO (Correspondent Member)

Bosnia and Herzegovina
Institute for Standards, Metrology and Intellectual Property of Bosnia and Herzegovina (BASMP)
http://www.bas.gov.ba (No English)
Membership: ISO (Member Body)

Botswana
Botswana Bureau of Standards (BOBS)
http://www.bobstandards.bw
Membership: ISO (Member Body)

Brazil
Brazilian Association of Technical Standards
Membership: COPANT (Active Member), ISO (Member Body)

Brunei Darussalam
Construction Planning and Research Unit (CPRU)
http://www.mod.gov.bn (No English)
Membership: ISO (Correspondent Member)

Bulgaria
State Agency for Standardization and Metrology (BDS)
http://www.bds-bg.org
Membership: CEN, ISO (Member Body)

Burkina Faso
Direction de la Normalisation et de la Promotion de la Qualité (FASONORM)
No Web site
Membership: ISO (Correspondent Member)

Burundi
Bureau Burundais de Normalisation et de Contrôle de la Qualité (BBN)
No Web site
Membership: ISO (Subscriber Member)

Camaroon
Division de la Normalisation et de la Qualité (CDNQ)
Membership: ISO (Correspondent Member)

Cambodia
Department of Industrial Standards of Cambodia (ISC)
http://www.isc.gov.kh
Membership: ISO (Subscriber Member)

Canada
Standards Council of Canada
http://www.scc.ca/en/index.shtml
Membership: COPANT (Active Member), ISO (Member Body)

Chile
Instituto Nacional de Normalizacion (INN)
http://www3.inn.cl/portada/index.php (No English)
Membership: COPANT (Active Member), ISO (Member Body)

China
China State Bureau of Quality and Technical Supervision (CSBTS)
http://www.sac.gov.cn/templet/english (English)
Membership: ISO (Member Body)

Colombia
Instituto Colombiano de Normas Tecnicas y Certificacion (ICONTEC)
http://www.icontec.org.co/Home.asp?CodIdioma=ESP (No English)
Membership: COPANT (Active Member), ISO (Member Body)

Congo, The Democratic Republic of
Instituto Colombiano de Normas Tecnicas y Certificacion (ICONTEC)
No Web site
Membership: ISO (Member Body)

Costa Rica
Costa Rica Institute of Technical Standards (INTECO)
http://www.inteco.or.cr/esp/index.html (No English)
Membership: COPANT (Active Member), ISO (Member Body)

Cote-d'Ivoire
Côte d'Ivoire Normalisation (CODINORM)
No Web site
Membership: ISO (Member Body)

Croatia
State Office for Standardization and Metrology (DZNM)
http://www.hzn.hr/english/indexen.html
Membership: ISO (Member Body)

Cuba
National Office of Standards (NC)
http://www.nc.cubaindustria.cu (No English)
Membership: COPANT (Active Member), ISO (Member Body)

Cyprus
Cyprus Organization for Standards and Control of Quality (CYS)
http://www.cys.org.cy/default.asp?id=175
Membership: CEN, ISO (Member Body)

Czech Republic
Czech Standards Institute (CSNI)
Membership: CEN, ISO (Member Body)

Denmark
Dansk Standard (DS)
http://www.en.ds.dk
Membership: CEN, ISO (Member Body)

Dominica
Dominica Bureau of Standards (DBOS)
Membership: ISO (Subscriber Member)

Dominican Republic
Main Directorate of Standards and Quality Control Systems (DIGENOR)
http://www.digenor.gov.do (No English)
Membership: COPANT (Active Member), ISO (Correspondent Member)

Ecuador
Ecuador Institute of Standardization (INEN)
http://www.inen.gov.ec (No English)
Membership: COPANT (Active Member), ISO (Member Body)

Egypt
Egyptian Organization for Standardization and Quality Control (EOS)
http://www.eos.org.eg/Public/en-us/Default
Membership: ISO (Member Body)

El Salvador
Nacional Council of Science and Technology of El Salvador (CONACYT)
http://www.conacyt.gob.sv (No English)
Membership: COPANT (Active Member), ISO (Correspondent Member)

Eritrea
Eritrean Standards Institution (ESI)
No Web site
Membership: ISO (Correspondent Member)

Estonia
Eesti Standardikeskus (EVS)
http://www.evs.ee/Esileht/tabid/111/language/en-US/Default.aspx
Membership: CEN, ISO (Correspondent Member)

Ethiopia
Quality and Standards Authority of Ethiopia (QSAE)
http://www.qsae.org
Membership: ISO (Member Body)

Fiji
Fiji Trade Standards and Quality Control Office (FTSQCO)
Membership: ISO (Member Body)

Finland
Finnish Standards Association (SFS)
http://www.sfs.fi/en
Membership: CEN, ISO (Member Body)

France
Association Francaise de Normalisation (AFNOR)
http://www.AFNOR.fr/portail.asp?Lang=English
Membership: COPANT (Adherent Member), CEN, ISO (Body Member)

Gabon
Centre de Normalisation et de Transfert de Technologies (CNTT)
No Web site
Membership: ISO (Correspondent Member)

Georgia
Georgian National Agency for Standards, Technical Regulations and Metrology (GEOSTM)
Membership: ISO (Correspondent Member)

Germany
DIN Deutsches Institut für Normung (DIN)
http://www.din.de/cmd;jsessionid=0BD1C972FAF90974F20F1EF06D388313.4?level=tpl-home&contextid=din&languageid=en
Membership: CEN, ISO (Member Body)

Ghana
Ghana Standards Board (GSB)
http://ghanastandards.org
Membership: ISO (Member Body)

Greece
Hellenic Organization for Standardization (ELOT)
http://www.elot.gr/home.htm
Membership: CEN, ISO (Member Body)

Grenada
Grenada Bureau of Standards (GDBS)
http://gdbs.gd/
Membership: COPANT (Active Member)

Guatemala
Guatemalian Commission of Standards (COGUANOR)
http://www.coguanor.org (No English)
Membership: COPANT (Active Member), ISO (Correspondent Member)

Guyana
Guyana National Bureau of Standards (GNBS)
http://www.gnbs.info (English)
Membership: COPANT (Active Member), ISO (Subscriber Member)

Honduras
Honduran Council of Science and Technology (COHCIT)
http://www.cohcit.gob.hn (No English)
Membership: COPANT (Active Member), ISO (Subscriber Member)

Hong Kong, China
Innovation and Technology Commission (ITCHKSAR)
http://www.itc.gov.hk/en/welcome.htm
Membership: ISO (Correspondent Member)

Hungary
Magyar Szabvanyugyi Testulet (MSZT)
http://www.mszt.hu/angol/index_eng.htm
Membership: CEN, ISO (Member Body)

Iceland
Icelandic Council for Standardization (STRI)
http://www.ist.is/english
Membership: CEN, ISO (Member Body)

India
Bureau of Indian Standards (BIS)
http://www.bis.org.in
Membership: ISO (Member Body)

Indonesia
Badan Standardisasi Nasional (BSN)
http://www.bsn.or.id
Membership: ISO (Member Body)

Iran, Islamic Republic of
Institute of Standards and Industrial Research of Iran (ISIRI)
http://www.isiri.org
Membership: ISO (Member Body)

Iraq
Central Organization for Standardization and Quality Control (COSQC)
No Web site
Membership: ISO (Member Body)

Ireland
National Standards Authority of Ireland (NSAI)
http://www.nsai.ie
Membership: CEN, ISO (Member Body)

Israel
Standards Institution of Israel
http://www.sii.org.il (No English)
Membership: ISO (Member Body)

Italy
Ente Nazionale Italiano di Unificazione (UNI)
http://www.uni.com/uni/controller/en
Membership: COPANT (Adherent Member), CEN, ISO (Member Body)

Jamaica
Jamaica Bureau of Standards (JBS)
http://www.bsj.org.jm
Membership: COPANT (Active Member), ISO (Member Body)

Japan
Japanese Industrial Standards Committee (JISC)
http://www.jisc.go.jp/eng
Membership: ISO (Member Body)

Overseas Standards 281

Jordan
Jordan Institution for Standards and Metrology (JISM)
http://www.jism.gov.jo
Membership: ISO (Member Body)

Kazakhstan
Committee for Standardization, Metrology and Certification (KAZMEMST)
http://www.memst.kz (No English)
Membership: ISO (Member Body)

Kenya
Kenya Bureau of Standards (KEBS)
http://www.kebs.org
Membership: ISO (Member Body)

Korea, Democratic People's Republic
Committee for Standardization of the Democratic People's Republic of Korea (CSK)
No Web site
Membership: ISO (Member Body)

Korea, Republic of
Korean Agency for Technology and Standards (KATS)
http://www.kats.go.kr/english/index.asp
Membership: ISO (Member Body)

Kuwait
Standards and Industrial Services Affairs (KOWSMD)
https://www.pai.gov.kw/portal/page/portal/pai/Home
Membership: ISO (Member Body)

Kyrgyzstan
National Institute for Standards and Metrology of the Kyrgyz Republic (KYRGYZST)
Membership: ISO (Correspondent Member)

Lao People's Democratic Rep
Department of Intellectual Property, Standardization and Metrology (DISM)
No Web site
Membership: ISO (Subscriber Member)

Latvia
Latvian Standard (LVS)
http://www.lvs.lv (No English)
Membership: CEN, ISO (Correspondent Member)

Lebanon
Lebanese Standards Institution (LIBNOR)
http://www.libnor.org
Membership: ISO (Member Body)

Lesotho
Standards and Quality Assurance Department (LSQAS)
No Web site
Membership: ISO (Subscriber Member)

Libyan Arab Jamahiriya
Libyan National Centre for Standardization and Metrology (LNCSM)
http://www.lncsm.org.ly
Membership: ISO (Member Body)

Lithuania
Lithuanian Standards Board (LST)
http://alpha.lsd.lt/en
Membership: CEN, ISO (Member Body)

Luxembourg
Service de l'Energie de l'Etat (SEE)
http://www.see.lu (No English)
Membership: CEN, ISO (Member Body)

Macau, China
Macau Productivity and Technology Transfer Center (CPTTM)
http://www.cpttm.org.mo/home_e.php
Membership: ISO (Correspondent Member)

Macedonia
Zavod za Standardizacija i Metrologija (ZSM)
No Web site
Membership: N/A

The former Yugoslav Republic of Macedonia
"Standardization Institute of the Republic of Macedonia" (ISRM)
http://www.isrm.gov.mk/ang_ISRM.aspx
Membership: ISO (Member Body)

Madagascar
Bureau de Normes de Madagascar (BNM)
No Web site
Membership: ISO (Correspondent Member)

Malawi
Malawi Bureau of Standards (MBS)
No Web site
Membership: ISO (Correspondent Member)

Malaysia
Standards and Industrial Research of Malaysia (SIRIM)
http://www.standardsmalaysia.gov.my (English)
Membership: ISO (Member Body)

Malta
Malta Standards Authority (MSA)
http://www.msa.org.mt
Membership: CEN, ISO (Member Body)

Mauritius
Mauritius Standards Bureau (MSB)
http://msb.intnet.mu/MSB/MSBHome.nsf?Open
Membership: ISO (Member Body)

Mexico
Nacional Office of Standards (DGN)
http://www.economia.gob.mx/?NLanguage=en&P=85
Membership: COPANT (Active Member), ISO (Member Body)

Moldova, Republic of
Department of Standards, Metrology and Technical Supervision
http://www.ssm.gov.md (No English)
Membership: ISO (Correspondent Member)

Mongolia
Mongolian National Centre for Standardization and Metrology (MNCSN)
http://www.pmis.gov.mn/masm (No English)
Membership: ISO (Member Body)

Montenegro
Institute for Standardization of Montenegro (ISME)
No Web site
Membership: ISO (Correspondent Member)

Morocco
Service de normalization industrielle marocaine (SNIMA)
http://www.mcinet.gov.ma/snima (No English)
Membership: ISO (Member Body)

Mozambique
Instituto Nacional de Normalização e Qualidade (INNOQ)
No Web site
Membership: ISO (Correspondent Member)

Myanmar
Myanma Scientific and Technological Research Department (MSTRD)
No Web site
Membership: ISO (Correspondent Member)

Namibia
Namibia Standards Information and Quality Office (NSIQO)
No Web site
Membership: ISO (Correspondent Member)

Nepal
Nepal Bureau of Standards and Metrology (NBSM)
No Web site
Membership: ISO (Correspondent Member)

Netherlands
Nederlands Normalisatie-instituut (NNI)
http://www2.nen.nl/nen/servlet/dispatcher.Dispatcher?id=ABOUT_NEN
Membership: CEN, ISO (Member Body)

New Zealand
New Zealand Standards
http://www.standards.co.nz (English)
Membership: ISO (Member Body

Nicaragua
Ministry of Development, Industry and Commerce Office of Technology, Standard and Metrology (MIFIC)
http://www.mific.gob.ni (No English)
Membership: COPANT (Active Member), ISO (Correspondent Member)

Nigeria
Standards Organisation of Nigeria (SON)
http://www.soncap.com/
Membership: ISO (Member Body)

Norway
Norges Standardiseringsforbund (NSF)
Norwegian Council for Building Standardization
http://www.standard.no/imaker.exe?id=5010
Membership: CEN, ISO (Member Body)

Oman
Directorate General for Specifications and Measurements (DGSM)
http://www.mocioman.gov.om/english/home.html
Membership: ISO (Member Body)

Pakistan
Pakistan Standards Institution (PSI)
http://www.psqca.com.pk
Membership: ISO (Member Body)

Palestine
Palestine Standards Institution (PSI)
http://www.psi.gov.ps
Membership: ISO (Correspondent Member)

Panama
Panamian Commission of Industrial and Technical Standards (COPANIT)
http://www.mici.gob.pa/index2.php (No English)
Membership: COPANT (Active Member), ISO (Member Body)

Papua New Guinea
National Institute of Standards and Industrial Technology (NISIT)
http://www.nisit.gov.pg
Membership: ISO (Correspondent Member)

Paraguay
National Institute of Technology of Standards (INTN)
Membership: COPANT (Active Member), ISO (Correspondent Member)

Peru
National Institute of Defense of the Competition and the Protection of the Intellectual Property (INDECOPI)
http://www.indecopi.gob.pe (No English)
Membership: COPANT (Active Member), ISO (Member Body)

Philippines
Bureau of Product Standards (BPS)
http://www.bps.dti.gov.ph
Membership: ISO (Member Body)

Poland
Polish Committee for Standardization (PKN)
http://www.pkn.pl/?lang=en&pid=en_strona_glowna
Membership: CEN, ISO (Member Body)

Portugal
Instituto Portugues da Qualidade (IPQ)
http://www.ipq.pt/backhtmlfiles/ipq_mei.htm (No English)
Membership: COPANT (Adherent Member), CEN, ISO (Member Body)

Qatar
Qatar General Organization for Standards and Metrology (QS)
http://www.qs.org.qa
Membership: ISO (Body Member)

Romania
Asociatia de Standardizare din Romania (ASRO)
http://www.asro.ro/engleza2005/default_eng.html
Membership: CEN, ISO (Body Member)

Russian Federation
State Committee of the Russian Federation for Standardization and Metrology (GOST R)
http://www.gost.ru (No English)
Membership: ISO (Member Body)

Rwanda
Rwanda Bureau of Standards (RBS)
http://www.rwanda-standards.org
Membership: ISO (Correspondent Member)

Saint Vincent and the Grenadines
St. Vincent and the Grenadines Bureau of Standards (SVGBS)
http://www.gov.vc/svgbs
Membership: ISO (Subscriber Member)

Santa Lucia
Santa Lucia Bureau of Standards (SLBS)
http://www.slbs.org.lc/
Membership: COPANT (Active Member), ISO (Member Body)

Saudi Arabia
Saudi Arabian Standards Organization
http://www.saso.org.sa (English)
Membership: ISO (Member Body)

Senegal
Association Sénégalaise de Normalisation (ASN)
http://www.asn.sn (No English)
Membership: ISO (Correspondent Member)

Overseas Standards 287

Serbia
Institute for Standardization of Serbia (ISS)
http://www.jus.org.yu
Membership: ISO (Member Body)

Seychelies
Seychelles Bureau of Standards (SBS)
http://www.seychelles.net/sbsorg
Membership: ISO (Correspondent Member)

Singapore
Singapore Productivity and Standards Board
http://www.spring.gov.sg/Content/HomePage.aspx
Membership: ISO (Member Body)

Slovakia
Slovak Institute for Standardization (SUTN)
http://www.sutn.org/?lang=eng
Membership: CEN, ISO (Member Body)

Slovenia
Standards and Metrology Institute (SMIS)
http://www.sist.si (No English)
Membership: CEN, ISO (Member Body)

South Africa
South African Bureau of Standards (SABS)
http://www.sabs.co.za
Membership: COPANT (Adherent Member), ISO (Member Body)

Spain
Asociacion Espanola de Normalizacion y Certificacion (AENOR)
http://www.aenor.es/desarrollo/inicio/home/home.asp
Membership: COPANT (Adherent Member), CEN, ISO (Member Body)

Sri Lanka
Sri Lanka Standards Institution (SLSI)
http://www.slsi.lk
Membership: ISO (Member Body)

Sudan
Sudanese Standards and Metrology Organization (SSMO)
http://www.ssmo.org (No English)
Membership: ISO (Member Body)

Suriname
  Suriname Standards Bureau (SSB)
  No Web site
  Membership: ISO (Subscriber Member)

Swaziland
  Swaziland Standards Authority (SWASA)
  No Web site
  Membership: ISO (Correspondent Member)

Sweden
  Swedish Standards Institute (SSI)
  http://www.sis.se
  Membership: CEN, ISO (Member Body)

Switzerland
  Swiss Association for Standardization (SNV)
  http://www.snv.ch/?en/home
  Membership: CEN, ISO (Member Body)

Syrian Arab Republic
  Syrian Arab Organization for Standardization and Metrology (SASMO)
  http://www.sasmo.org/en/index.php
  Membership: ISO (Member Body)

Tajikistan
  Agency of Standardization, Metrology, Certification and Trade Inspection (TJKSTN)
  No Web site
  Membership: ISO (Correspondent Member)

Tanzania, United Republic of
  Tanzania Bureau of Standards (TBS)
  http://www.tbstz.org
  Membership: ISO (Member Body)

Thailand
  Thai Industrial Standards Institute (TISI)
  http://www.tisi.go.th
  Membership: ISO (Member Body)

Togo
  Conseil Supérieur de Normalisation (CSN)
  No Web site
  Membership: ISO (Correspondent Member)

Trinidad and Tobago
Trinidad and Tobago Bureau of Standards (TTBS)
http://www.ttbs.org.tt
Membership: COPANT (Active Member), ISO (Member Body)

Tunisia
Institut National de la Normalization et de la Propriete Industrielle (INNORPI)
http://www.inorpi.ind.tn/en/inorpi.asp?langue=english
Membership: ISO (Member Body)

Turkey
Turk Standardlari Enstitusu (TSE)
http://www.tse.org.tr (No English)
Membership: ISO (Member Body)

Turkmenistan
The Major State Service "Turkmenstandartlary" (MSST)
No Web site
Membership: ISO (Correspondent Member)

Uganda
Uganda National Bureau of Standards (UNBS)
http://www.unbs.go.ug
Membership: ISO (Correspondent Member)

Ukraine
State Committee of Standardization, Metrology and Certification of Ukraine (DSTU)
http://www.dssu.gov.ua/control/en
Membership: ISO (Member Body)

United Arab Emirates
Directorate of Standardization and Metrology (SSUAE)
http://www.esma.ae/en/esenmain.html
Membership: ISO (Member Body)

United Kingdom
British Standards Institute (BSI)
http://www.bsigroup.com
Membership: CEN, ISO (Member Body)

Uruguay
Uruguayan Institute of Technical Standards (UNIT)
http://www.unit.org.uy (No English)
Membership: COPANT (Active Member), ISO (Member Body)

Uzbekistan
Uzbek State Centre for Standardization, Metrology and Certification (UZGOST)
http://www.standart.uz/?x=news&lang=eng&PHPSESSID=d9390a7ed03a13ea9ad1af9a2d129628
Membership: ISO (Member Body)

Venezuela
Standards and Quality Certification Fund (FONDONORMA)
http://www.fondonorma.org.ve (No English)
Membership: COPANT (Active Member), ISO (Member Body)

Viet Nam
Directorate for Standards and Quality (TCVN)
http://www.tcvn.gov.vn/en/index.php
Membership: ISO (Member Body)

Yemen
Yemen Standardization, Metrology and Quality Control Organization (YMSO)
No Web site
Membership: ISO (Correspondent Member)

Zambia
Zambia Bureau of Standards (ZABS)
No Web site
Membership: ISO (Correspondent Member)

Zimbabwe
Standards Association of Zimbabwe (SAZ)
http://www.saz.org.zw
Membership: ISO (Member Body)

Notes:

COPANT—Pan American Standards Commission (http://www.copant.org/English/index.asp)
CEN—European Committee for Standardization (Comite European de Normalisation)
ISO—International Organization for Standardization

EXHIBIT 7.2
Approaches to Development of Standards

Consensus means that there is substantial agreement by directly and materially affected interest categories. Generally, this means concurrence of more than a simple majority (usually two-thirds). Unanimity is not required. Consensus requires that all views and objections be considered, and that an effort is made to resolve them.

There are degrees of consensus, and consensus standards can be developed in several different ways.

FULL CONSENSUS METHOD

Full consensus is the most open and balanced approach to standards development. For that reason, it can also be the most time consuming and labor intensive. Full consensus requires the following elements:

Due Process

Due process requires that any person or entity with a material interest in the subject of the standard has a right to participate in the process of development and approval. Participation includes the right to:

1. Express a position and its basis
2. Have that position considered
3. Appeal

For this reason, full consensus requires making the draft document public for a reasonable period to permit all interested parties the opportunity to review and comment on it.

Transparency (Openness)

Participation is open to all entities directly and materially affected by the subject of the document. There cannot be onerous financial or unreasonably restrictive technical requirements precluding the involvement of interested parties. The right to vote cannot be contingent upon membership in any organization.

Timely notice of actions regarding actions of a committee or document must be provided to known materially affected interests. Notice must include a clear and meaningful description of the purpose of the activity and include a readily available source of additional information.

The affiliation and interest category of each member of the committee must be made available to interested parties on request.

Balance (Lack of Dominance)

The committee must have a balance of interests to preclude one interest from dominating and controlling the outcome of the process. Dominance means a position of undue influence by reason of superior leverage or representation that inhibits fair and equitable consideration of other viewpoints.

There are several methods of achieving and maintaining balance. In the case of ASTM, no more than 50% of the membership of a technical committee can be classified as producers of the subject of the standard. NFPA divides membership so that a technical committee has no more than 25% of any voting interest classification. Traditional guidelines are that no single interest category constitutes:

- More than one-third of the membership in a committee dealing with safety
- A majority of the membership in a committee dealing with product standards

Not Infallible

While the full consensus method is the most rigorous means of standards development, it is by no means immune from manipulation. To illustrate how even the full consensus method can be subverted, consider the following example using ASTM's processes and procedures.

A committee of 200 members is on the large side for most committees. The rules state that 60% of members must return ballots, and that a ballot item must obtain a 90% affirmative at the main committee. This is calculated by dividing the number of all votes (minus abstentions) by the number of affirmative votes. Consider also that a substantial percentage of members on a given committee do not vote, and many that do, abstain. Of the 200 voting members, 60% (or 120 members) vote. Of those 120 votes, one-third (40 members) abstain, not an uncommon occurrence.

To secure a 90% affirmative, of the 80 votes cast either affirmative or negative, only 8 can be negative votes. As we have seen, ASTM allows up to 50% of voting members (in this case, 200) to be producers/manufacturers. One can see what an easy matter it is for this interest group to stop any standard from moving forward—they can have 100 votes and only need 9 to stop a ballot item.

This is why it is argued by some that even the most rigorous standards process is weighted strongly in favor of the industry for which such standards are being developed.

Canvass Method

The canvass method is often seen as an abbreviated version of the full consensus methods of standards. In it, certain aspects of full consensus are delayed in order to increase flexibility and potentially reduce the time and effort needed to complete the standards process.

It appears that the canvass method was originally developed in order to have a mechanism for trade associations to be able to document existing industry practices and to have them recognized nationally. It is now used to initiate, develop, and publish standards from the ground up. The canvass method has been used to develop more than 1,000 ANSI standards. About half of all ANSI-accredited organizations are approved to use this method.

The canvass method is used when a level of agreement among a drafting group has been obtained, and wider consensus is sought through a "canvass" (or polling) of interested parties. Once a draft standard has been developed, the standards developer identifies, to the extent possible, entities directly and materially affected by the subject of the standard. The standards developer then conducts a letter ballot of those interests to determine consensus on a document. While the committee developing the standard need not be balanced, the standard must be approved by a canvass of balanced interests. Thus, participation in the process by a balance of interests does not occur while a standard is in development, but when a fully drafted document is out for approval.

The canvass list is held to the same requirements as a committee would be in the full consensus method. Procedures must be approved by ANSI, there must be a roster, and interest categories must be assigned. Typically, a majority is required to return the ballot from the canvass list, and two-thirds must approve the ballot, although these ratios vary depending on the procedures of the organization. Organizations on a canvass list have the same right of appeal as a committee member does in the full consensus method.

However, development of the organizations on the canvass list can be selected and assigned an interest category by a committee dominated by a single interest; it is more readily manipulated than the full consensus methods. In addition, true consensus depends on broad participation by canvass list organizations in returning the ballot. Thus, a single-interest committee could select organizations that are nominally interested in the subject of the standard and less likely to respond to the ballot (thus fulfilling the balance requirement), with the remaining canvass organizations with the same interest as the committee returning ballots in the affirmative.

INDUSTRY METHOD

Unless full consensus or canvass methods are used, standards developed by an industry association or trade group may have no requirements for due process, transparency, or balance. Because the mission of these organizations is generally self-serving (i.e., to promote and protect their industry), these standards can be replete with inherent biases. Although an industry standard may be the product of the dictionary definition of consensus, it is the consensus of a single interest group, generally providing no opportunity for comment from the public and other interested parties. More appropriately, documents from these sources should be considered "voluntary standards" instead of "voluntary consensus standards."

8 Footwear

Organizational Responsibilities

Component	Potential Responsibility
Footwear Specification, Selection	Safety, Purchasing
Maintenance	Line Supervisor, Employees
Footwear Program Design	Safety, Operations, Purchasing, Human Resources, Legal
Footwear Program Implementation	Operations, Human Resources

The fear's as bad as falling.

William Shakespeare

8.1 INTRODUCTION

Falls are responsible for a significant portion of employee injuries. Aside from the obvious costs, employee injuries also interrupt customer service and can affect profitability. Although it is unlikely that you will influence the general public, management can certainly exercise a high degree of control over the footwear worn by employees.

This chapter provides information on the elements of design and construction of slip-resistant footwear, labeling and testing issues, advertising traps to avoid, general guidelines on selection of footwear in general, other protection features to consider, maintenance, and issues to consider when establishing a footwear policy. It also provides a brief overview of key international standards related to footwear slip resistance.

8.1.1 INDUSTRY CONDITIONS

In 2005, imports supplied virtually the entire U.S. footwear market in many categories, including athletic, juvenile, women's, and slippers. Plastic/protective footwear and men's work shoes, where import penetration in 2005 was 76.2% and 78.2%, respectively, represent the only categories where U.S. production still maintains a significant share of the market (AAFA, 2006b).

In 1985, China had a 2.8% share of the U.S. footwear market. In 1995, however, China had a 58.3% share, and by 2005 it had climbed to 84.2%, which translates to over 1.9 billion pairs of shoes. By 2005, U.S. shoe production had reached a record low of less than 50 million pairs (AAFA, 2006a). As such, the standards to which shoes are manufactured overseas are of substantial concern.

Footwear industry experts indicate that, because China has such a substantial presence in U.S. and EU markets, most China imports are manufactured to U.S. standards for protective footwear (e.g., ASTM F2413 and F2412) and U.S. and overseas test methods for slip resistance (e.g., EN13287, ASTM F1677). See Chapter 4, U.S. Tribometers, and below for more information on these test methods and standards.

8.1.2 Potential Impact of Footwear

Clear and compelling evidence exists that using slip-resistant footwear reduces accidents. In 2001, St. Lawrence University (Canton, New York) completed a pilot study by providing slip-resistant footwear to its employees. The reduction in injury rates and workers' compensation claims of this small school resulted in savings of an estimated $100,000. A food service operation employing 250 people at a major airport also reduced slips and fall accidents to zero for over a year by providing slip-resistant footwear.

Jack in the Box established a mandatory slip-resistant shoe program. The company buys slip-resistant overshoes and requires employees to use either these or slip-resistant shoes. They report that using the overshoe has been the vehicle to get the employees into the shoes, and that with a voluntary program only about 40% tend to participate in such programs (Slip Slidin' Safety, 2005).

At The Cheesecake Factory, 90% of staff wear slip-resistant shoes. Since the inception of the program, the company reports that the frequency of slip-and-fall related claims has reduced by 72%, and associated expenses have reduced by 81% (Shoes for Crews).

Friendly's Restaurants used a cost-sharing program to help employees purchase slip-resistant safety shoes, which resulted in a 30% reduction of slip and fall injuries in the first year. That year, cuts replaced slips and falls as the number one cause of injury, saving the company an estimated $750,000 (Hedden, 1997).

However, one make of footwear promoted as "slip resistant" may be significantly more or less slip resistant than another. Also, footwear that is slip resistant in one environment may not be in another. Slip-resistant shoes may be tested on different floor types. Environmental factors can significantly affect the performance of a shoe, so be sure the shoe selected is suited for the environment in which it will be worn.

The effectiveness of slip-resistant footwear depends not only on the efficacy of the design, but also the environmental conditions, type of floor surface and finish, shoe soling materials and compounds, and other factors. Beyond selection, maintenance and administration of footwear programs is needed in order to achieve success.

8.2 FOOTWEAR DESIGN FOR SLIP RESISTANCE

The types of footwear that employees are permitted to wear should be limited. In general, footwear with hard plastic or leather soles or heels should not be permitted, nor should soles with little or no tread pattern (e.g., smooth). Instead, footwear with slip-resistant soles should be required, designed according to the guidelines in Section 8.2.3.

8.2.1 Sole Compounds

The material of the heel and sole of footwear is a major factor in its performance as slip resistant (see Chapter 4, U.S. Tribrometers, Section 4.8).

In general, softer compounds are more slip resistant than harder materials because they grab the floor surface more effectively. For example, women's high-heeled shoes are often made from a very hard plastic, known as PVC (polyvinyl chloride). Even with a tread design, this material against a relatively smooth, hard walkway surface (such as marble, terrazzo, or ceramic) provides little slip resistance.

The consistency and life of materials should also be considered. Leather is an inconsistent material—its properties change over time, with wear, and when saturated with water, oil, or other such materials. As such, leather is a poor choice for footwear bottom material.

Some footwear bottom materials have a finite time during which they perform consistently. After that period (which differs with the extent of use and the environment), the properties of the material change with wear, and may not perform in the same way. Styrene-butadiene rubber (SBR), nitrile-butadiene rubber (NBR), and polyurethanes are some of the most commonly used footwear bottom materials. Similar to many other soling materials, they can be formulated in a wide range of hardness.

8.2.2 Outsole Tread Patterns

Tread patterns can significantly impact slip resistance performance, due to issues such as directionality, contaminant runoff, contact area, and impact of wear.

Tread patterns that run in the direction of travel are ill advised because they tend to accentuate rather than retard forward motion (like skis and skates). Random patterns and patterns perpendicular to the direction of travel are more effective in providing slip resistance.

Patterns that produce enclosed areas are also not recommended. These areas can collect and trap water or other liquid contaminants. Having no path to disperse, the liquid is squeezed, and thus unable to compress, which may result in hydroplaning (see Chapter 3, Principles of Slip Resistance).

Contact area is the percentage of the footwear bottom that makes contact with the walkway surface. In general, a larger the contact area increases the slip resistance. At the same time, large unpatterned areas (especially the heel), while providing greater contact area, do not allow for liquid runoff, nor do they provide a means of gripping the surface.

For improved traction, select an outsole with small incisions (or snips) that divide the tread shape into multiple movable parts. This design channels more liquid to the outer portion of the outsole, increasing slip resistance.

If the shapes and patterns are situated to closely together, there may not be sufficient width to channel liquid to the outer edges of the outsole. Recommended are shoes with at least 2 millimeters of space between tread pattern shapes. With regard to tread depth, there should be about 2 to 3 millimeters between the sole of the shoe and the bottom of the tread (Douglas, 2004).

8.2.3 General Guidelines for Shoe Design and Selection

Verification of the effects of tread groove depth is significant in assisting designers in designing proper footwear for slippery floor conditions. In one study, COF was measured using the Neolite test pads on terrazzo, steel, and vinyl floors under three liquid-contaminated conditions. Footwear pads with tread grooves 3 and 9 mm wide were used with a Brungraber Mark II Slipmeter. The depth of tread grooves ranged from 1 to 5 mm.

The results showed that tread groove depth had a significant effect on traction. Higher friction was noted for footwear pads with deeper tread grooves on wet and water-detergent contaminated floors. The averaged COF gain per tread groove depth increase in millimeters under these conditions ranged from 0.018 to 0.108, depending on the tread groove width, floor, and contaminant (Li et al., 2006).

Few published guidelines regarding shoe design and construction are available. The following are some recommendations published by SATRA Technology Centre, U.K.:

Heel
- Round and patterned.
- Square heel breast (acts as leading edge) as opposed to a rounded edge.

Sole
- Flat, flexible bottom construction—Consider a low-density midsole that conforms to ground and maximizes contact area.
- Material that wears evenly and smoothly.
- Consider wedge sole (rather than a rounded edge) for indoor occupational footwear (e.g., catering, hospitals, sports footwear). This helps to provide the maximum contact area with the floor.
- A raised tread pattern extending over whole sole and heel area, a pattern with leading edges in many directions (e.g., a crosshatch or similar design).

Cleat patterns
- Cleat width between 3 and 20 mm.
- Channel width at least 2 mm.
- Well-defined square leading edges.
- Radius internal corners (prevents sole cracking).
- Minimum depth of at least 2 mm.

8.3 LABELING

8.3.1 Labeling for Usage

Although many factors may contribute to any injury, many have a common contributing factor: improper footwear. Shoes are made with specific characteristics to provide the best attributes for a particular activity. For example, a sprinter's shoes are very lightweight, have almost no heel, and include a sturdy sole for flat-out speed. A linesman's boot has heavy, ankle-high leather uppers to prevent twisted ankles and

Footwear

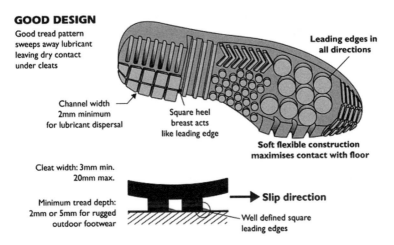

FIGURE 8.1 Attributes of "good sole design" taken from SATRA slip resistance soling design guidelines. Slip-resistant footwear bottom design guidelines based on the SATRA Chart ("Slip Resistant Sole Design Diagram"). (Reproduced with the permission of SATRA. For more information, contact SATRA at http://www.satra.co.uk.)

a steel sole shank to ease pressures on the foot while standing on narrow telephone pole pegs.

Many specific attributes also create hazards for other types of activities. A tennis shoe provides good traction, but does not provide puncture resistance or toe protection. Wrestling shoes may be lightweight and grip well, but have insufficient padding or shock resistance for activities like outdoor running.

So, the first step is to determine the activities to be performed and the types of contaminants present in your facility, such as petroleum-based oils (e.g., auto oil, axle grease, and other lubricants), solvents/acids, or food-based oils (e.g., cooking oil, shortening, grease). Then, be sure the label indicates that the manufacturer recommends the footwear for such conditions.

8.3.2 Labeling for Slip Resistance Testing

Some footwear on the market is promoted as "slip resistant." Although slip-resistant footwear can be tested, most evaluations contemplate clean, dry floor surface conditions. Test method ASTM F489 (James Machine, see Chapter 5, U.S. Standards and Guidelines) is one of the most commonly performed and cited tests for footwear slip resistance, but it tests only under dry testing conditions.

The SATRA STM 603 Slip Resistance Tester™ is preferred by many footwear industry professionals and is used extensively in Europe and increasingly in the U.S. and worldwide. The ISO has adopted such a method (ISO 13287:2007).

Some manufacturers of slip-resistant footwear may also use ASTM F1677 (Portable Inclineable Articulated Strut Tester, or Brungraber Mark II) and F1679 (*Variable Incidence Tribometer, or English XL*, see Chapter 5, U.S. Standards and Guidelines). These methods do contemplate wet and contaminated conditions. One concern, however, is that only a portion of the shoe can be tested on these instruments. Thus, the overall performance of all design aspects of the footwear cannot be fully engaged.

8.4 ADVERTISING

Caution must be exercised when evaluating product claims regarding the performance of slip-resistant footwear. In some cases, vendors have cited wet or contaminated test results using a dry only test method (e.g., ASTM F489 James Machine), which usually yields unrealistically higher results due to stiction (see Chapter 3, Principles of Slip Resistance, Section 3.9.2).

Self-serving slip resistance scales are sometimes used in an attempt to show superior performance of footwear products. In one example, a vendor indicated that:

> 0.35–0.50 was considered "acceptable" performance, and 0.50–0.60 was "good" for dry and wet conditions; 0.05–0.10 was considered "fair," and 0.10–0.20 was "good" for oily and oily/wet conditions.

In addition, the vendor implied that these were government standards, but they were not.

In other instances, numbers have been distorted through improper averaging methods. In one example, a manufacturer showed results of 0.44 (oily) and 0.45 (oily/wet) against a combined average for four different competitors of 0.34 (oily) and 0.34 (oily/wet). In reality, two of those competitors could have had poor results (significantly lower than the vendor's), bringing the two with far superior results (significantly higher than the vendor's) down to the combined average cited.

Other advertising methods use wording such as "restaurant tested and approved," which is too vague to be considered an endorsement. "Patent pending design" indicates only that the design may be the subject of a patent application. Patent protection is no guarantee of effectiveness of any design.

8.5 OTHER SELECTION GUIDELINES

Comfort is an important factor in the selection of footwear. For the program to be successful, employees must actually wear the footwear. Footwear that is uncomfortable will make this less likely to happen, therefore:

- Employees should try on the footwear first.
- Their foot should be properly measured. If one foot is slightly larger than the other, the larger size should be selected.
- If heavy socks or liners will be worn, they should be worn while trying on the footwear to ensure a proper fit.
- Because feet are the most swollen at the end of the day, employees should wait until then to shop for footwear.
- Footwear should not be selected on style. Comfort is more important.

8.6 OTHER PROTECTIVE FEATURES

In 1994, the Occupational Safety and Health Administration (OSHA) launched its first formal hazard assessment program, which required employers to institute personal protective equipment (PPE) programs. According to statistics from the Bureau of Labor Statistics (BLS), the result has been a reduction in foot and toe injuries range from 32% to 39%.

Depending on the environment in which the footwear is to be used, other protection features may also be needed. OSHA 29 CFR Part 1910.136 requires that protective footwear purchased after July 5, 1994, comply with ANSI Z41-1991—*Personal Protection–Protective Footwear* or be demonstrated by the employer to be equally effective. In 2005, ASTM assumed responsibility for Z41 (which has since been withdrawn) and produced two standards that replace it. Footwear that meets these ASTM standards meets and exceeds the ANSI Z41 standards, and so it meets OSHA standards and is compliant.

ASTM F2412—*Standard Test Methods for Foot Protection* is under the jurisdiction of F13.30 Footwear. This standard establishes test methods for impact resistance, and compression resistance of protective toes, metatarsal protection, conductive footwear, electric shock (EH/ESR) resistance, static dissipative (ESD) footwear, puncture resistance, chain saw cut resistance, and dielectric insulative footwear (e.g., high voltage).

ASTM F2413—*Standard Specification for Performance Requirements for Foot Protection* (also under ASTM F13.30 Footwear) specifies minimum requirements for design, performance, testing, and classification for protective footwear. While ASTM F2412 establishes the methods of testing, F2413 specifies fit, function, and performance criteria for protective footwear. It also provides measurable criteria for several hazards. Impact and compression requirements measure the footwear's ability to protect against falling or rolling objects. Impact- and compression-resistant shoes are constructed using a steel or nonmetallic toe cap for foot protection from objects that could crush or break toes.

Although these topics are beyond the scope of this chapter, protective features may include:

- Toe caps to guard against the impact of falling objects.
- Puncture-resistant steel-plated or other composite midsoles to protect the foot bottom from sharp objects (e.g., nails, screws, scrap metal) and worksites such as landfills where there is potential for coming into contact with needles.
- Electric shock–resistant footwear to insulate from electrical shock.
- Static dissipative footwear to minimize the buildup of static electricity where sensitive equipment is present. These are generally worn when working with sensitive computer equipment and (to help prevent potentially damaging static charges being transferred to the materials) in explosive atmospheres.
- Conductive footwear for grounding when working with hazardous materials to help prevent dangerous sparks.

- Metatarsal guards to protect the top of the foot from falling/rolling objects. F2413 requires that such guards be integrated into the shoe, which is optimal. However, strap-on guards made of aluminum, steel, fiber, or plastic that attach to the outside of the shoe to protect the top of the foot from impact and compression are still sold.
- Ankle support for wet or sloping environments.
- Anti-fatigue qualities, providing shock absorption, cushioning, and lightweight materials.
- Leggings protect the lower legs and feet from heat hazard injuries, such as molten metal or welding sparks.
- Thermal insulation to keep feet warm in cold temperatures.
- Chemical resistance to protect from contact with certain caustic chemicals.
- Water resistance in wet environments.

8.7 MAINTENANCE

8.7.1 Keeping Clean

Inspection and maintenance is part of any effective program, and footwear is no different. Footwear should be checked for cleanliness and condition daily, prior to the start of work. The presence of liquid contaminants or solid matter wedged in tread patterns can reduce slip-resistant properties. Disposable slip-resistant shoe covers are available, but they are generally designed to last for one workday and should be discarded thereafter.

8.7.2 Wear and Inspection

Wear impacts the effectiveness of slip-resistant footwear, but several factors influence the degree of wear. In any event, it is just a matter of time before slip resistance begins to deteriorate and the footwear needs to be replaced. The primary consideration is the effect of environmental and floor conditions on the footwear bottom material and tread pattern design. In general, walking on smooth surfaces tends to "polish" or smooth the footwear bottom, whereas walking on rougher surfaces tends to result in uneven wear or damage.

Shoe tread is like car tire tread—it thins and wears away over time. Because tread depth is gradually reduced by wear, monitoring the extent of outsole wear is important. Upon inspection, signs of wear are usually clear: the rear of the heel is worn away, and the sharp peaks of treads are shorter and flatter.

Failing to properly inspect footwear may result in failure to promptly identify and replace excessively worn footwear caused by the grinding and leaching of contaminants into the footwear bottom. This not only reduces the useful life of the footwear, but more quickly degrades their slip-resistant qualities.

Some employee education (and a spot check process) should be implemented to inspect and clean footwear bottoms, just as floors are regularly cleaned to ensure effectiveness. It can be helpful to develop and distribute a brief set of guidelines for staff as a payroll stuffer and/or a toolbox talk to include a discussion of:

Footwear 303

- Proper cleaning of footwear bottoms to extend their life and keep them effective
- A warning not to wear them outside of work
- A recommendation to have multiple pairs and rotate them
- A suggestion to use lockers (if available) to store shoes

Figure 8.2 shows an example of the same sole design showing a relatively new, intact pattern and an older, worn pattern.

8.7.3 Replacement

Since slip-resistant footwear is softer, it wears more quickly than conventional street footwear. It is often subject to substantial contamination. For these reasons, a system to track shoe replacement is essential. The scheduled frequency of replacement should reflect the expected conditions—sometimes as frequently as semiannually.

Consider that wear and degrading of footwear is accelerated because many employees wear slip-resistant shoes intended for work elsewhere. Traveling to and from work and walking on rough concrete sidewalks and parking lots subject footwear to conditions that puncture, gouge, scrape, abrade, and scuff them.

Depending on the size of the company, coordinating a tracking program at the corporate level may be a challenge, especially if turnover is high. Assigning responsibility for footwear replacement to the local manager helps to divide that labor and place responsibility at the most appropriate level of management, as unit management is most intimate with their workforce. Maintain information on the date each employee last purchased shoes to track when new shoes are obtained and prompt employees to purchase replacements.

FIGURE 8.2 An example of the same sole design showing a relatively new, intact pattern and an older, worn pattern.

There are no set guidelines concerning how long a shoe will retain slip resistance, because this depends on the extent and type of wear, as well as the outsole material and design. In the absence of obvious signs of sudden damage, such as punctures, gouges, or cuts, some experts suggest that the life of average footwear is six months to a year. A practical approach is to monitor the snip grooves and replace the footwear before they are worn away.

8.8 FOOTWEAR PROGRAMS

8.8.1 MANDATE OR RECOMMEND

Before instituting a slip-resistant footwear program, it is important to seek legal counsel. Although a mandatory policy is likely to maximize footwear usage, it could also present potential legal exposures if improperly drafted or implemented. It may be helpful to request that footwear vendors provide sample footwear policies for you and counsel to examine when designing your own. Employee orientation is an opportune time to sign up employees for this program.

8.8.2 SPECIFICATIONS

There are often requirements that "slip-resistant safety shoes" be worn, but rarely is this definition clear enough for employees and managers. If a single vendor is not selected for slip-resistant footwear, it is prudent to clearly specify (or pre-approve) what constitutes slip-resistant footwear suitable for the environment for which it is intended.

8.8.3 PURCHASE OPTIONS

It is advisable to review the legal and practical implementations of each option for footwear purchase. Issues such as labor and union agreements, the impact on employee morale, and record-keeping requirements need to be considered and balanced.

> Purchase for employees—It could be argued that slip-resistant footwear, for the purposes of employee safety, is no different from any other type of personal protective equipment. In general, employees are not asked to purchase and maintain their own face masks, hard hats, safety goggles, or respirators. Clearly, the downside of providing footwear for employees is the purchase price, but the benefits can certainly outweigh this consideration. Purchasing the footwear for your employees allows control over the kind of footwear worn. If consistency in look/style is important to your operations, this can also be specified. In addition, a purchase plan makes it easier to track the age and condition of footwear, permitting a system for scheduling replacing them in a timely fashion. Some companies have found that the cost-benefits of slip-resistant footwear favor purchasing them for employees. This approach can eliminate many excuses and is

more likely to be perceived as required and part of the uniform, and much easier to ensure daily use.
- Employee purchase—If employees are required to provide their own footwear, this should be clear at the time of hiring. The cost, along with criteria for acceptable footwear, should be provided in writing. It is often possible to negotiate discounts for employees to purchase selected types and styles of footwear through a specific vendor.
- Payroll deduction—Offering a voluntary payroll deduction option can make it easier for employees to purchase slip-resistant footwear. The benefit is the ease with which footwear can be purchased, helping to make implementation more expedient. Provisions should be in place for handling reimbursement of costs when employees leave the company.

8.8.4 Enforcement

Having clearly outlined what is considered acceptable footwear, company policy should clearly state that any employee arriving to work in footwear other than that specified in the policy will be sent home. There should be a progressive disciplinary system to address repeated failure of employees to wear and maintain the specified footwear.

While it may be difficult to do so in some cases (such as when the unit is short-handed), compliance is unlikely to take hold unless it is uniformly enforced. After an initial period of frequent discipline, it is likely to drop off as workers understand that management is committed to enforcing its footwear policy. Alternatives to sending the individual home include: (1) giving him or her a less desirable job for the day in an area/operation that does not require slip-resistant footwear, or (2) having on hand a small supply of overshoes or shoes he or she must sign out, use, and return at the end of the shift (optimally the least attractive variety).

8.9 FEDERAL SPECIFICATION—USPS NO. 89C

USPS No. 89C is the U.S. Postal Service Specification for Footwear. A 1993 standard, this specification applies to the footwear worn by U.S. Postal Service employees. For slip resistance, it specifies the use of ASTM F489, Static Coefficient of Friction of Shoe Sole and Heel Materials as Measured by the James Machine, and requires a threshold of no less than 0.5.

8.10 CONSENSUS STANDARDS

8.10.1 ASTM F08 Sports Equipment and Facilities

ASTM's Committee F08 on Sports Equipment and Facilities was organized in 1969 and as of this writing has a membership of 600 who participate in any of 25 technical subcommittees. Committee F08 has developed 125 standards for sports equipment, surfaces, and facilities to reduce inherent risk of injuries and promote knowledge as it relates to these standards.

8.10.1.1 ASTM F2333 Standard Test Method for Traction Characteristics of the Athletic Shoe–Sports Surface Interface

The responsibility of subcommittee F08.54 on Athletic Footwear, this test method was developed for use by three groups:

- Athletic footwear manufacturers to characterize the traction of the athletic shoe–sports surface interface and to develop athletic shoe outsoles
- Researchers in determining the effect of sport surface conditions on traction characteristics of the athletic shoe–sports surface interface
- Sports surface manufacturers to characterize the traction of the athletic shoe–sports surface interface, and to develop sports surfaces

The method was also intended to be used to research relationships between traction at athletic shoe–sports surface interfaces and athletic performance or injury, which could ultimately lead to recommended levels of traction.

The method specifies two generic apparatuses for performing tests in the field and in the laboratory. However, because these tests are not to be performed simultaneously, separate test devices are specified for measuring linear traction and rotational traction.

In order to avoid duplication or overlap with the activities of ASTM F-13 and other ASTM committees dealing with slip resistance, F-08 described their test method in terms of static traction, dynamic traction, and traction ratios.

8.11 INTERNATIONAL FOOTWEAR STANDARDS FOR SLIP RESISTANCE

Literally hundreds of domestic and international footwear-related standards are in existence. Although some slip resistance-related specifications are contained within some general footwear standards, the following sections outline standards specific to footwear slip resistance.

For information on U.S. Standards, see Chapter 5, U.S. Standards and Guidelines. For details on each of the standards organizations mentioned next, see Chapter 7, Overseas Standards.

8.11.1 ISO

8.11.1.1 ISO TC 94 Personal Safety—Protective Clothing and Equipment

Standards Australia is secretariat for this ISO technical committee. Work related to foot protection is done under SC 3, for which BSI is secretariat (http://www.iso.org/iso/iso_catalogue/catalogue_tc/catalogue_tc_browse.htm?commid=50592).

8.11.1.1.1 ISO 13287

ISO 13287-2006 is the standard for Personal protective equipment—Footwear—Test method for slip resistance. This standard was first published in 2004 as a European standard, and it is now recognized as a unified EN ISO standard test method and

Footwear

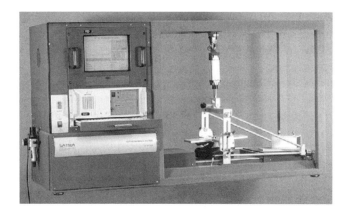

FIGURE 8.3 The SATRA Slip Resistance Tester. (Photograph courtesy of SATRA.)

allows the incorporation of slip resistance testing into the certification process for personal protective equipment (safety, protective, and occupational) footwear sold in the European market.

In this standard, the kinetic friction is measured by the movement of the shoe under test against a floor surface or vice versa. Although the apparatus is also manufactured by others, the standard describes a protocol and instrument of considerable resemblance to the SATRA Slip Resistance Tester (STM 603), which costs an estimated $50,000 (Figure 8.3 and Figure 8.4). It is not applicable to special-purpose footwear containing spikes, metal studs, or similar components. The standard includes discussion of the apparatus, test conditions, sole preparation, procedure, and test report requirements.

The method fixes the shoe at a seven-degree angle and places it on a floor material that is in motion at 0.3 meters per second. Vertical force is applied to make contact,

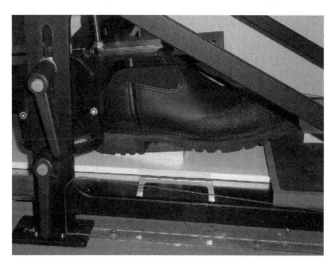

FIGURE 8.4 The SATRA Slip Resistance Tester in operation.

then 0.1 second later, horizontal force is applied. On glycerin-coated stainless steel, the minimum friction standard is a coefficient of friction of 0.18, and for water on stainless steel, the minimum is 0.23. Measurements are taken at 0.3 seconds and 0.6 seconds of contact, which are averaged to arrive at a final determination. Three sets of measurements are obtained: one at heel contact, one with flat contact, and one at forefront contact.

If industrial footwear is CE marked and claimed to be slip resistant, it must have been tested, and the coefficient of friction (COF) test values must be available. The CE mark is the official marking required by the European Community for products sold in the EU. It proves to the buyer—or user—that the product fulfills all essential safety and environmental requirements as they are defined in European Directives.

According to research by the HSE, the test may not be capable of differentiating between footwear with differing levels of slip resistance under some test conditions. For this reason, it may be of limited use in selecting slip-resistant shoes for a particular environment (HSE, Relevant Laws and Standards).

As of this writing, there is a proposal for the United States to annex this standard.

8.11.1.1.2   ISO/NP 20350

ISO/NP 20350 is the proposed standard for *Personal protective equipment—Footwear for the food industry*. As of 2007, this standard was listed as "under development" with an estimated completion date of March 2010.

8.11.1.1.3   ISO/TR 1120-1993

There are no relevant standards under development or in force, as of 2003 when ISO/TR 11220:1993 was withdrawn. Determination of Slip Resistance was the standard for Footwear for Professional Use—Determination of Slip Resistance. Placing the footwear on the testing surface with glycerin, applying a given load, and either moving the footwear horizontally in relation to the surface or moving the surface in relation to the footwear determines slip resistance. The frictional forces are measured, and the dynamic coefficient of friction (DCOF) is calculated. For practical purposes, the replacement for this standard is ISO 13287 (see below).

8.11.1.1.4   Related ISO TC94 Standards

Non-slip-resistance standards from TC94 include the following. (Each was originally promulgated in 2004 and reissued unchanged in 2007.)

- ISO 20344:2004 (A1: 2007) PPE—Footwear–test methods for footwear. This standard defines all the test methods used for certifying personal protective equipment footwear. Annex A defines the calibration procedure for ceramic tiles used in ISO 13287.

The following standards contain the specifications for footwear testing, using the methods defined in ISO 20344. Slip resistance is now a requirement in each of the following standards. Annex A of these standards defines the three marking symbols available (SRA, SRB, and SRC, depending on the tests performed) and associated slip resistance specifications.

- ISO 20345:2004 (A1: 2007) PPE—Footwear–safety footwear
- ISO 20346:2004 (A1: 2007) PPE—Footwear–protective footwear
- ISO 20347:2004 (A1: 2007) PPE—Footwear–occupational footwear

8.11.1.2 ISO TC 216 Footwear

AENOR is the secretariat for this ISO technical committee (http://www.iso.org/iso/standards_development/technical_committees/list_of_iso_technical_committees/iso_technical_committee.htm?commid=54972). Eleven countries participate. Of the current 60 published standards, none address slip resistance or coefficient of friction, although there are two proposed standards.

8.12 GERMAN—DIN 4843–100

DIN 4843–100, Safety, Protective and Occupational Footwear; Slip Resistance, Metatarsal Protection, Protective Insert and Thermal Behaviour; Safety Requirements, Testing, was published in August 1993 (in German only). Until 2004, it served as the method for measuring the slip resistance of shoes in Germany by means of "light-footed" walking over a ramp. The angle of inclination that is borderline acceptable to the test person serves as the measure for the slip resistance. In 2004, DIN 4843-100 was replaced by DIN EN 13287 (see ISO 13287 above).

9 Food Service Operations

If your foot slips, you may regain your balance; if your mouth slips, you cannot recall your words.

Proverb

9.1 INTRODUCTION

Business insurance costs are driven by two primary factors: exposures (e.g., sales or payroll) and past losses or claims. Falls can represent as high as 45% of all workers compensation claims dollars in food service operations.

The Bureau of Labor Statistics (BLS) annual census of occupational injuries and fatalities reports that 3.4 of every 100 food service workers were injured from a slip or fall severe enough to require at least a day off work in 2002, versus 8.5 per every 100 workers in 1992.

According to BLS, injured servers, kitchen workers and food preparation workers miss approximately five days of work following an accident. And 22% of slip and fall injuries resulted in more than 31 days away from work.

Slips and falls in restaurant/hospitality occupancies account for 34% of all restaurant worker injury cases (Hedden, 1997).

Restaurant Insurance Corp. (RIC), a specialty insurance company, conducted a study of over 3,700 claims filed by its insured restaurants since 2002. Slips, trips, and falls by patrons and others were the most frequent type of claims from full-service restaurants, with each resulting in an average cost of $3,550 (Restaurants, 2004). At modest profit margins, a full service restaurant must serve hundreds of additional meals to make up $3,550 in additional costs.

According to another study, slips and falls account for 31% of restaurant worker injuries. Same-level falls in restaurants is six times the number of claims of the next highest industry, nursing homes in Washington (Li et al., 2006).

Customer falls intensify the seriousness of the issue. Slip and fall accidents accounted for 57% of food service general liability insurance claims (Heavyweight Solutions, Inc.). Of foodservice premises liability verdicts, 41% were between $10,000 and $99,999, and 47% were between $100,000 and $5 million. The verdict mean was $85,840, and the average was $397,000 (Heavyweight Solutions, Inc.).

One lasting effect of kitchen injuries is their impact on insurance premiums. According to one commercial insurer, the national average for workers' compensation premiums in the restaurant industry is $7 per $100 of payroll. For those with a poor safety record, the premium is higher still (Hedden, 1997).

Leamon and Murphy (in McCabe, 2004) reported that the incidence rate of falls on the same level over a two-year period was 4.1 per 100 full-time-equivalent restaurant employees, and calculated that workers' compensation losses due to falls in restaurants resulted in an industry-wide cost of $116 per restaurant worker (Courtney et al., 2006).

It is clear from these statistics that reducing claims and their associated costs is one way that restaurant owners and managers can positively impact overall costs.

9.2 EXPOSURE OVERVIEW

CHEF SLIPS AND SUFFERS SEVERE BURNS FROM HOT OIL

A chef working in a hotel kitchen was walking by a deep fat fryer carrying a box of potato peelings when his foot slipped. As he reached out to steady himself, he plunged his arm into hot oil. He sustained burns to his hand, arm, and face, and was out of work for five months.

An investigation found that there had been daily problems with water pooling on the floor around the dishwasher, vegetable preparation area, and steamers. They were due to leaks and improper operation of the dishwasher.

The ongoing problem with water on the floor was before and following the accident, and the issue had never been addressed. The company was charged with failing to undertake a suitable risk assessment and put in place a safe system of work. The company pleaded guilty, were fined £14,000 and had to pay costs of £2,000.

Since the incident, the company made several improvements; a clean as you go policy has been set up; the floor has been replaced with one that will remain slip resistant when wet; and risk assessments are now reviewed twice a month and results are reported to management.

Source: UK HSE, Preventing Slips and Trips in the Workplace

Food service is a fast-paced industry where customer service is determined by two factors: food quality and speed of service. Conditions in a restaurant kitchen can include slippery floors, knives and other sharp tools, hot surfaces, heavy pieces of movable equipment, oversized food packages, and congested quarters. The impact of these hazards is amplified during peak times when the pace is taken up a notch.

Restaurant, bar, and other food service operations present a high potential for slips, trips, and falls. Patrons are often unfamiliar with the layout of the facility. Dim lighting for mood can mask the unique architectural and decorative features that are aesthetically pleasing but function poorly for the purposes of pedestrian safety: oversized handrails, unmarked changes in levels, mixed types of floor coverings, and glossy flooring. The presence of alcoholic beverages and commonly experienced food and drink spills contribute further to conditions that heighten the potential for falls.

Slips and falls can occur on wet or contaminated surfaces, and where there are transitions in floor types (e.g., from the carpet in a dining area to the ceramic tile in a kitchen area). Common sources of slippery floors include dishwashing over-spray or runoff, leaking equipment or pipes, food debris, and spillage from transport of open containers such as those holding fryer grease and food wastes (Courtney et al., 2006).

Employees not only face most of the hazards encountered by customers, they are also subject to the dangers in the back of the house. Wet and greasy floors are often a potential hazard. Water in the dishwashing area, oil and grease residue, spills in

Food Service Operations

the food preparation area, and ice-covered floors located in the walk-in freezers each present a substantial slip hazard. Supplies and equipment stored in the walkways, trash, and other similar housekeeping issues are not uncommon. High employee turnover creates ongoing training and awareness challenges.

9.3 PEDESTRIAN FLOW AND SLIPS, TRIPS, AND FALLS

Traffic patterns can be defined as the time and location of where people walk. People do not always walk where you expect or intend them to, so traffic patterns may not follow the way in which the facility was designed. Traffic patterns are important to understand, because where people walk can increase the potential for falls and result in accidents, especially in times of considerable activity. Knowing the traffic patterns in the facility can reveal opportunities to reduce exposure to slip, trip, and fall accidents.

In order to truly understand traffic patterns, time must be spent observing employees from a distance as they perform their daily work activities. High-traffic-flow areas are especially conducive to trips and falls. These areas, such as transition areas between the front and back of the house, should be observed closely, and traffic patterns should be determined for critical times of the day. The staff should be observed as a system, not as individuals. The objective is to identify the common paths taken when moving between the front and back of the house.

Supplement observations with an analysis of where and when falls are occurring to better pinpoint areas of concern. It may at times be appropriate to ask employees why they walk where they do, but since choosing a path is not always a conscious decision, they may not have a ready explanation.

After this information has been assembled, work to redirect traffic flows away from undesirable paths and onto preferred, safer paths, or to make the undesirable paths safer. It may be determined that the traffic flows are appropriate, but that some controls such as floor cleaning may need to be improved or conducted more frequently.

9.3.1 Making the Undesirable Path Safer

Because the shortest and quickest path is favored, it is often best to try and adopt the path created by human behavior. This reduces or eliminates the potential for pedestrians to veer into unintended and possibly hazardous areas, because they are already taking the most expedient route. In short, a determination of where people walk should be made. From there, a design should be crafted around this behavior. Options to consider include:

- *Increase the focus.* Remove or limit distractions that take employees' attention away from the task at hand. Avoid vivid floor patterns and obstructions where practical, especially those below knee height that can be easily missed in the fast pace of food service operations.
- *Increase lighting.* Insufficient lighting, or glare in the path of travel can increase the potential for falls and can cause employees to change their travel paths to avoid such conditions. Providing substantial, glare-free lighting

makes it easier to anticipate unexpected and undesirable floor conditions and gives employees a chance to avoid or more gingerly traverse the area.
- *Upgrade the housekeeping.* One of the greatest challenges in food service is maintaining good housekeeping during times of peak activity. Grease, liquids, and food can be readily tracked from the kitchen to the front of the house. Increasing monitoring of floor or mat conditions may be what is needed to make an unsafe path significantly safer.
- *Use good basic design features.* Proper workspace design is one of the fundamentals of slip/trip/fall prevention. Ramps that are too steep, uneven steps, unmarked changes in levels, and missing handrails are major contributors to falls. Installing and maintaining a slip-resistant floor surface is essential. If possible, allow for elbow room in kitchen areas, because greater congestion often leads to increased potential for accidents.

9.3.2 Redirect Traffic onto Preferred, Safer Paths

- *Accentuate the positive.* Emphasize the desired path with barriers to limit access to the undesired path, lighting to highlight the desired path, and/or other visual cues such as floor arrows or mat designs to guide the way.
- *Redesign paths or workflow.* Should other measures fail to achieve the desired effect, the remaining solution may involve a redesign of paths by creating an entirely new arrangement for the back and/or front of the house. While this may require more effort, it also provides an opportunity to optimize the space not only to minimize slip/trips/falls, but to improve efficiency.

9.3.3 Customer Falls

The same methods used to put employees on the right path can also help customers. The flow of how customers move through the facility should be noted, and customer slip, trip, and fall accidents should be reviewed to identify areas with high potential. The ways to redirect customers to the desired path or adapt customer-preferred flows in a safer way should be explored.

As each change is implemented, watch employees and customers and monitor accident experience to determine if there has been improvement. Several adjustments may be required to minimize exposures to slip, trip, and fall accidents. Change is a continuous process and an inevitable step to improving efficiencies and creating a safer workplace.

CASE STUDY: Fractured Skull from a Hazardous Floor Surface

The worker experienced an uncontrollable slip and hit her head on the hard tiled floor. She was rushed to hospital, where she drifted in and out of consciousness, suffered seizures, and spent a lengthy period in the hospital's high dependency unit. The employer was prosecuted and ordered to pay over £36,000 including prosecution costs.

A local authority health and safety inspector investigating the event observed that the floor surface in the kitchen was slippery with even the smallest amount of water or grease on it.

In addition, some areas were sloped, increasing the hazard. Kitchen staff was seen walking with a peculiar gait, trying to avoid slipping. Cleaners had removed mats at the time of the accident, leaving workers to walk on the floor that quickly became contaminated with food waste, water, and oily residues.

The company's safety records showed there had been other slip incidents in the area (including four in the prior 12 months), but the response had been to provide the most heavily contaminated areas with matting that was slippery, especially when wet. It was determined that the floor surface was not fit for the purpose.

After a number of options were considered, the company elected to replace the floor surface with one that was suitable for use in an area where elimination of floor contaminants was not possible. The new floor was specified to provide sufficient traction, even in wet or contaminated conditions.

Source: UK HSE, December 2008

9.4 FLOOR SURFACES/HOUSEKEEPING

9.4.1 Keep Floors Dry

To reduce the potential for food and beverage spills and leaking equipment:

- Arrange tasks to reduce carrying. When less is carried, the likelihood that a spill occurs decreases.
- Steps that can take a bit of extra time, but can avoid a much larger problem:
  - Use lids on portable containers.
  - Avoid carrying too much at one time.
  - Use hot mitts when carrying hot items.
- Basic kitchen design features that can keep contamination off the floor include providing:
  - Lips around table edges.
  - Curbs around equipment (e.g., ice machines).
  - Drip trays under taps.
- Follow manufacturer preventative maintenance schedule, especially for water-producing equipment such as drink stations and ice machines.

To mitigate the impact of wet floors:

- "Clean as you work." Promptly cleaning spills as they occur is fundamental. The longer a spill remains on the floor, the greater the odds that an accident will result.

- Immediately mark and section off spills from all directions until area is cleaned up and dry.
- Identify areas where spills routinely occur and increase inspection during peak activity time.
- Inspect equipment for leakage daily, and report problems to the appropriate contact for repair. If equipment cannot be shut off, use place mats and/or increase frequency of cleanup until repaired.
- Inspect floor drains to make sure they are clear and free flowing. Failure to maintain floor drains makes the accumulation of water and oil not a matter of if, but of when.
- Clean floors
  - In times of low activity
  - In sections, leaving a dry path to reach needed areas
  - By marking and sectioning off wet areas from all directions until dry

9.4.2 Keep Floors Clean

- Use separate, color-coded mops and buckets for the front and back of the house. This will ensure that the equipment used for the back of the house, which is significantly greasier and contaminated, does not spread contamination to the front of the house.
- Follow the manufacturer instructions for concentration/floor area of cleaner.
- Follow the cleaning schedule, but increase frequency if indicated based on the condition of the floor, complaints, or accidents.
- Inspect cleaning equipment daily and report to the appropriate contact for repair.
- Use scrubber or deck brush in grease-laden areas when possible.
- Change mop heads and scrub pads daily or more frequently if indicated based on the condition of this equipment.
- Use hot water and change water when dirty. Otherwise, all that is accomplished from mopping is spreading the contamination to less contaminated areas.
- Identify areas of high grease accumulation (e.g., fryer area) and increase spot cleaning as needed.

Figure 9.1 shows a quarry tile kitchen floor in relatively pristine condition, while Figure 9.2 and Figure 9.3 illustrate the smoothing out and contamination (or fouling) to which quarry tile is subject due to inadequate cleaning. Note the condition of the grout, a strong indicator of the effectiveness of floor cleaning.

Figure 9.4 shows a kitchen floor that has been poorly maintained. Note the variety of replacement tiles used, the eroding/damaged condition of the grout, and the pooling of liquids due to uneven surfaces.

9.4.2.1 Common Cleaning Scenario

Typically, major floor-cleaning jobs are performed at night when traffic is the lightest. Unfortunately, the low priority placed on floor cleaning results in cost cutting, which impacts the quality of the job. Whether the work is performed in-house or by

Food Service Operations 317

FIGURE 9.1 A quarry tile kitchen floor in relatively pristine condition.

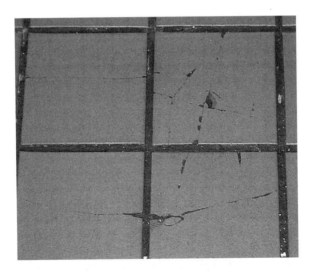

FIGURE 9.2 The smoothing out and contamination (or fouling) to which quarry tile is subject due to inadequate cleaning.

FIGURE 9.3 The smoothing out and contamination (or fouling) to which quarry tile is subject due to inadequate cleaning. Note the condition of the grout, a strong indicator of the effectiveness of floor cleaning.

FIGURE 9.4 A kitchen floor that has been poorly maintained. Note the variety of replacement tiles used, the eroding/damaged condition of the grout, and the pooling of liquids due to uneven surfaces.

an independent contractor, a common scenario involves workers that are hired at low wages with minimal supervision using the least expensive cleaning supplies and poorly maintained equipment.

Typically, floors are cleaned by dust mopping (or dry mopping) first and then wet mopping; however, this process works better in theory than in practice. In theory, a clean, freshly treated dust mop is first used to remove dry contaminants including loose debris, grit, and soil. In reality, the condition of the mop often deposits more contaminants than it removes. Disposable dust mops are available, but there is a tendency to overuse them due to the expense involved. Mop cleaning should be available, optimally by either an open hose on a vacuum or preferably in a washing machine.

In theory, the worker then obtains a new or cleaned mop head with a fresh pail of water/cleaner and begins mopping up to the square foot area specified by the floor cleaning product cleaner. Then, the mop is cleaned or changed, and the pail is cleaned and filled with fresh water and cleaner. In practice, workers usually fill a bucket once, use the water mixed with cleaning solvent to swab an area far larger than recommended by the manufacturer of the cleaner, and use a dirty mop. The result is the solvent loses effectiveness and becomes diluted and increasingly contaminated. The end result is that dirt, grease, and contaminants that may have been confined to one small section are spread over a much wider area.

It has been demonstrated that within about a month of continuous mopping in oil or grease-laden environments (e.g., restaurants, supermarkets), a hard, clear, shiny layer forms. Although the floor is still slip resistant under dry conditions, traction becomes lower when the floor is wet. This phenomenon is known as polymerization, in which the oils break down into fatty acids that attach to the surface, making them difficult to dislodge through cleaning by mopping (non-agitation) methods.

Polymerization is defined as the process of chaining together many simple molecules to form more complex molecules, with different physical properties, which combine and harden. This results in a film that is impervious to detergents at warm and moderately hot temperatures.

Although mopping is the most commonly used method of cleaning kitchen floors, it is simply not a viable method for removing a polymerized surface. An effective solution is to use a deck brush or automatic scrubbing equipment, with water of 160°F (334 K) or higher, followed by wet vacuuming or comparable removal of residue prior to rinsing. Although it may require more effort, and perhaps more time, the agitation created by scrubbing cleans the walkway surface far more thoroughly than does mopping. This approach is strongly recommended for high-traffic areas that become extremely slippery when wet or frequently become contaminated. An additional benefit of automatic scrubbing is that fresh water and soap is continuously applied and moves aside the used solvent, providing optimal cleaning conditions.

CASE STUDY: Optimal Cleaning Depends on the Type and Condition of the Flooring Being Cleaned

This study shows that, in many restaurants, floor friction can be improved by simple changes in cleaning procedures. Food safety officers visited 10 restaurants in London to identify floor cleaning procedures including cleaning methods, the concentration of floor cleaner, and temperature of the wash water. The onsite cleaning procedures were repeated in the laboratory to remove olive oil from new and sealed quarry tiles, fouled and worn quarry tiles, and new porcelain tiles. For each site, the cleaning efficiency was optimized by changing the parameters of the cleaning procedures.

Recently installed flooring such as porcelain tiles or sealed quarry tiles are relatively easy to clean; however, as the roughness and the sealer wears off, they become smoother and fouled, and less slip resistant and more difficult to clean. As such, damp mopping with a neutral floor cleaner becomes inadequate. Simple changes, such as using a two-step cleaning method with a degreaser diluted as recommended with water at 24°C, can result in substantial improvement of the floor friction. It is essential that the cleaning method be adapted to the condition of the flooring as it ages.

Note: Two of the three degreasers gave very good results, while the cleaning efficiency of the neutrals was generally lower. It is fair to conclude that degreasers are better suited than neutrals for cleaning kitchen floors exposed to olive oil.

Source: Quirion et al., 2007

9.4.2.2 The Chemistry of Fat and Flooring

9.4.2.2.1 Impact of Fats on Flooring Materials

Fat penetrates more deeply into porous floorings such as quarry tile, the most commonly used flooring material in commercial kitchens, and stripped vinyl, while it remains at the surface of impermeable floorings such as finished vinyl and glazed ceramic. Permeable floorings are more difficult to clean because fat is trapped in the floor material. Impermeable floorings are easier to clean because fat is accessible on the surface. Floor cleaners can be classified into six categories based on the chemical nature of the main ingredients:

Neutral Anionic (NA = 1)
Neutral Non-ionic (NN = 2)
Degreaser Anionic (DA = 3)
Cationic (C = 4)
Degreasers based on Glycol Ether (DG = 5)
Degreasers based on Limonene (DL = 6)

Cleaning vegetable oil with a type NA floor cleaner removes twice as much fat from the ceramic than the quarry tile. Aggressive degreasers such as DG and strongly alkaline C may remove part or all the acrylic finish, leaving the flooring unprotected against fat penetration.

Common fats typically found in the kitchen may be vegetable, such as shortening and vegetable oil, or animal, such as pork and chicken fat. Shortening is more viscous than vegetable oil, and cooked chicken fat falls between the two. On impermeable

Food Service Operations 321

> **Online Resources: Floor Cleaning**
> 1. **Stop slips in kitchens**—An initiative aimed at people working in and managing kitchens from the UK Health and Safety Executive (HSE), http://www.hse.gov.uk/slips/kitchens/.
> 2. **Floor cleaning training program and support materials**—Francois Quirion of Q Inc has carried out several experimental field and laboratory studies to determine the optimum conditions for using floor cleaners to reduce the risks of slipping. The suggested training scenarios are based on training tools (CD and printable documentation) developed by the Institut de Recherche Robert-Sauvé en Santé et en Sécurité du Travail (IRSST). They last between 15 and 75 minutes and are adapted to the availability or not of computers and audio-visual aids.
>   - Interactive training program
>   - Training scenarios
>   - Game sheets
>   - Quiz sheets
>   - Information sheets
>   http://www.qinc.ca/entretien/index1.html
> 3. **Floor cleaning as a preventive measure against slip and fall accidents** by François Quirion—Technical Guide RF-366 (IRSST).
>   http://www.irsst.qc.ca/files/documents/PubIRSST/RF-366.pdf

flooring, fat remains on the surface, and the ability to be cleaned is comparable. For porous floorings, cleaning efficiency depends on the penetration of the fat into the pores. Shortening will remain on the surface longer and will generally be easier to clean than vegetable oil that penetrates rapidly due to lower viscosity.

Normal cleaning of quarry tile with a type NN cleaner left half as much shortening as did the vegetable oil. If cleaning is not performed frequently, fats on the surface will migrate deeply into the flooring. In that case, the higher the viscosity of the fat, the more difficult removal is. Vegetable oil penetrates rapidly into pores, but low viscosity helps dispersion and removal with an adequate cleaner and method. Shortening, with its higher viscosity, becomes difficult to disperse once trapped in porous flooring.

Cooked chicken fat has a viscosity low enough for rapid penetration into the porous floorings and high enough to be difficult to disperse and remove. In general, it was found that cooked chicken fat was more difficult to clean from porous surfaces than shortening and vegetable oil (Quirion, 2004).

9.4.2.2.2 Impact of Fats on Cleaning

When floor cleaning begins, the cleaning solution contains water and floor cleaner. As dirt is transferred to the cleaning solution, some remains are left in the mop. The dirtier the mop and the cleaning solution become, the more solution left on the floor will be dirty. As the solution dries, nonvolatile components like surfactants, alkaline salts, dirt particles, and fat remain on the floor and become cleaning residues. Adding

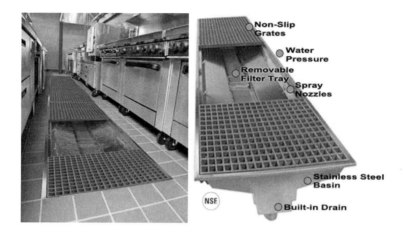

FIGURE 9.5 An example of a self-washing floor system. (Photograph courtesy of Sanifloor.)

cleaner to a dirty cleaning solution does not make it cleaner. In reality, overdosing the floor cleaner generates more cleaning residues that may make the floor more slippery.

To reduce the amount of residue, it is important to use adequate dosage of the floor cleaner, to change frequently the cleaning solution, and to change or clean the mop as it becomes dirty. Rinsing the floor with a damp mop is an effective way to remove most cleaning residues.

For similar reasons, tools used to clean should not also be used to pick up food spills. A scraper and shovel and towels should be used to pick up the most of the spill. Only at this point should a mop be used to clean the floor.

Cleaning residues and floor contaminants not removed during cleaning remain trapped in the floorings, where they may react with air and other materials and eventually fill and clog the pores. With time, the fouling action, combined with the wear of the flooring, will even the surface and make it more slippery (Quirion, 2004).

9.4.2.3 Self-Washing Floors

An innovation in the quest to maintain floors is the self-washing floor system. This involves the installation of a basin and drain into the floor equipped with built-in water nozzles and a slip-resistant fiberglass grate on top. The primary connections are a simple cold water supply and cold water drain.

Liquids and debris fall through the grates into the basin below, allowing the floor to stay drier and debris-free, minimizing cross contamination. The automatic high pressure water nozzles spray debris off the basin sides and onto the filter tray, and the stainless steel filter tray lifts out for debris disposal (see Figure 9.5).

The manufacture estimates savings in several areas including the following:

- Reduction of slip and fall accidents
- Reduced labor
- Less water usage
- Use of fewer cleaning products and related equipment

Food Service Operations

- Elimination of rubber mats and associated cleaning, change out, and replacement

(Sani-Floor, 2007)

For a detailed discussion of floor cleaning procedures and equipment, see Chapter 6, Flooring and Floor Maintenance.

9.4.3 KEEP WALKWAYS CLEAR

- Make sure there are enough trash cans, placed where waste is created.
- Clean as you work. Remove trash when done.
- Remove trash and materials from walkway paths when observed.
- Increase inspection in areas where trash accumulates and storage is not put away.
- Organize the work area.
- Remove hoses and electrical cords from walkway paths.
- Mark and section off any hazards.

9.4.4 PERCEPTIONS OF FOOD SERVICE WORKERS

Some research has suggested that worker self-reports may be a reasonably good indicator of floor slipperiness. To test this theory, 126 workers employed in 10 fast food restaurants participated in a study in which participants' ratings of floor slipperiness and occupational slip history within the prior four weeks were collected through written questionnaires. Additional factors collected by questionnaire included age, gender, shift length, and shoe type. Shoe wear and contamination were visually assessed by researchers, and floor friction was measured.

Lower restaurant mean coefficient of friction and the presence of contamination on workers' shoe soles were environmental factors significantly associated with workers reporting more slippery conditions. A recent workplace history of slipping with or without a subsequent fall was also significantly associated with workers reporting more slippery conditions. The results suggest that worker ratings of slipperiness are influenced not only by the actual level of friction, but also by the other individual and environmental factors as noted above (Courtney et al., 2006).

9.4.5 SPILL CLEANUP

One tactic used by the Cheesecake Factory to prevent customer slips and falls is the use of collapsible wet floor signs hidden behind planters or other fixtures. These signs can be quickly deployed when a hazard is observed.

One alternative to cones is an over-the-spill station kit. These are pads that can be dropped over a spill and absorb a substantial amount of liquid. Because they are brightly colored, usually yellow, they also serve as an alert to employees that clean up is required and to customers to exercise caution (See Figure 9.6).

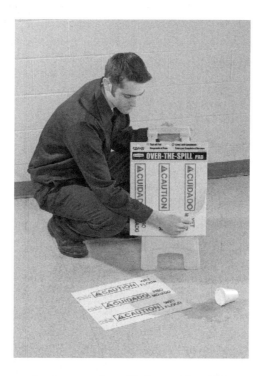

FIGURE 9.6 An example of an over-the-spill cleanup kit, which serves as a warning sign and an absorbent. (Courtesy of New Pig Corporation, www.newpig.com)

Another option is to have managers or hostesses stand over a wet spot until it is cleaned. If available, a headset–intercom system could be used to call bussers to the spot quickly and without leaving the area unattended.

Also see Exhibit 9.2, Kitchen Slip/Fall Hazard Assessment.

9.5 FLOOR MATS

Just as in entryways, mats for food service operations are intended to supplement, not supplant good housekeeping. Mats in commercial kitchens should serve as a short-term measure of protection from food and liquid spills, not an attempt to excuse inadequate floor cleaning.

Like entry mats, kitchen and dining room matting should never be overlapped and should be sized appropriately for the area (see Chapter 1, Physical Evaluation, Section 1.12).

9.5.1 Rubber Mats

Rubber, anti-slip, fatigue-fighting, grease-proof mats are designed for use in commercial kitchens, particularly in areas such as dishwashing, freezers, ice machines, and in front of deep fat fryers where wet or oil-contaminated floors are inevitable

Food Service Operations 325

FIGURE 9.7 Drink Station. Anti-fatigue mats are being used, and they only cover a portion of the counter area. These do not provide absorptive qualities, so spills will tend to bead up and roll off onto the hard flooring. These should be replaced with one or two carpet mats extending the length of the drink station.

(Figure 9.7). They are also commonly used where food (nonliquid) spills are likely, such as in food prep areas and along salad bars.

Rubber mats are most often black or red and ½ to 1 in. (13 to 25 mm) in thickness. Older styles are thicker, heavier, and not beveled, which makes them difficult to move and clean, and hard to move carts over. Other characteristics of these mats are shown below:

- Heavy rubber construction helps keep mats in place.
- Provides comfort and relief for extensive standing (anti-fatigue properties).
- Raised rib design enhances traction in wet or dry conditions.
- Heavy-duty rubber and large circular pattern is intended to allow easy drain-through of liquids and debris.

Desirable features available for these mats include the following:

- A jigsaw-style interface so they can be seamlessly interlocked when longer runs are needed
- Nitrile rubber mats that are resistant to many water, oils, greases, and animal fats
- Treated with antimicrobial agents to resist growth of contaminating organisms
- Low-profile mats to accommodate cart traffic (e.g., ⅜ in. (10 mm)

Although normally provided with beveled edges, the thickness of rubber mats can be an issue when carts such as those used to move clean dishes to the dining room are used. In such cases, proper placement (e.g., outside of the cart path) or design (such as thinner mats of the same construction) are options to consider.

Rubber mats are commonly cleaned by placing them into a commercial dishwasher, which should normally be done daily.

9.5.2 Olefin Fiber Mats

Olefin fiber mats are appropriate for locations such as backroom and patron drink stations and serving lines where spills may occur, the area is not a habitually contaminated area—and appearance is important. Olefin mats have the following characteristics:

- Available in a wide variety of colors
- Normally range from ⅜ to ½ in. (10 to 13 mm) thick
- Can hold up to a gallon of water per square yard
- Dries three times faster than nylon or polyester

9.5.3 Wiper/Scraper Mats

For mats leading from the kitchen into the dining room, a wiper/scraper mat is generally recommended. For further detail, see Chapter 1, Physical Evaluation, Section 1.12.4.

For more information on care of mats, see Exhibit 9.1, Floor Mat Cleaning Requirements.

CASE STUDY: Cafeteria/Kitchen Floors

Serving Line Mat. A mat is provided at the beginning of the serving line where trays are stored. It collects any residual water from the trays so it is not tracked onto the hard surfaces. The mat currently in use is unsecured and too small such that it must continually be reset into place. To resolve this issue, this mat should be replaced with a larger one that covers the width of the aisle way. For additional securement, the mat can also be slipped under the wire racks located on the wall in front of the mat.

9.6   FOOTWEAR

The impact of appropriate footwear in food operations has been well established by substantial industry experience. In addition to the guidance provided in Chapter 8, Footwear, the following are issues to consider specific to food service.

In kitchens, a sensible shoe will not be enough to reduce slips; slip-resistant footwear is needed. Sole tread needs to be kept clear of waste or debris. If they constantly clog, the effectiveness of the sole design to interrupt a slip will be impeded.

Food Service Operations

FIGURE 9.8 Serving line mat.

Suitable food service footwear generally has a well-defined tread pattern with more edges.

Shoe comfort is also an important aspect of safety for the hospitality industry. Comfort increases energy. Since uncomfortable footwear is unlikely to be used, it is also a practical way to improve the likelihood of it being worn.

Some experts believe ideal footwear for the food service industry is a three-quarter top, high enough to protect the ankle bone and to also protect against falling objects (Torres, 2007).

Finally, consider how easy it is to clean and maintain and how long it will last.

CASE STUDIES: Slip-Resistant Footwear Program for Commercial Kitchens

Friendly's implemented a cost-sharing program to help employees purchase slip-resistant safety shoes. The investment paid off by reducing slip and fall injuries by 30% in its first year. Friendly's had already tried slip-resistant mats, experimented with various cleaning products and employee training, but slip and fall injuries still remained troublesome.

Friendly's implemented a payroll deduction copayment plan. Every quarter, in each of its 725 restaurants, the company contributes $10 toward shoes for employees. The balance is deducted over four paychecks. While it is mandatory that employees wear slip-resistant shoes, they are not required to purchase them through Friendly's.

In the first full year of the program, cuts replaced slips and falls as the number one cause of injury, and the company saved an estimated $750,000 (Hedden, 1997).

The Bailey Company, which operates 50 franchise restaurants, bought slip-resistant footwear for every employee—more than 1,000 pairs.

Slip and fall injuries have been virtually eliminated, and correspondingly, the company's workers' compensation premium has dropped significantly. The company now sponsors an employee footwear program. Newly hired employees are given certificates to purchase safe footwear; the cost of the certificate is then deducted from employees' paychecks. The company reports the small deduction is hardly noticeable, and they have received no complaints about the cost (Pinnacol Assurance, 2006).

9.7 MULTIPLE INTERVENTION STUDY

In a single study by the Ohio Bureau of Workers' Compensation, eight different methods of improving slip resistance in restaurants were tested. Each intervention was tested at a minimum of four restaurant locations and included a control group where no interventions were initiated. The following interventions were implemented and evaluated:

- *Nonslip shoes.* Each worker in this group was provided a pair of nonslip shoes worn through the follow-up test period.
- *Shoe covers.* Shoe covers consisted of a rubber cover with treaded soles that fit over employees' shoes.
- *Nonslip floor treatment.* The floor treatment consisted of a one-time treatment.
- *Floor grit strips.* Grit strips consisting of a rough surface with an adhesive backing were placed in the center of each tile.
- *Floor treatment and grit strips.* A combination of the floor treatment and grit strips.
- *New floor cleaning product.* This product was designed to reduce the buildup of grease and starch on the floor and was used nightly to clean the floor.
- *New floor cleaning equipment.* Consisting of a weighted scrub brush, the intent is to reduce the push forces required to scrub floors, which could lead to cleaner floors due to reduced fatigue.
- *Contracted floor cleaning service.* An outside vendor was contracted to clean the floors.

The injury rate was calculated by taking the number of OSHA-recordable injuries and normalizing the number based on hours worked. Data were collected for one year prior to (baseline) and one year after (follow-up) implementation. The following impacts on injury rates by intervention type were achieved:

Type of Intervention	Injury Rate Before	Injury Rate After	Injury Rate Change
Shoes	11.6	0	−100%
Shoe covers*	17.9	4.66	−73.96%
Floor treatment	4.23	19.78	367.80%
Grit strips*	16.73	33.80	102.04%
Floor treatment & grit strips*	45.77	38.87	−15.08%
New cleaning product	15.65	16.67	6.56%
New cleaning equipment	6.64	8.43	27.10%
Contracted cleaning service	33.82	59.66	76.40%
Control group (no intervention)	11.13	4.33	−61.10%

* Discontinued test.

Nonslip shoes provided the greatest reduction in injuries; no slips or falls occurred in the one-year follow-up period.

Although the only other intervention to perform better than the control group was shoe covers, this test was discontinued at the advice of a physician. Since covers did not allow the feet to breathe, and moisture resulted in incidents of rashes, these were not deemed appropriate for use.

Floor treatment and grit strips also resulted in a decreased incidence rate, although not more than the control group. However, this test and grit strips were discontinued, because the strips made cleaning difficult. Mop fibers became entangled, making mopping difficult. Entangled fibers also created an unattractive floor. The strips also came up after a relatively short period of time, particularly when exposed to water.

EXHIBIT 9.1

Floor Mat Cleaning Requirements

Area	Cleaning
Entry (vestibule)	**Vacuum daily.**
	Surface clean. Use water-based liquid or foam cleaner and gently scrub with hand-held brush to remove surface soil. Do not use solvent based cleaners or rotary scrubbers.
	Drying. Remove excess water by hanging. Allow to dry completely before placing back in use.
	Tufted mats. Use power washer or carpet extractor. Do not use solvents, alkalis, Ph detergents, or rotary scrubbers.
Entry (walk-off)	**Vacuum daily.**
	Spot/Stain removal. Hand wash or extract stains with a carpet spotter using water-based carpet detergent or shampoo.
	Deep clean. Periodically clean with carpet extractor using water-based detergent or shampoo. Follow manufacturer's instructions for operation and chemical use. Do not use solvent-based cleaners.
	Drying. Remove excess water by extracting or hanging. Allow to dry completely before returning to use.
Salad bar	**Mat-loc or rug-hugger backing.** Do not drag to prevent damage to backing. To move, roll up and pick up entire mat.
Dishwasher	**Weekly**
Fryer/cooking	Use water-based detergent and hot, soapy water with a mop or pressure washer.
Freezer	Do not put in dishwasher
Bar (behind)	
Serving line	*Note: Failure can occur if mats are not cleaned with detergent for extended periods.*
Checkout	**Daily**
	Shake dirt and debris by flipping the mat over or gently sweep with a broom.
	Sweep underneath before replacing.
	For fingertip type mats, flip the mat over to remove sand, dirt and debris collected inside.
	Weekly
	Wash off mats with hose or pressure washer or gently scrub by hand with a soft brush and nonabrasive, water-based detergent.
	Let it fully dry before returning to use.
	Do not use rotary scrubbers, vacuums, or other mechanical cleaners. Do not use solvents, pH detergents, or alkalis.

EXHIBIT 9.2

Kitchen Slip/Fall Hazard Assessment

	Problem Area(s)	Corrective Action(s)	Responsible Person	Target Date
WALKWAYS IN GOOD CONDITION	A sloped floor/ramp or stairway that feels too steep			
	Lack of handrails where stairs or ramps rise more than 30" above the floor			
	Handrails that are difficult to grasp, loose, or in poor condition			
	Poor lighting			
	A single step that is not clearly marked			
	Uneven steps			
	Floors with cracks, holes, dips, unevenness, or similar conditions			
KEEP FLOORS DRY	Arrange tasks to reduce carrying			
	Use lids on portable containers			
	Avoid carrying too much at one time			
	Provide lips around tables, curbs around equipment, and drip trays under taps			
	Use hot mitts when carrying hot items			
	"Clean as you work"—promptly clean up own spills			
	When seen, immediately mark and section off spills from all directions until cleaned up and dry			
	Identify areas where spills routinely occur, and increase inspection during peak activity time			
	Follow manufacturer/client preventative maintenance schedule, especially drink stations, and ice machines			
	Inspect equipment for leakage daily and report problems to the appropriate contact for repair			
	If equipment cannot be shut off, place mats and/or increase frequency of cleanup until repaired			
	Clean floors in times of low activity			
	Clean floors in sections, leaving a dry path to reach needed areas			
	Mark and section off wet areas from all directions until dry			
	Inspect floor drains to make sure they are clear and free-flowing			

Continued

EXHIBIT 9.2 (*Continued*)
Kitchen Slip/Fall Hazard Assessment

	Problem Area(s)	Corrective Action(s)	Responsible Person	Target Date
KEEP FLOORS CLEAN	Use separate (color code) mops/buckets for front and back of the house			
	Follow manufacturer instructions for concentration/floor area of cleaner			
	Follow the cleaning schedule, but increase frequency if needed			
	Inspect cleaning equipment daily, and report to the appropriate contact for repair			
	Where possible, use scrubber or deck brush in grease-laden areas			
	Change mop heads and scrub pads daily or more frequently if needed			
	Use hot water, and change water when dirty			
	Identify areas of high grease accumulation (e.g., fryer area), and increase spot cleaning as needed			
KEEP WALKWAYS CLEAR	Make sure there are enough trash cans, placed where waste is created			
	"Clean as you work"—remove trash when done			
	When seen, remove trash and materials from the walkway path			
	Increase inspection in areas where trash accumulates and storage is not put away			
	Organize work area so there is a place for everything, and everything is kept in its place			
	Return material and equipment to its place when done			
	Remove hoses and electrical cords from walkway paths			
	Mark and section off any hazards until they are corrected			

Food Service Operations

EXHIBIT 9.2 (*Continued*)

Kitchen Slip/Fall Hazard Assessment

	Problem Area(s)	Corrective Action(s)	Responsible Person	Target Date
PROPER FOOTWEAR	Purchase only from an approved vendor			
	Footwear must match the floor and environment where the employee works			
	Slip-resistant footwear is required for:			
	• Nonoffice employees working on hard surfaces			
	• Employees working in wet or potentially wet outdoor environment			
	Slip-resistant overshoes may be used:			
	• For those who enter these areas but do not normally work there			
	• For new employees and others waiting for proper footwear			
	• Employees who come to work without proper footwear			
EMPLOYEE INVOLVEMENT	Employees are trained on slip/fall hazards and what to do			
	There is ongoing communication about slip/fall hazards (handouts, talks)			
	Employees understand their role in preventing slips and falls, including:			
	• "Clean as you go"			
	• Reporting hazards			
	• Wearing and maintaining proper footwear			
	• Keeping three points of contact when using ladders and similar equipment			
	• Following rules and safe work practices			

10 Healthcare Operations

Whenever I see an old lady slip and fall on a wet sidewalk, my first instinct is to laugh. But then I think, what if I was an ant, and she fell on me. Then it wouldn't seem quite so funny.

Jack Handey
Saturday Night Live

10.1 INTRODUCTION

The healthcare industry, comprised of approximately 13 million workers, is the largest employer in the United States and ranks second in percentage of slip, trip, and fall (STF) accident severity. In 2002, the hospital industry had the most injuries, over 296,000, in the United States. Even with a large workforce, hospitals have a higher than average rate of STFs. Bureau of Labor Statistics (BLS) data has shown that the incidence rate of lost-workday injuries from STFs in hospitals was 38.6 per 10,000 full-time employee (FTE) and 85% greater than the average rate for all other private industries combined (Bell et al., 2007).

A 2007 study performed by the UK National Health Service (NHS) calculated that the overall direct healthcare cost to the NHS of patient falls is estimated to be £15 million annually. This represents a cost of £92,000 a year for an 800-bed acute hospital. Additional costs are more difficult to quantify. Falls can result in patients needing extra healthcare, social care, or residential care after discharge from hospital. Even minor injuries can require extended, costly treatment (Healey and Scobie, 2007).

Nursing homes and other healthcare occupancies present a unique set of slip, trip, and fall hazards due to the nature and needs of the residents. They also pose unusual circumstances that increase the exposure of staff and visitors. Although it would appear that the elderly or infirm residents are the greatest risk, many serious falls at healthcare occupancies are associated with employee activities in their daily interactions with the residents, and to visitors to the facility, many of whom are often elderly individuals visiting loved ones. Special controls are needed to meet the unique exposures of patients, staff, and visitors in healthcare settings.

Patients and residents of healthcare facilities are impaired by age and/or health problems. They can injure themselves by simple errors in judgment or as a result of diminished faculties. Vision problems can make it difficult to distinguish changes in floor coverings or surface levels, and the glare of highly waxed floors may cause disorientation. Loss of physical body strength and the need for assistance can lead to problems if hallway handrails and bathroom grab bars are not properly designed and maintained. Placement and operation of light switches can become critical if a resident is forced to use the bathroom in the dark because he or she cannot locate the switch and cannot or will not wait for assistance. Patients lacking control of bodily functions can unintentionally create significant slip hazards; the prompt identification and cleanup of bodily fluids and other spills is critical.

From an employee standpoint, the physical job demands associated with the daily living activities of patients/residents regularly place employees at risk of slips, trips, and falls. The transfer of residents (e.g., from bed to chair, wheelchair to toilet) requires employees to maintain firm footing while lifting from awkward positions, moving the residents while often off balance, and dodging the personal belongings and room furnishings inevitably underfoot as they go about their tasks. Many employee activities take place in environments where liquids on the floor are the norm. Assisting residents in bathing or showering activities, working in the kitchen and dining room, and handling the laundry operations place the employee in direct and frequent contact with wet floor surfaces.

Finally, the needs of visitors and guests cannot be overlooked. Many guests are the elderly spouses or loved ones of patients/residents, who are often weak and infirm themselves. A higher degree of care is required for these individuals, who may fall making the short trek from their cars into the building. Once inside, they are often prone to the same physical challenges as the patients/residents they are visiting. Hazards to other visitors to these facilities must be considered. This can range from groups of children who come to sing Christmas carols to paid or volunteer providers of entertainment.

10.1.1 Impact of Age

A key concern is not only the high frequency of falls in older persons, but the combination of high incidence and a high susceptibility to injury. This propensity for fall-related injury in elderly persons stems from a high prevalence of comorbid diseases (e.g., osteoporosis) and age-related physiological decline (e.g., slower reflexes) that make even a relatively mild fall potentially dangerous.

In addition to physical injury, falls can also have psychological and social consequences. Recurrent falls are a common reason for admission of previously independent elderly persons to long-term care institutions. One study found that falls were a major reason for 40% of nursing home admissions. Fear of falling and the postfall anxiety syndrome are also well recognized as negative consequences of falls. The loss of self-confidence to ambulate safely can result in self-imposed functional limitations (American Geriatrics Society, 2001).

10.2 KNOWN PARAMETERS OF PATIENT FALLS

10.2.1 Type of Units

Some care delivery areas appear to experience a higher rate of patient falls than normal. One stroke rehabilitation unit reported that 39% of all patients fell, while a geriatric department of an acute care hospital reported a fall rate of 26%. Acute care hospitals have reported fall rates of 1.6%, 1.7%, and 6%. An unpublished Australian study conducted in five hospitals found a benchmark range of 5.90 to 17.78 falls per 1,000 bed days. Unfortunately, due to differences in reporting, sound conclusions are

TABLE 10.1
Fall Rates per 1,000 Patient Days for Selected Units

Unit	Fall Rate
Geropsychiatry	13.1–25
Rehabilitation	7.6–12.6
Geriatric medical	7.8
Neurology	5.2
Psychiatry	4.1
Oncology	3.5
General medical	3.0
Surgery	2.2
Ophthalmology	2.2
Obstetrics-gynecology	1.8

Source: Mahoney, J.E., *Clinics in Geriatric Medicine*, 14(4), 699–726, 1998.

difficult to reach. Research suggests that patients in rehabilitation units or geriatric departments of acute hospitals may be at greater risk of falling (Evans et al., 1998).

Comparisons to external organizations with similar populations may help provide context and establish realistic reduction objectives. These are available from performance measurement systems such as the Maryland Hospital Association Quality Indicator Project (www.qiproject.org) and other published research studies. The fall rates in Table 10.1 were developed in various research studies.

10.2.2 Location of Falls

In a review of study series that reported the location of falls, the patient's bedside and ward area is the most commonly identified area. One unpublished report noted that 43% of all falls occurred from, or near, the patient's bed. Other common locations include the bathroom, toilet, and corridors (Evans et al., 1998).

10.2.3 Time of Day of Falls

According to an analysis of incidents reported to the National Reporting and Learning System (NRLS), the most common time for falls is midmorning when patients are most active. There are few falls during mealtimes and in the early hours of the morning. There are slightly more falls during weekdays than weekends, when there are more patients in hospital (Healey and Scobie, 2007).

10.2.4 Activity at Time of Fall

Patient transfer from one location to another is the most commonly cited. Transferring into or out of bed and moving about in bed have been identified in many studies.

> **Online Resource: National Patient Safety Agency (U.K.)**
>
> The NPSA is a U.K.-based organization that has produced a comprehensive national picture on inpatient falls in a third report of the Patient Safety Observatory (2007). The study, entitled, "Slips, Trips, and Falls in Hospitals," examines research, evidence, and information on falls including over 200,000 incident reports from acute and community hospitals and mental health units. More information can be found by visiting the NPSA website at *http://www.npsa.nhs.uk/patientsafety/alerts-and-directives/directives-guidance/slips-trips-falls/*.

Transferring in or out of a chair is commonly cited as well. Other activities associated with falls include walking, toileting, and sitting in a chair, commode, or wheelchair. One study in a rehabilitation setting found that wheelchairs were involved in 57% of all falls (Evans et al., 1998).

According to an analysis of incidents reported to the NRLS, most falls tend to happen when patients are moving from a bed or chair, walking, or using a toilet or commode (Healey and Scobie, 2007).

10.2.5 LENGTH OF STAY

Regarding the relationship of length of stay to falls, the results are contradictory. While some studies suggest the first week is associated with a higher incidence of falls, others suggest falls are more likely in the later period of hospitalization. It has been speculated that issues such as an unfamiliar environment, hesitancy in asking for assistance, or weakness following recovery from illness and hospital treatment are contributing factors (Evans et al., 1998).

10.3 CAUSES OF PATIENT FALLS

The reasons that patients fall are complex because they are impacted by factors such as physical illness, mental health, medication, and age, as well as environmental factors. A fall can be the result of a single factor, such as tripping or fainting which affects an otherwise healthy person. However, most falls are generally the result of multiple factors.

10.3.1 CATEGORIZING CAUSES OF PATIENT FALLS

10.3.1.1 Morse Fall Scale

Nursing fall risk assessment, diagnoses, and interventions are based on use of the Morse Fall Scale (MFS). The MFS is used widely in acute care settings in hospital and long-term care inpatient settings. The MFS is intended to be done upon admission, fall, change in status, and discharge or transfer to a new setting (Table 10.2). Consisting of six variables which have been shown to be a valid predictor of falls, a majority of nurses (82.9%) rate the scale as quick and easy to use, and 54% estimated that it took less than three minutes to rate a patient. MFS suggests that falls be classified by the following:

- *Accidental falls*—when patients fall unintentionally (e.g., trip or slip due to a failure of equipment or by environmental factors such as spilled water on the floor)
- *Unanticipated physiologic falls*—when the physical cause is not reflected in the patient's risk factor for falls (e.g., due to fainting, a seizure, or a pathological hip fracture)
- *Anticipated physiologic falls*—in patients whose score on risk assessment scales (e.g., Morse Fall Scale) indicates they are at risk due to characteristics

TABLE 10.2
Morse Fall Scale Components, Scale, and Criteria

Component	Scale	Criteria
1. History of falling	00 (No) 25 (Yes)	25 if a fall has occurred during current hospital admission or if there were physiological falls within the last three months (e.g., from seizures or an impaired gait).
2. Secondary diagnosis	00 (No) 15 (Yes)	15 if more than one medical diagnosis is on the patient's chart.
3. Ambulatory aid	0 (Bed rest, nurse assist) 15 (Crutches, cane, walker) 30 (Furniture)	• 0 if walking without a walking aid (even if assisted), uses a wheelchair, or is on a bed rest and does not get out of bed at all. • 15 if using crutches, a cane, or a walker. • 30 if walking requires holding onto the furniture for support.
4. IV/heparin Lock	00 (No) 20 (Yes)	20 if an intravenous apparatus or a heparin lock is inserted.
5. Gait/transferring	00 (Normal) 10 (Weak) 20 (Impaired)	• 0 if Normal—Walks with head erect, arms swinging freely at the side, and striding without hesitance. • 10 if Weak—Stooped, but is able to lift the head while walking without losing balance. Steps are short and the patient may shuffle. • 20 if Impaired—Difficulty rising from the chair, attempting to get up by pushing on the arms of the chair/or by using several attempts. Head is down, and watching the ground. Due to poor balance, grasps onto furniture, a support person, or a walking aid for support and cannot walk without assistance.
6. Mental status	00 (Oriented to own ability) 15 (Forgets limitations)	Measured by checking the patient's own self-assessment of ability to ambulate. Ask "Are you able to go the bathroom alone or do you need assistance?" • 0 if the response is consistent with the ambulatory order. • 15 if the response is not consistent with the nursing orders or if unrealistic (considered to overestimate own abilities and forgetful of limitations).

Source: Morse, J. M. *Preventing Patient Falls.* Newbury Park, CA: Sage, 1997.

TABLE 10.3
Morse Fall Scale Sample Risk Levels

Risk Level	MFS Score	Interventions Required
No risk	00–24	None
Low risk	25–50	Standard fall prevention measures
High risk	>50	High-risk measures

such as a prior fall, weak or impaired gait, use of a walking aid, intravenous access, or impaired mental condition

According to Morse, the vast majority of patient falls (78%) are anticipated physiologic falls.

The score recorded on the patient's chart and risk level and recommended actions (e.g., no interventions needed, standard interventions, high-risk interventions) are identified. Morse Fall Scale risk levels are intended to be tailored to each healthcare setting so that fall prevention strategies are targeted to those most at risk (Table 10.3).

10.3.1.2 Tideiksaar Classification Method

Another frequently used fall classification method is based on the assumption that they result from a complex interaction of intrinsic and/or extrinsic risk factors.

- Intrinsic risk factors (e.g., integral to the patient's condition):
  - Previous fall
  - Reduced vision
  - Unsteady gait
  - Musculoskeletal system—impacts the ability to maintain balance and proper posture
  - Mental status—affected by confusion, disorientation, inability to understand, and impaired memory
  - Acute illnesses—rapid onset of symptoms related to seizures, stroke, orthostatic hypotension, and febrile conditions
  - Chronic illnesses—conditions such as arthritis, cataracts, glaucoma, dementia, and diabetes

- Extrinsic risk factors (e.g., relating to the physical environment):
  - Medications—those affecting the central nervous system, such as sedatives and tranquilizers, and the number of administered drugs
  - Bathtubs and toilets—equipment without support (e.g., grab bars)
  - Design of furnishings—chair and bed heights
  - Condition of walkway surfaces
  - Poor illumination conditions—intensity or glare
  - Type and condition of footwear

- Improper use of devices—bedside rails and mechanical restraining devices that may increase fall risk
- Inadequate assistive devices—walkers, wheelchairs, and lift devices

10.3.2 Personal Risk Factors of Patient Falls

Several studies have shown that the risk of falling increases dramatically as the number of risk factors increases (American Geriatrics Society, 2001).

10.3.2.1 Medical Conditions

Specific diagnoses that may be associated with a higher risk of falling include anemia, neoplasms, and general medical disease, congestive heart failure, and cerebrovascular accident. Stroke patients have been singled out as a patient group at greater risk of falling. Studies have addressed areas specific to stroke patients. One study identified postural sway (movement of the body during standing) as a significant factor in patient falls. Findings from other studies suggest that impulsive behavior and response time may also influence the stroke patient's risk of falling (Evans et al., 1998). Common medical conditions that contribute to the potential for falls include the following:

- Diabetes—Among those who are 60 years and older, women with diabetes are 1.6 times more likely to have fallen in the previous year and twice as likely to have fall-related injuries as women without diabetes.
- Parkinson's disease (PD)—38% to 68% of PD patients experience falls as a serious complication of gait disturbances. Compared to age-and-sex matched non-PD community subjects, PD patients had a 2.2 fold increased risk of fractures and a 3.2 fold greater risk of hip fracture.
- Depression—Older people with a symptom of depression have an approximately 2.2 fold increased risk of falls. However, depression could be the result of a fall rather than a causal or risk factor (e.g., resulting from fear of falling or from self-imposed functional limitations).
- Alzheimer's disease—People with Alzheimer's disease have twice the risk of falling as those of the same age without this disease. Contributing factors may include defects in attention and visual/spatial abilities (Yoshoda, 2006).

10.3.2.2 Physical Conditions

Factors directly or indirectly related to mobility have been identified as being associated with a risk of falling. Identified risk factors include a weak or impaired gait, weakness, decreased mobility of lower limbs, and poor coordination and balance. One study found that patients that fell were more likely to have been using a mobility aid such as walking frame, cane, or wheelchair. A study by Morse reviewing multiple fallers identified impaired gait as a significant difference between multiple fallers and nonfallers (Evans et al., 1998). Common medical conditions that contribute to the risk of falling include the following:

- Muscle weakness—A decline in muscle strength is frequently reported among older people and can interfere with balance. Those with muscle weakness are almost five times more likely to fall. Persons with lower extremity weakness have a 1.8 fold increased risk.
- Visual impairment—A decrease in visual acuity has been shown to increase the risk of multiple falls. Older people with impaired depth perception have a threefold increased risk of multiple falls. Slower reaction time and increased body sway were associated with falls.
- Cognitive impairment—Even relatively modest cognitive impairment and confusion can increase the risk of falling. One study found an increased risk of 1.8; others have reported increases ranging from 2.0 to 4.7.
- Foot problems—Foot problems, such as severe bunion, toe deformity, and deformed nails, are associated with a twofold increased risk of falling.
- Body mass index (BMI)—Weight loss and low body mass index are associated with low bone mineral density and increased risk of fall-related fractures (Yoshoda, 2006).

10.3.2.3 Medication

Some studies have found that taking more than four medications of any type increases the risk of falling and is associated with fear of falling and a ninefold increased risk of cognitive impairment. The most common drugs associated with falls are those acting on the central nervous system. Changes associated with aging, such as decreased lean body mass, increased body fat, and decline of kidney and liver function affect the absorption, distribution, metabolism, and elimination of medications. Medications that increase risk of falls and fractures include the following:

- Benzodiazepines
- Antidepressants
- Antipsychotics
- Antihypertensive: beta blockers, ACE inhibitors, thiazide diuretics, and loop diuretics
- Cardiac medications: cardiac glycosides, antiarrhythmics, calcium channel blockers, and nitrates
- Analgesics: nonsteroidal anti-inflammatory agents (NSAIDs), opioid analgesics, anticonvulsants, antihistamines, and gastro-intestinal-histamine antagonists (Yoshoda, 2006).

According to a 2008 study completed by the University of North Carolina at Chapel Hill, some prescription drugs may increase the risk of falls among the elderly. This study found seniors taking four or more medications are at a greater risk, perhaps two or three times greater, of falling than those taking none. Allergy pills, sleep aids, and other popular prescription medications along with several treatments that result in dizziness and loss of balance are on the list of drugs that appear to increase the risk of falling. The complete list can be obtained by visiting the

University of North Carolina's Web site at http://uncnews.unc.edu/images/stories/news/health/2008/drugslist.pdf (Leach, 2008).

10.3.2.4 Other Factors

Behavioral factors—Fear of falling is common among older people, occurring in approximately 30% of those who have never fallen and 60% for those who have fallen previously. Fear of falling is significantly associated with changes in balance, mobility, and muscle weakness, and with increased spontaneous sway, decreased one-leg stance time, and reduced gait speed.

Sedentary behavior—Muscle function is strongly associated with physical activity. Those who fall tend to be less active, which causes muscle atrophy. Those who are inactive fall more often than those who are moderately active or very active.

Alcohol misuse—Research has demonstrated a link between alcohol consumption and falls. When compared with nondrinkers, those with a monthly ethanol intake of more than 1000 grams have a threefold increased risk of serious falls. Long-term alcohol use combined with the age-related decline in posture and balance can increase postural instability and resulting increased potential for falls (Yoshoda, 2006).

Risk factors identified only in single studies include intravenous therapy, dizziness, type of nursing unit, substance abuse, postoperative conditions, admission to an intensive care unit, sleeplessness and the length of hospital stay (Evans et al., 1998).

10.3.2.5 Specialty Units

There are unique risk factors for falls for specialty units. Potential triggers for patients in these units are listed in Table 10.4.

TABLE 10.4
Risk Factors of Specialty Units

Specialty Unit	Risk Factors
Critical Care:	Awakening from coma, overdose, hypoxia, environmental psychosis
Obstetrics:	Prolonged labor, excessive blood loss, first and second time out of bed after delivery, within 24 hours after epidural
Psychiatry:	Unpredictable affect, antidepressants, orthostatic side effects, polypharmacy, wandering, acting out
Rehabilitation:	Incontinence, polypharmacy, stroke disability requiring assistance with activities of daily living, and mobility deficits
Geripsychiatric:	Confusion, polypharmacy, antidepressants, hypotension, anxiety, age-associated conditions
Pediatrics:	Age (developmental vs. fall), within 24 hours after surgery, head injured patients and other neurological disorders, respiratory disorders

Source: Hendrich, A., et al. *Appl. Nurs. Res.*, 8(3):129–139, 1995.

> **Online Resource: National Center for Patient Safety**
>
> Maintained by the Department of Veterans Affairs, this resource recommends a team approach to initiate fall prevention interventions after an assessment. If the patient is at risk for falls, this guide suggests interdisciplinary interventions that include medical, nursing, and rehabilitations management.
>
> This Fall Prevention and Management aid is intended to prompt clinical staff such as nurses, physicians, rehabilitation therapists, and others to consider a systematic assessment for determining patients' risk for falling and to recommend interventions.
>
> Postfall management guidelines are also provided that include postfall assessment, fall risk level, interventions, and documentation. If a patient is not at risk for falling based on your assessment, interventions should still be implemented to protect the patient from extrinsic fall risk factors such as the presence of clutter, spills, and electrical cords.
>
> *http://www.va.gov/NCPS/CogAids/FallPrevention/index.html*

10.4 CALCULATING FALL RATES

10.4.1 Number of Patients at Risk Rate (commonly used in long-term care facilities)

$$\frac{\text{Number of patient falls}}{\text{Number of patients at risk}} \times 1,000$$

10.4.2 Number of Patients Who Fell Rate

$$\frac{\text{Number of patients who fell}}{\text{Number of patients at risk}} \times 1,000$$

In this formula, repeated falls experienced by the same person are only included once in the numerator.

10.4.3 Number of Falls per Bed

$$\frac{\text{Number of patient falls (for a given time period)}}{\text{Number of beds}}$$

10.5 INTERVENTIONS/CONTROLS

Patient safety should be balanced against their right to make their own decisions about risk along with their dignity and privacy. In wards where there are no falls, patients are more likely to regain their independence and return home.

Healthcare Operations

10.5.1 Flooring

- Use low-gloss waxes and buff floors lightly to avoid a high-sheen finish. Highly waxed floors can produce a glare that can interfere with the eyesight of those with poor vision. In addition, the reflected glare can give the floor an appearance of being slippery which can lead to patient fear of ambulation and gait changes. Employ shades and curtains where needed to block sunlight from streaming in and reflecting off the floors.
- Avoid floor coverings with bold and loud colorful patterns and colors. These can be very disorienting to individuals with depth perception issues.
- Distinguish changes in floor coverings clearly. Moving from tile to carpet or vice versa can pose a significant slip or trip hazard to the elderly.
- Consider slip-resistant flooring and slip-resistant adhesive strips in areas such as the floor next to the sink and toilet.
- Use contrasting colors at floor level to make it easier to differentiate steps and edges.
- Observe environment for potentially unsafe conditions such as loose carpeting and water on the floor. Clean up spills immediately. For more information see Chapter 2, Management Controls.

10.5.2 Lighting

- Provide proper illumination in the environment and adjust to the individual needs of patients and residents.
- Extra (strategic) lighting may be needed in certain locations such as the bedroom, path from the bed to the bathroom, and toilet area. Consider using motion-sensor bathroom lighting.
- Use floor-level night lights that do not create shadows and glare.
- Use full-spectrum fluorescent light as it is more effective than incandescent lighting for overall illumination.

10.5.3 Beds/Bedside

- Ensure patient can reach necessary items including assistive devices (e.g., walker or cane) and the call light and other personal care items (Figure 10.1).
- Minimize obstacles and bedside clutter.
- Provide nightlights at bedside and toilet.
- Ensure beds and bedside furniture is secure.
- Consider use of cordless phones, because they eliminate the hazard of the cord.
- Ensure bed on lowest setting except when giving care.
- Use height adjustable hi–low beds or fixed low–deck height beds where applicable.

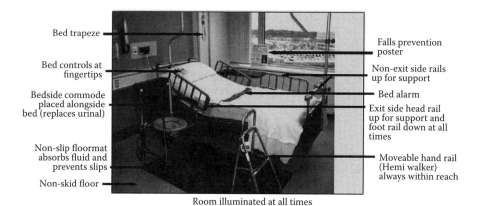

FIGURE 10.1 An optimal bed and bedside design and arrangement developed by one Veteran's Administration facility (VHA NCPS Toolkit, May 2004).

- Use bed wheel-locking systems such as a combination swivel-and-wheel brake. Consider placing nonslip adhesive strips underneath the bed wheels.
- Consider providing a bed footboard to assist patients and residents use in transferring in and out of bed or in ambulating about the bedroom.
- Ensure mattresses are firm enough to support necessary safe bed transfers.
- Demonstrate the use of call bell to patients and ensure it is within reach of patient.
- Provide appropriate armchair with wheels locked at the patient's bedside.
- Arrange furniture to maximize open walkways. Ensure that the pathway to the restroom is free of obstacles and properly lighted.

See Section 10.5.8, Assistive Devices, for information on bed side rails and bed alarm systems.

Online Resource: Hospital Bed Safety

The Hospital Bed Safety Workgroup (HBSW) is a partnership including the FDA, the medical bed industry, national healthcare organizations, patient advocacy groups, and other federal agencies. Its goal is to reduce the risk of side rail entrapment in hospital beds.

The work of the HBSW resulted in a number of FDA references including the "Hospital Bed System Dimensional and Assessment Guidance to Reduce Entrapment." This guidance provides recommendations for manufacturers of new hospital beds and for facilities with existing beds.

Access more information at the FDA's Web site: http://www.fda.gov/cdrh/beds/index.html.

10.5.4 BATHROOMS

- Ensure bathroom toilets and showers have well-maintained, secure grab bars mounted vertically for use by the patient in performing their daily living activities.
- Mark light switches in patient rooms and bathrooms with tapes or paints in colors contrasting to the wall background so that they do not blend into the background and are clearly visible. They should be readily accessible and function easily.
- Place patients with urgency near toilets.
- Check patients receiving laxatives and diuretics.
- Toilet at-risk patients routinely.

10.5.5 HALLWAYS

- Avoid clutter (e.g., medicine, laundry, food carts, cleaning equipment, wheelchairs, etc).
- Because patients and residents may become fatigued in ambulating long hallways on the way to reach the nurses' station, dining rooms, etc., consider providing rest stops or chairs strategically placed every 20 to 30 ft. (6 to 9 m).
- Provide handrails for support and ensure proper height from floor (26–36 in./66 cm–91 cm) and design (e.g., round grip).

10.5.6 FOOTWEAR

Evaluate residents' footwear for fit and construction to assure that it is appropriate for the exposure. High heels, soft-bottom slippers, etc. are generally inappropriate for healthcare environments and can lead to serious slips and falls. Avoid long robes or loose-fitting slippers.

For more information, see Chapter 8, Footwear.

Online Resource: Registered Nurses Association of Ontario

Registered Nurses Association of Ontario, Prevention of Falls and Fall Injuries in the Older Adult, *Nursing Best Practice Guideline,* **Revised 2005.**

The purpose of this guideline is to increase nurses' ability to identify adults in healthcare facilities at risk of falling and to define interventions for prevention of falling. The guideline is relevant to acute and long-term care and is intended to assist nurses in applying the best available research to clinical decisions.

This and other useful resources are available on the RNAO Web site: http://www.rnao.org/Page.asp?PageID=924&ContentID=810.

10.5.7 OTHER POLICIES AND PROCEDURES

- Orientation—Provide residents with a thorough orientation of the facility. Walk them through their room and the building so they can gain familiarity with the layout. If residents have access to outdoor areas such as enclosed patios, they should have a similar orientation to these areas.
- Assessment—Evaluate residents as to their physical capabilities and their fall risk using an assessment tool such as the Morse Fall Scale.
- Mobility—Provide physical/occupational therapy, instruct patients to rise slowly, assist high-risk patients and in patient transfers, repeat activity limits to patient and family, and lead patients in corridor once or twice per shift.
- Mental state—Reorient confused patients, transfer confused patients near nurses station, use family members to sit with confused patients, and nurse confused patients in low bed.
- Medications—Review medications frequently, check for patients receiving laxatives and diuretics, minimize combinations of medications, consider peak effect for prescribed medications that affect level of consciousness, gait, and elimination.
- Wheelchairs—Use safety straps or seat belts in chairs and wheelchairs, geriatric chairs, and latex mesh in chairs to prevent patients slipping, select chairs with arm rests and of the proper height for rising and sitting.
- Use colored identification arm bands and stickers for doors and charts of patients at risk of falling.
- Involve the patient's family in care. Include them in the development of an individualized safety plan considering age-specific criteria and patient cognition when planning care. Collaborate with the family to provide assistance as needed while maintaining the patient's independent functioning.
- Communicate the patient's at-risk status during shift report and with other disciplines as appropriate.
- Reassess staffing needs in relation to high-risk patients.
- Do not leave at-risk patients or residents unattended in diagnostic or treatment areas. Ensure patients or residents being transported by stretcher/bed have side rails in the up position during transport, or if left unattended briefly while awaiting tests or procedures.
- Consider one-to-one nursing for high-risk patients.

RESEARCH STUDY: 10-Year Evaluation of Integrated Interventions

A 10-year study to evaluate the effectiveness of a comprehensive slip, trip, and fall (STF) prevention program was conducted in three acute care hospitals. The prevention program included onsite hazard assessments, changes to housekeeping procedures and products, introduction of products and procedures, slip-resistant footwear, general awareness campaigns, and other prevention measures.

The study involved almost 17,000 employees working in excess of 80 million hours. Of the 472 STF incidents, 85% were classified as same-level falls.

Approximately one of every four accidents involved workers slipping on water and slick spots such as grease, or other fluids (e.g., beverages, body fluids) on the floor.

STF hazard assessments were conducted including the areas cited below. Recommendations addressed walkway repairs, degreasing kitchen floors, keeping floors clean and dry, clutter and cord containment, employee training, and products and procedures to assist in preventing STF incidents.

- Interior—Entrances, stairs, ramps, operating rooms, emergency room, scrub sink areas, nursing stations, pharmacy, histology lab, hallways, kitchen, dishwashing areas, cafeteria, patient rooms, bathrooms, instrument decontamination areas, engineering and carpentry shops, and the morgue
- Exterior—Parking garages, ramps, sidewalks, and employee shuttle bus stops

Other interventions consisted of slip-resistant shoes for food service workers and housekeeping staff, recommendations for slip-resistant flooring to be added during renovations, revised housekeeping procedures, ice cleats for home health nurses, and a hazard awareness campaign. Housekeeping managers wore beepers with numbers advertised through e-mails and posters, notifying all staff to promptly report spills and other contaminants on the floor. Outside, snow and ice removal were enhanced by strategically located containers with ice-melt chemicals to be used by any employee who noticed icy conditions.

The analysis identified a cluster of STF incidents at a shuttle bus stop with heavy pedestrian traffic. Three downspouts were releasing rain water onto a sloped sidewalk, creating icy patches in freezing temperatures. Redirecting the downspouts under the sidewalk eliminated this exposure.

Total STF workers' compensation claims rate declined by 58% from the pre-intervention to the postintervention period, suggesting that implementation of a broad-scale prevention program can significantly reduce STFs in hospitals.

(Bell et al., 2007)

10.5.8 ASSISTIVE DEVICES

Studies of multifactorial interventions that have included assistive devices such as bed alarms, canes, walkers, and hip protectors have demonstrated benefit. However, there is no direct evidence that the use of assistive devices alone will prevent falls. The fundamental aspect of the effectiveness of these devices is that nurses and aides have prompt and easy access to them.

10.5.8.1 Restraints

The use of physical restraint is a controversial method to minimize the risk of falls through limiting mobility of patients. There is a range of physical restraint devices including jackets and vests, limb restraints, mitts, wristlets, anklets, and wheelchair restraints.

Restraints have been traditionally used as a falls prevention approach; however, they have serious drawbacks and can contribute to serious injuries. There is no experimental evidence that widespread use of restraints or, conversely, the removal of restraints, reduces falls (American Geriatrics Society, 2001).

10.5.8.2 Bed Alarm Systems

A decline in the use of mechanical and chemical restraints to prevent high-risk patients and residents falls has encouraged the use of bed alarm systems designed to warn nursing staff that at-risk patients or residents are attempting to leave their bed unassisted (Figure 10.2).

Sensor alarms are pressure-sensitive pads that are placed on a bed or chair. If a patient rises from the pad, an alarm sounds, alerting the patient and staff. The alarm reminds the patient that they require assistance to mobilize, which often encourages

FIGURE 10.2 An example of a bed alarm system. (Images provided courtesy Posey Company, Arcadia, California.)

them to wait for assistance. In addition, it informs the ward staff that the patient is mobilizing, and they can then go to their aid before an incident occurs.

There is no rigorous evidence currently available. As such, no recommendations can be made on the effectiveness of alarm systems in preventing patient falls (Evans et al., 1998).

10.5.8.3 Bed Side Rails

Routine use of bed side rails is still a standard practice in many hospitals and long-term care facilities. However, studies have shown that patients fall from beds despite bedrails being raised. The only study identified that looked at the falls in relation to bedrails was a review of 181 incident forms which challenges the effectiveness of bedrails. While bedrails come in varying lengths and heights, there is no information on which is the most effective in stopping falls. For example, half-length bedrails may stop accidental rolls from bed while not creating an obstacle for patients who would otherwise climb over the top of the rail (Evans et al., 1998).

Indiscriminate use of bedrails has been shown to increase the risk of falling and there have been three U.K. safety warnings regarding the use of bedrails (Barnett, 2002).

10.5.8.4 Identification Bracelets

A study involving 134 patients found that bracelets were of no benefit in preventing falls among patients at high risk of falling. Due to the lack of rigorous evidence, no recommendations can be made on the effectiveness of bracelets or other methods of identifying high-risk patients in preventing patient falls (Evans et al., 1998).

10.5.8.5 Hip Protectors

One approach to reduce the impact of falls by dispersion is hip protectors (Figure 10.3). Various types of hip protector have been developed. Most consist of plastic shields which are kept in place by pockets within specifically designed underwear (Barnett, 2002).

A number of studies, including three randomized trials, strongly support the use of hip protectors for prevention of hip fractures in high-risk individuals (American Geriatrics Society, 2001). Parker et al. (2006) conducted a review of the evidence regarding hip protectors and concluded that the incidence of hip fractures is reduced when hip protectors are worn by those vulnerable to fracture. They are a costly intervention, and each patient would require two or three pairs for hygiene purposes. A thorough needs assessment is required when a patient may benefit from hip protectors. When warranted, the financial advantages of providing hip protectors clearly outweigh the disadvantages. Other benefits include allowing patients more freedom to move with decreased risk of serious injury and reducing ethical dilemma of restraint (Barnett, 2002).

The available evidence for hip protectors demonstrates the following:

- Hip protectors can substantially reduce hip fractures in older people, but compliance and adherence remains low.
- Hip protectors do not reduce the incidence of a second hip fracture.

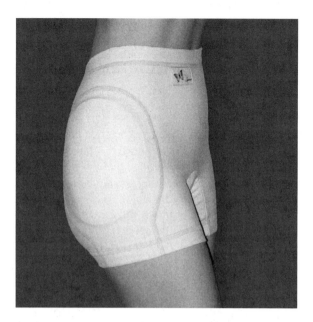

FIGURE 10.3 An example of a hip protector. (Courtesy of HipSaver, http://www.hipsaver.com.)

- Hip protectors do not appear to affect the risk of falling or reduce injury to other parts of the body.
- Adherence to hip protector wear is improved by staff education.
- Investigation of osteoporosis risk and appropriate interventions is an important complementary strategy to reducing fall incidence.

10.5.9 Coordination and Strength Training

A review of the literature on prevention of trip and falls among the elderly found that changes in gait pattern, reduced vision, and cognitive impairments seem to increase the probability of tripping. Recovery of balance after a trip is probably limited because forward placement of the recovery leg is slower. Balance recovery is also impaired because joint movements in the stance leg are generated more slowly. These results suggest that coordination and strength training can help prevent falls among the elderly (van Dieën et al., 2005).

RESEARCH STUDY: Implementation of a Comprehensive Fall Prevention Program

In 2002, Northeast Health System (NHS) launched a comprehensive fall prevention program at its two acute care hospitals. After the first six months of the project, there was a dramatic overall reduction in the rate of falls. NHS used a

Healthcare Operations 353

multifaceted approach to implementing changes, and interventions included the following:

- **Using a reliable and valid instrument to predict and identify prone-to-fall-patients.** NHS developed a risk assessment tool used to assess patients. The tool is based on the Morse scale and is recorded in an electronic log along with the appropriate risk-reduction strategies and interventions. Nurse managers receive daily reports of at-risk patients and post them on the units. High-risk patients are identified with a gold star on the unit.
- **Developing a system to track incidence and type of falls institution wide.** The team revised the falls report to include more information on factors that contribute to falls. An update to the administrative database allows better unit-specific information for trending and the ability to develop interventions appropriate to the patient population.
- **Maintaining a safe environment.** The team worked with plant operations to examine potential environmental fall factors and performed checks on components including beds, wheelchairs, walkers, handrail placement, and bathroom call bells. The falls prevention program coincided with the purchase of many new beds equipped with bed alarms.
- **Developing and targeting interventions for those likely to fall.** A multifaceted approach including administrative, direct care, environmental, and equipment initiatives included identifying patients with a high risk fall score by placing a gold star on the unit, then regularly toileting those patients, ensuring they had adequate lighting at night, placing patients near the nursing station, and equipping the beds and chairs with alarms.
- **Reducing the risk of those likely to fall.** NHS created a falls committee and assigned a clinical educator to provide ongoing falls education to staff. The committee conducts falls rounds, during which they provide direct education regarding current fall assessments.
- **Constantly monitoring patients who have fallen using a postfall protocol.** NHS developed an assessment and reporting flow sheet for nurses and physicians to provide standardized monitoring, treatment, and physician/family notification after a fall. The flow sheet outlines responsibilities and steps for staff to follow after a patient experiences a fall.

(QI Project, 2004)

10.6 REDUCING EMPLOYEE FALLS

Many of the controls implemented to minimize patient falls also benefit employees. In addition to those measures, the following precautions should be taken, based on the area/activities in which employees are engaged.

10.6.1 ALL EMPLOYEES

- Require employees to wear slip-resistant footwear and to maintain footwear in good condition because floors can be wet in many places in a healthcare operation. See Chapter 8, Footwear, for more detail.
- Use only properly maintained ladders to reach items. Do not use stools, chairs, or boxes as substitutes for ladders.
- Reduce the presence of cramped working spaces to avoid awkward positions and promote safe work.

10.6.2 DIETARY

- Arrange traffic flow to minimize overlap between workers entering and leaving the kitchen.
- Clean kitchen floors regularly, using a degreaser/neutral cleaner applied by deck brush or floor machine (see Chapter 6, Flooring and Floor Maintenance).
- Use thin, beveled rubber floor mats in areas including dishwashing, adjacent to deep fat fryers, in front of ice machines, and other high-risk areas. Carts can more readily roll over thinner mats.
- Inspect floor drains regularly and keep them unobstructed and free flowing.
- Run the kitchen as a "clean-as-you-go" operation, requiring that once a task is done, the employee cleans the designated area and puts away equipment before moving to the next task.
- Keep storage well organized, easy to reach, and out of walkway paths.

See Chapter 9, Food Service Operations, for more detail.

CASE STUDY: Hospital Employee Loses Leg after Two Slip Accidents

Alison was an occupational therapist in a large hospital when she had two slip accidents at work. The first was a slip on rotting leaves on the concrete steps of the hospital. The leaves were supposed to be cleared daily, but had not been, due to staff shortages. She fell heavily on her right knee and was still in pain three months later, when she was told she needed surgery.

Six years later, she slipped again, this time on a wet vinyl floor. Someone had mopped the floor and failed to dry it or put out barriers or warning signs. She slipped and fell directly onto her right ankle. Over the next few years, she underwent 30 operations. Eventually she was told that the only solution was to have her foot amputated. In her 32nd operation, her leg was amputated at the knee.

HSE Information Sheet
Slips and Trips: The Importance of Floor Cleaning

Healthcare Operations

10.6.3 Housekeeping

- Train housekeeping staff on inspection, maintenance, and cleaning procedures.
- Inspect hallways, aisles, and stairwells regularly. Keep clear of storage, debris, or other obstructions that pose a tripping hazard or otherwise impede traffic flow.
- Implement a formal spill control and cleanup program to quickly identify and eliminate fluids from the floor surfaces. Identify, section off, and report damaged or contaminated walkways.
- Use beveled floor mats with a slip-resistant backing during inclement weather seasons. Inspect mats regularly to assure edges lie flat and the mat does not buckle, ripple, or creep. Change out more frequently during inclement weather so they do not become saturated.
- Provide enough trash containers close to points of generation of waste; otherwise, waste materials are more likely to end up on the floor.
- Provide umbrella sleeves/bags at main entrances to avoid dripping water onto the floor.
- Allow an adequate walkway path through the area, especially in corridor areas when cleaning and waxing floors. Section off and clean/wax half of the floor at a time.
  - Use signs and barriers at all approaches/entrances to the wet area. Do not remove until the floor is completely dry. When mopping a large area, place wet floor signs at least every 15 feet.
  - Do not use the same mop head for more than four patient rooms. If the head is heavily soiled, change it more frequently.

For further detail, see Chapter 2, Management Controls.

10.6.4 Laundry

- Vacuum to remove lint from floors continuously.
- Pick up dropped laundry immediately.
- Assure floor mats are clean, in good repair, and lying flat.
- Assure there are enough trash receptacles and that they are emptied frequently enough.
- Immediately clean up spills. To minimize spills, use premeasured packets of detergent.
- Arrange workflow to minimize employees continuously crossing each other's path.

10.6.5 Crocs™ Footwear

There has been an increased use of Crocs footwear in hospital situations. While many healthcare workers who use them find comfort and relief from foot and back problems, this footwear presents two primary concerns: the poor slip resistance of the footwear bottom and the potential for exposure to infectious fluids due to the perforated uppers.

OSHA's Bloodborne Standard, 1910.1030, section (d)(3)(i) states that employees should wear shoes that protect from potential blood or other potentially infectious materials (OPIM) splashing. This would seem to prohibit the use of footwear with openings in the uppers (Partnership for Health and Accountability, 2007). In August of 2007 on its e-mail forum, OSHA stated that Crocs are not appropriate in hospital settings where there is a reasonable expectation that an employee's feet could be exposed to blood or other potentially infectious materials. Such exposures are likely to occur in the OR, ER, and labs, for example (Wallask, 2007).

Hospitals are also concerned with slower reaction times in responding to emergencies, since clogs are more difficult to run in than conventional hospital footwear.

Many hospitals, including all Canadian hospitals, have banned the use of this footwear (White, 2007). In response to the infection issue, Crocs Inc. began manufacturing a new version called the Specialist in the fall of 2008 which is like the original but without holes on top (Greene, 2008).

However, standard slip-resistant booties appear most appropriate for such environments because they address all the issues—infectious exposure, slip resistance, and response time.

EXHIBIT 10.1

Guidelines for a Fall Prevention Program

Risk	Intervention	Rationale/Key Points
Psychosocial Aspects of Falls (Healthcare)		
Denial of Aging Process		
1. Denial of aging process	1.1 Counseling initiated. A. Explanation of decline in sensory, neurological, and musculoskeletal function that accompanies normal aging.	1.1 Aging process includes physical limitations including decline in visual acuity, speed of walking, reaction time and balance. A. Acceptance of restrictions imposed
2. Refusal to ask for or accept assistance in getting out of bed or using bathroom	2.1 Instruct residents on reasons and consequences of falls and use of assorted assistive devices.	2.1 Grab bars, shower, chairs, walkers, canes, other assistive devices. A. Use of call light/button.
3. Failure to take corrective action of environmental fall hazards	3.1 Assist resident in assessing environment for fall hazards. A. Explain corrective actions to take. 1. Adequate illumination.	3.1 Maintains image as a capable and functional individuals. A. Guard against risk of falling. 1. Does not attempt to accomplish tasks in areas of inadequate lighting.
4. Refuses to restrict or accept assistance in activities that resident may no longer perform safely	4.1 Advise resident to allow for more time to complete specific task and to identify times when risk is lowest. 4.2 Explain alternative measures resident should take to maintain present level of functioning. A. Assistive support should, while being acceptable, also bolster self-esteem.	4.1 Examples: A. Ambulating from point A to point B. B. Reaching for objects from high places. 4.2 At the same time will guard against risk of falling. A. Target modification of fall-related activities.

Continued

EXHIBIT 10.1 (Continued)
Guidelines for a Fall Prevention Program

Risk	Intervention	Rationale/Key Points
Fear of Falling		

1. Risk of falling increased because of preoccupation with falling	1.1 Environmental correction of falling hazards.
	  A. Adequate illumination.
	  B. Proper handrails.
	  C. Nonslip step surfaces.
	  D. Proper stair climbing.
	  E. Wearing of individual alarm devices.
		1.1 Less cognitively alert to potential environmental fall hazards.

2. Avoidance of functional activities that may lead to a fall	2.1 Minimize resident's anxiety at performing feared activity:
	  A. Behavior resumption of feared activity.
	  B. Psychotherapeutic intervention.
	  C. Environmental correction.
	  D. Elimination of hazards.
		2.1 Fear of falling basis for avoidance.
		  A. Perform activity with resident several times to gain confidence.
		  B. Panic attacks not indicative of underlying medical problems.

Cognitive Impairment (Psychological Dysfunction)		

1. Residents who experience cognitive impairment due to anxiety, depression, or dementia	1.1 Treat underlying cause of anxiety or depression.
	  A. Look for reversible causes.
	
	1.2 Remove environmental hazards.
	  A. Provide appropriate furniture.
	  B. Utilization of assistive devices to support ambulation.
		1.1 Less alert to environmental fall hazards.
		  A. Dementia, adverse drug affects, depression, visual or hearing deficits, electrolyte disturbances, hypoglycemia, dehydration, anemia, or systemic diseases.
		1.2 Less able to effectively prevent a fall in progress.

Healthcare Operations 359

Environmental Relocation

1. Environmental relocation may place residents at risk

    1.1 Ensure clear understanding about need for relocation.

    1.2 Observe residents for ability to function safely in new environment.
       A. Knowledge of new environment.

   1.1 Cannot be maintained in own home.
      A. Need for protective care.
      B. Emotional reactions.

   1.2 Unfamiliar with new environment.

Attention-Seeking Behavior

1. May fall in an effort to gain attention

    1.1 Recognize the need for attention and the need for increased assistance.
       A. Psychotherapy.

    1.2 Resolve presence of relational conflicts.
       A. Meet with family or staff members.
       B. Social service involvement.

    1.3 Provide community support and activities.
       A. Participation in activities.

   1.1 Residents may be lonely due to feeling abandoned in a nursing home.
      A. Falling episodes may be reflective of conscious or unconscious gesture to receive help or ensure care and involvement of family.
      B. Helps to deal with feelings.

   1.2 Family and friends emotionally disengaging or withdrawing or decreasing contact with resident.
      A. May represent focus of attention-seeking falls.

   1.3 Enhances feelings of accomplishment and self-sufficiency.
      A. Helps cope with changes in health status.

Continued

EXHIBIT 10.1 (Continued)

Guidelines for a Fall Prevention Program

Risk	Intervention	Rationale/Key Points
Poor Compliance		

1. Residents are noncompliant to therapeutic recommendations by physicians	1.1 Outline treatment plan in simple language and involve family in implementing.
	  A. Write out all recommendations.
	  B. Monitor progress consistently.
	1.2 Discuss lifestyle changes and modifications in resident's environment.
	1.3 Build rapport with resident and family.
	1.4 Coordinate medical care with community services.
		1.1 Residents with cognitive dysfunction may not understand or remember recommendations.
		1.2 Residents may not accept underlying problems as being responsible.
		1.3 Relationship may be impaired.
		  A. Distrust and failure to implement recommendations.
		1.4 May improve compliance in residents lacking social supports.
		  A. Community services may include visiting nursing homes, home healthcare, and social work agencies.

©ESIS, Inc. All rights reserved.

11 Profiles of Other High-Risk Industries

If you ever fall off the Sears Tower, just go real limp, because maybe you'll look like a dummy and people will try to catch you because, hey, free dummy.

Jack Handey
Saturday Night Live

11.1 INTRODUCTION

Fall hazards and exposures can vary widely depending on the type of facility. Some operations that historically involve high potential for falls by visitors, patrons, residents, customers, and employees include the hospitality (lodging), mercantile, theater, and trucking industries. While not a thorough discussion, these profiles attempt to cover the commonly observed conditions contributing to falls. Hazards particular to the food service and healthcare industries are discussed in their respective chapters.

11.2 ALL OCCUPANCIES

11.2.1 EXTERIOR CONTROLS

1. Parking lots and other exterior walking surfaces must be kept well maintained and in good repair. Even small cracks or imperfections can pose a significant trip hazard to an elderly or impaired visitor.
2. Storm drains should be inspected on a scheduled basis to ensure they are clear and free flowing. This will keep water from accumulating on the surface where it can freeze.
3. Prompt snow and ice removal is essential. Gutters and downspouts should be regularly inspected to make sure they are clear of debris and do not allow water to drip or pass over any walking surfaces. Whereas water from gutters and downspouts might not pose a problem during warmer summer months, in the winter it freezes into ice and presents a serious slip and fall exposure.
4. Exterior lighting should be adequate throughout and should be inspected and maintained on a scheduled basis to ensure uniform coverage. Timers or photoelectric sensors should be used to ensure that exterior lights come on at dusk.
5. In the fall, leaves should be kept off of parking lot and sidewalk surfaces. Wet leaves can present a slippery surface.

11.2.2 INTERIOR CONTROLS

1. Good housekeeping practices are critical. For example, hallways, aisles, and stairwells should be kept clear of all unnecessary storage and materials that might pose a tripping hazard or otherwise impede the flow of foot traffic.
2. Responsibility should be assigned for specific individuals to check the floors on a regular basis for dropped items, spills, damaged mats, and housekeeping concerns.
3. A formal spill control and cleanup program should be implemented to quickly identify and remove fluids from walkway surfaces.
4. Slop rugs and floor mats should be used during inclement weather. These should be inspected regularly to ensure that edges do not curl up and present a tripping hazard and that they do not become waterlogged. Waterlogged rugs can increase the hazard because people will track water off of the rug and onto hard flooring as they enter the building.

11.3 HOSPITALITY (LODGING)

In hotels, slips, trips, and falls are the leading cause of injury for employees and guests, consistently accounting for approximately 42% of all accidents for each group. Falls most often occur in stairways, balconies/landings, ramps, parking lots, and bathtubs or showers (Kohr, 1991).

By their very nature, hospitality operations present a high potential for slips, trips, and falls:

- Continuous foot traffic by individuals unfamiliar with the parking lot arrangement or the interior building layout is the norm.
- Guests range in age from newborn children to the elderly and also include the physically and mentally challenged.
- In many cases, language and other ethnic barriers may pose a problem.
- Businesses operate 24/7, which increases the exposure.

11.3.1 OUTSIDE HAZARDS

Depending on the size and scope of the operation guests and employees are exposed to a wide range of slip, trip, and fall exposures. Parking lots are sometimes large and confusing, and incoming guests often have to maneuver through them after dark. Luxury hotels may have elaborate exterior landscaping with planted specialty walkways, bridges traversing man-made or natural streams, and unique overhead skyways. Exterior playground equipment for younger guests may also be provided.

11.3.2 INSIDE HAZARDS

Inside, visitors may be exposed to still more slip, trip, and fall hazards. Large, open lobbies and common areas with sunken seating groups and changing floor surfaces increase the exposure as soon as one steps inside. Meeting rooms often have temporary electrical cord tangles and tightly packed seating conditions on poorly maintained folding chairs. In many hotel rooms, the tub is raised and the unexpected step-down presents potential for serious injury from a trip. Many facilities have full-service restaurant and lounge bar operations with multiple levels, raised furniture, highly waxed dance floors, and trendy floor coverings. Dim lighting, the presence of alcoholic beverages, and food and beverage spills increase the hazard (also see Chapter 9, Food Service Operations).

Swimming pools and exercise rooms present other concerns. Water-covered tile floors in the shower rooms and along the pool edges can present a significant slip hazard if they have not been properly treated. Improperly designed or installed exercise equipment can also create the potential for a tumble. Slip and fall/trip and fall accidents also occur frequently in exercise rooms because of excess water on floors and workout gear left where people can trip over it. The following procedures should be utilized to reduce risk:

- Check bathroom floors and areas around water fountains and drink machines frequently.
- Use absorbent, nonskid mats at the base of water fountains to absorb excess moisture.
- Require clients to store gym bags in lockers (if provided).
- Remove unused towels and weight equipment, and place them in appropriate storage racks.
- Document inspection activities with a log.
- If you contract such services, obtain a certificate of insurance.

11.3.3 Employees

Employees are subject to the same exposures and more. Many of the behind-the-scenes operations present increased hazards for the average staff member. Kitchen operations occasionally have water and grease on the floor, and stock, equipment, and trash can often block or impede the walkway surface. Laundry areas also have the potential for water-slick floors and poor housekeeping. Maintenance staff is at risk while performing routine activities. Climbing ladders, performing snow and ice removal operations, and making plumbing repairs in water-soaked areas can be a part of their daily chores. High employee turnover creates ongoing training and awareness issues.

11.3.4 Guest Room Bathrooms

> When Aimee Fitzgerald stepped out of the shower at her hotel, she didn't notice the small amount of water that had accumulated on the floor because the white marble floor combined with the bathroom's bright lights made it difficult to see. Ms. Fitzgerald slipped, fell, and broke her sacrum. The water wasn't the only issue. The shower curtain didn't extend below the rim of the bathtub, and the rubber backing on the bath rug was worn. Settlements for injury-related medical expenses range from $1,800 into

six figures. In 2000, a jury awarded $189,000 to a plaintiff who slipped and fell in the bathtub at a Milwaukee hotel.

According to hotel industry statistics, slips and falls consistently account for about 42% of guest accidents; bathrooms are among the top five places they occur.

Glossy bathroom floors are increasingly popular at hotels. Hotel-industry experts project that proliferation of marble and other smooth, easy-to-clean materials, combined with a lack of federal standards for hotel bathtub safety, means an increase in accidents, especially as baby boomers age.

In addition, existing safety standards, which are consensus and voluntary, do not apply to bathroom floors, only to the interior of the tub or shower. And even the safety threshold for the bathtub standard (ASTM F462) was not developed based on the exposure to slips, but to bathtubs manufactured at the time (see Chapter 4, U.S. Tribometers). Bathtubs should have factory-installed slip-resistant etching; regular cleaning is the best way to maintain that etching. Another essential element is the use of rubber mats, something that some hotels hesitate to use because they detract from the look of the bathroom.

Grab bars are one safety feature that can be easily added to a bathroom, and they have become more commonplace. *ASTM F446 Standard Consumer Safety Specification for Grab Bars and Accessories Installed in the Bathing Area* specifies requirements for grab bars and accessories designed to decrease the probability of slips and falls. It covers performance requirements, test methods, and levels of performance to ensure satisfactory functioning of the grab bars and accessory items during reasonable use to assist a person entering, leaving, or moving in the bathing area. The standard requires that grab bars be installed horizontally to withstand an applied downward load of 250 lbs (113 kg), applied to a 3½ in. (9 cm) area in the center of the bar for five minutes.

But safety features must also be properly installed and maintained. In November 2001, a jury awarded a plaintiff $1.75 million for an injury when the grab bars in a tub at a hotel in Chattanooga detached from the wall.

11.3.5 Swimming Pool/Whirlpool Areas

According to the Association of Pool and Spa Professionals, slips and falls constitute the greatest number of accidents involving pools (Figure 11.1).

- *Decking*—Be sure the deck has a slip-resistant surface with adequate and free-flowing drainage. Decks are generally graded away from the pool to facilitate water runoff from overspill and should be monitored for unevenness that could result in pooling. Keep the deck or patio clean and clear of debris.

  Lifeguards need to conduct a visual inspection of the poolside prior to the activity commencing and remove potential trip hazards. Trip hazards that cannot be removed must be made clearly visible, sectioned off, and/or edges softened.

- *Pool*—Ladders with nonslip treads should be on at least one side of the deep section to access the swimming area. Diving boards should have nonslip tread surfaces.

Profiles of Other High-Risk Industries 365

FIGURE 11.1 Commercial swimming pools present a range of slip and fall hazards which require effective recognition and appropriate physical and management controls.

> **Online Resource**
>
> The Sensible Way to Enjoy Your Aboveground/Onground Swimming Pool, The Association of Pool and Spa Professionals (APSP), http://www.apsp.org/clientresources/documents/apsp_aboveground_v2.pdf.

- *Rules*—Rules should include prohibition of alcohol and designated areas away from the pool for food consumption, because spills on the deck can create a slip hazard. Rules should also specify acceptable or unacceptable footwear and require swimmers to dry off in the pool area so water is not tracked through the facility. Running and horseplay must also be prohibited.

Signs stating pool rules should be posted conspicuously at poolside. To be considered an effective management tool, rules must be enforced, a task which usually falls to the lifeguard.

11.3.6 PLAYGROUNDS

According to the Consumer Product Safety Commission (CPSC), 79% of playground injuries are caused by falls. Most fall injuries are associated with climbing—56% of playground injuries to children between the ages of 5 and 14 occur while climbing. Over 200,000 children ages 14 and younger are treated in U.S. hospital emergency departments for injuries from playground equipment. Brain injury is one of the top 10 diagnoses in emergency departments for playground-related injuries. Nearly 20 children die each year from playground-related injuries; about one-third result from falls.

Fall-related playground injuries can be reduced by decreasing the maximum height of play equipment and by improving protective surfacing (Figure 11.2).

FIGURE 11.2 Though playgrounds by nature are a common source of slip and fall injuries to children, there are a wide range of measures that can be taken to minimize the risk.

The top 12 problems found at playgrounds by the National Recreation and Park Association's National Playground Safety Institute (NPSI) is also known as the "Dirty Dozen." Most of these have a direct impact on the potential for falls:

1. Improper protective surfacing, including the surface or ground under and around playground equipment.
2. Inadequate use zone. The use zone should extend at least 6 feet in all directions from the edge of equipment.
3. Protrusion and entanglement hazards, hardware that could impale or cut a child or snag on clothing.
4. Entrapment in openings, including areas where an opening is large enough for the lower body, but could entrap a child's head or upper body.
5. Insufficient equipment spacing, which can cause overcrowding of a play area, resulting in an unsafe environment.
6. Trip hazards, including exposed concrete footings, abrupt changes in elevation, tree roots or stumps, and rocks.
7. Lack of supervision. A play area needs to be designed so a parent or caregiver can observe.
8. Age-appropriate activities. Equipment must be suitable for the age of intended users.
9. Lack of maintenance. Preventive maintenance must be performed.
10. Pinch, crush, shearing, and edge hazards. Components should be checked to make sure moving parts cannot crush children's fingers, and there should be no sharp edges.
11. Platforms with no rails. Elevated surfaces must have guardrails or barriers to prevent falls.
12. Equipment that is not recommended for public playgrounds, including heavy swings, free-swinging ropes, and swinging exercise rings or trapeze bars.

> **Online Resource**
> Consumer Products Safety Commission
>
> http://www.cpsc.gov/CPSCPUB/PUBS/playpubs.html
>
> This site includes excellent guidance on proper design and maintenance on playgrounds, including:
>
> - Public playground safety checklist
> - Handbook for public playground safety
> - Soft contained play equipment safety checklist
> - Playground surfacing materials

Although there may be local statutes, ASTM and the CPSC are the source for the most recognized national standards on playgrounds, including:

- ASTM F1487 Standard Consumer Safety Performance Specification for Playground Equipment for Public Use
- ASTM F2223 Standard Guide for ASTM Standards on Playground Surfacing
- CPSC Public Playground Safety Handbook

11.4 MERCANTILE (RETAIL)

The retail industry, regardless of the nature of the operation, faces many of the usual issues associated with the slip, trip, and fall exposure. Although the products and the size of operations may differ, the basic premise exposures remain the same for customers and employees.

Slip and fall exposures are often encountered in the large parking lot areas often associated with retail operations. Customer foot traffic is usually heavy, and the customer makeup ranges from the very young to the elderly, from the physically fit to the unsteady and infirm. Parking lots and sidewalks in disrepair, inadequate snow and ice removal, and insufficient lighting are leading causes of slip and fall injuries for retail operations. Other parking lot considerations that contribute to the exposure include speed bumps, tire stops, and uncollected shopping carts.

The interior design and layout of the operation can play a major role in limiting slip and fall exposure. Floor surfaces in poor repair and changes in floor coverings (e.g., carpet to tile) increase exposure to slips and falls. Fallen merchandise or spilled liquids or food items on an otherwise clean floor likewise magnify the hazard. Highly stacked stock items can create a fall exposure for employees using stools or ladders to reach them. Additionally, poorly placed merchandise displays often present tripping hazards, as do protruding signs, table legs, and crates (Figure 11.3).

One concern expressed is that of stacking displays of product on pallets. As the product is purchased, the stack becomes lower and less visible to customers. In

FIGURE 11.3 Keeping product aisles clear to reduce congestion and trip hazards is a fundamental part of fall prevention. This is an ongoing challenge for operations open to the public with a high volume of foot traffic.

addition, the pallet sometimes extends beyond the product display, creating a tripping hazard. Since raising the platform would require more frequent stocking, this is not usually a practical alternative. One solution is to design displays with the use of large bins or baskets with sides at least waist high. This would not require more frequent stocking, and it would provide sufficient visual notice to customers even when stock is low.

The National Safety Council provides the following additional guidance on retail displays:

- Remove empty racks, unused displays, and stock containers from the sales floor, and immediately close stock drawers after use.
- Sales items under 12 in. high should not be displayed on the floor unless in racks or large groups, and preferably not in aisles.
- Display bases and platforms for mannequins or other displays should be at least 12 in. high. For visibility, the color of the top and sides should contrast with that of the floor, and the top edge should have no overhang or lip so the side surface will be smooth.
- Bases should not be designed with unexpected extensions that present a tripping hazard. Bases should have rounded corners. A display should not be placed at a counter end or at a column when the diameter of the base exceeds the width of the counter or column.

11.5 THEATERS

The primary areas most susceptible to slip/falls for theaters are the parking lot, main entrance doors, lobby/concession area, restrooms, corridors, stairs, ramps, and seats.

Main entrance doors—Ideally, a grate system with catch basin should be used in entrances to allow ice, snow, and water to be removed from patrons' shoes upon entry. If this is not possible, heavy mats should be used to reduce tracked-in moisture. Entrance vestibules should not be mopped prior to customer traffic in freezing weather, as a thin film of ice can result. This area is also a common site for changes in flooring type. Changes in elevation should be clearly marked with a warning tape, contrasting colors, or a warning sign.

Lobby and concession areas—The lobby area is usually carpeted, with a tile surface adjacent to the concession area. The flooring inside the concession area is usually tiled (Figure 11.4). While some tile floors may have adequate slip resistance, it can be dramatically decreased by water or oil on the floor surface. Slip-resistant coatings consisting of an adhesive base material with slip-resistant substances mixed in or sprinkled on after the base is applied may be used in specific areas to increase traction on slippery surfaces.

In terms of slips, carpet is one of the best walking surfaces that can be employed, with the following limitations:

- Carpets should be checked for wear and looseness and repaired as necessary.
- Busy carpet patterns should be avoided, because they tend to reduce the ability to judge stair heights and other changes in walkway surfaces.

Restrooms—The flooring in restrooms is usually ceramic or vinyl tile, which can become slippery when wet. Therefore, it becomes essential for management to patrol the restrooms frequently to promptly identify and correct wet areas. The condition of restroom fixtures is also important. If they are slowly leaking water onto the floor, even the most diligent employees may not be able to keep the floor dry. Additionally, the grab bars installed in stalls for the disabled should be solidly anchored.

FIGURE 11.4 A typical movie theater concession area setup.

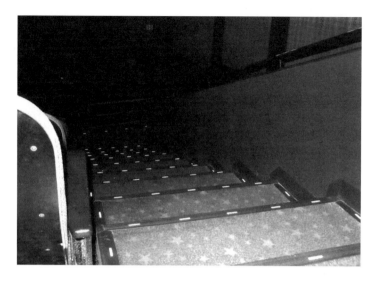

FIGURE 11.5 Strip lighting is frequently used to make viewing room stairs more visible to patrons.

Corridors—The corridors leading to the projection rooms are another potential area for customer falls. Lighting should be adequate but not too bright. Keep in mind some customers, especially those over 40 years old, are beginning to experience the normal physical loss of their visual threshold.

Stairs and ramps—Stairs are probably the leading fall sites within theaters. Stairs with nonstandard geometry are not uncommon. Controls to reduce the potential for a fall on a nonstandard stair may include enhanced marking, lighting, addition of handrail(s), nonskid treads, and increased housekeeping and maintenance (Figure 11.5).

As with stairs, a nonstandard ramp can be the cause of slips and falls. One of the best controls available is the installation of strip lighting along either side of the ramps in the viewing rooms. The lighting presents a point of reference for depth perception and marks the ends of the seating aisles. Newer style strip lighting is easy to maintain, since the bulbs can be replaced with relative ease. Strategies for dealing with nonstandard ramps are the same as those for nonstandard stairs: signage, improved lighting and maintenance, and the addition of center handrails or nonskid surfaces.

Seats—Seats are also a common site for theater accidents, because many falls occur as patrons are sitting down. Pay particular attention to the rows of seats.

- Look for seats that do not fully retract, and those with torn upholstery or loose armrests.
- The floors beneath the seats should be clean and free from spills. The theater should have a spill patrol in place, where the public areas are checked at least hourly for spills.
- Viewing rooms should be checked promptly as the patrons are leaving. Sitting-down accidents often involve the patron's heels sliding out from

Profiles of Other High-Risk Industries 371

under them while in the act of sitting down, potentially resulting in bruised or broken bones.

Temporary seating is not permanently bolted to the floor and can result in an unstable block of two or three seats. These seats should only be used if they can be temporarily secured to the floor.

Many theaters now install seats with built-in cup holders on the armrests. One company has reported success in reducing soda spills and has experienced no fall losses in viewing rooms where seats were so equipped.

Management Controls—It is especially important for the theater management to understand that any loss prevention program (e.g., spill patrol) must have an enforcement procedure. Documentation of spill patrol procedures, usually in the form of a log, and resolution of the concerns identified in completed logs must be part of the program. A progressive disciplinary procedure for ignoring spills should be developed to reinforce the need for employees to take slip, trip, and fall prevention seriously.

Also see Exhibit 11.1, Stadium-Style Seating (Theaters) and Exhibit 11.2, Theater Inspection Log: Common Areas.

11.6 TRUCKING

Spectacular collisions make the evening news, but a common cause of injury to truck drivers is their ongoing exposure to slips, trips, and falls. Injuries can often be severe, requiring long recovery periods. Truck drivers face the daily challenge of simply getting in and out of their vehicles. This is often attempted while they are loaded down with paperwork, coffee cups, and other extraneous materials, making it easier to misstep and fall. Poor training and improperly equipped or modified vehicles often contribute further to this exposure. The lack of adequate grab bars as well as improperly designed steps that do not provide enough traction or actually create a hazard under adverse weather conditions are additional contributory factors. A similar exposure exists for those truck drivers who are responsible for securing the loads they carry and must climb on and over the trailers, securing chains or coverings and inspecting tie-downs. Falls from the trailer are commonplace and often more severe due to the size, shape, and positioning of the load, which can put the driver at a higher risk of injury. An additional hazard is improperly dismounting from the trailer to the ground.

Once out of the truck, drivers are, at some point, faced with walking across customer delivery sites covered with snow and ice, slick with rain or sleet, mired in mud or muck, or simply damaged beyond repair. When the truck driver is handling a standard delivery route, they are exposed to many aspects of the industry they are serving. Falls from docks are common, as are falls inside of trailers while climbing over stock looking for delivery items. Once inside a customer building, the driver is subject to any of a number of exposures that are beyond their control, including poor housekeeping, slippery floors in walk-in coolers, and oil and grease-covered environments. Proper footwear protocols, better management controls, focused training, and the strategic use of additional equipment can greatly assist in the reduction of the slip, trip, and fall exposures associated with the trucking industry.

11.6.1 Falls from Cabs

Entering and exiting the cab repeatedly throughout the day can lead to complacency in following recommended and established safety procedures. Every year, thousands of drivers are injured as a result of falls from their cabs and trailers. Some of these injuries are serious, causing drivers to miss work days and perhaps sacrifice income. Studies have shown that 20% of all lost-time injuries relating to mobile equipment occurred when the operator was mounting or dismounting.

The following steps are actions that the operator can take to protect from serious injury while entering and exiting the cab of the truck and while mounting and dismounting equipment.

1. **Wear and maintain proper footwear.** Most companies that work in environments that contribute to falls have some footwear requirements. Use of sturdy work boots with slip-resistant soles is advisable to help prevent falls. However, even the best boots cannot prevent a fall if they are not properly worn and kept free of excessive dirt and mud. Also, boots with excessively worn soles provide little protection. See Exhibit 11.3, Sample Footwear Policy for Trucking Fleets.
2. **Use and inspect your equipment.** Most vehicles have safety handles and slip-resistant steps for climbing onto and off of equipment and trucks. Use these safety aids and keep them free of dirt, grease, oil, mud, or any other substance that may impair security while using them. Inspect these items daily and report excessively worn or damaged handles or steps.
3. **Enter and exit in the proper direction.** Always face the truck or equipment when entering or exiting a truck cab or when climbing on or off of mobile equipment. Access ladders and steps are designed to work with the operator facing the truck or equipment.
4. **Plan your exit and know where you are going.** Where possible, park your vehicle where stable, firm footing is available. If not available, extra caution should be used when getting on and off of the truck or equipment.
5. **Use the three-point system for access on and off of trucks and equipment.** Almost all slips and falls from truck cabs and trailers can be prevented with the use of three-point contact. The three-point system means that you have at least three points of contact with the truck or equipment at all times—either one foot and two hands or both feet and one hand. Having these three points of contact give you additional stability and balance as well as security if one point of contact should slip (Figure 11.6).
6. **Keep both hands free when climbing on or off your vehicle or equipment.** If you do not have both hands free, you cannot maintain three secure points of contact. When entering or exiting a truck cab, place objects in the truck seat before entering or exiting. When climbing on equipment, place tools or other items needed on a flat secure surface before climbing.

Profiles of Other High-Risk Industries 373

FIGURE 11.6 Unlike this example, in which the driver is not holding on to any part of the vehicle, entering and exiting a truck cab should be done using the three-point system.

11.6.2 FALLS FROM LOADING DOCKS

Loading docks and ramps are frequently congested, heavy-traffic areas, and are often wet. Dock plates can wear and become slippery, and the edge of a dock plate invites trips and falls.

An accidental backward step can result in a fall from the dock. Portable railings, which can be easily removed from the edge of the dock, could prevent many dangerous falls. They are removed when a truck or tractor is at the dock and replaced as soon as the truck or trailer leaves.

Proper housekeeping, well-designed traffic patterns, and the use of abrasive, slip-resistant surface coatings reduce the risk. The slopes of ramps and gangplanks should be as gradual, wide, and dry as possible. They should also have slip-resistant surfaces.

11.6.2.1 Dock Plates and Dock Boards

Between the rear of a trailer or truck and the edge of a loading dock is a space about 15 in. (33 cm.) wide. It is a space in which people are injured or killed with alarming frequency.

Dock plates and dock boards are removable metal ramps used to bridge the gap between a truck parked in a loading bay and the warehouse floor. Dock plates are normally constructed of aluminum, while dock boards are made of steel. Aluminum dock plates are recommended only for nonmotorized use, such as dollies, pallet trucks, and handcarts, while steel dock boards are suitable for motorized equipment.

Steel and aluminum dock plates do not have a curb, while dock boards have a bolted or welded curb.

The lowest capacity of a dock board is 6,000 pounds, and the highest standard board has a capacity of 20,000 pounds; however, 30,000-pound capacity and more is available. Consider the maximum height differential below and above the dock. Dock boards are designed with fixed leg lengths and a lip. A typical differential is approximately 10 inches, although this varies by the type of equipment in operation. Hitting the dock plate is the biggest jolt a lift truck driver will experience. To ease the transition from the dock to the trailer, manufacturers have designed dock levelers. Levelers range in length from 6 to 15 ft., and the lip crown can also be varied to capture the variety of trailer bed heights.

Many accidents occur when trucks pull away from docks when workers are still working within the truck. This unfortunate and dangerous problem can be minimized by always chocking wheels and/or using dock locks, which grab onto a truck's ICC bar and will not release. The trucker cannot physically leave the dock until the dock operator releases the lock.

EXHIBIT 11.1
Stadium-Style Seating (Theater)

Stadium-style seating is becoming a popular form of customer seating in theaters. Along with its benefit of improved visibility for customers, stadium seating may also pose some added slip, trip, and fall exposures. Listed next are some measures to help minimize slip, trip, and fall loss exposures associated with stadium-style seating.

DESIGN AND USE OF HANDRAILS

Steps should be of normal design in height of each riser and the width of each step, including nosings that protrude less than 1.5 in. There should be a minimum of deviation from step to step.

Steps between the seating levels or landings should not leave gaps or holes between the side end of the step and the nearby seats. Gaps and holes will increase the potential for patrons to misstep while ascending or descending the steps.

Consideration should be given to providing a short hand rail (12 to 16 in.) at each level of seats. The rail can be attached to the concrete floor at the edge of the upper seating level and then attach to the concrete of the next lower seating level at the point near the end of the seat armrest. These handrails will help give support to patrons who need this, plus the handrails will not block or protrude into the aisle area of the seats.

LIGHTING AND IDENTIFICATION

Lights or lighting strips need to be used to illuminate the steps within the auditorium. Lighting strips need to be on the top of each step and landing so that patrons can easily see the step outline as they descend. Lighting strips or louvered lighting should be used on the vertical face of each step so that each step will be illuminated while a patron is ascending the steps. There are also lighting strips inside special stair nosings that provide illumination to the patron from either direction.

In conjunction with the lighting strips, the nosings, and sides of the steps should be of contrasting colors or materials to again identify the boundaries of each step.

ADDITIONAL LIGHTING AND TIMING OF LIGHTING

In general, if the steps within the auditoriums have strip lighting on both the horizontal and vertical faces of each step and landing, the steps should be readily visible and identifiable for most patrons. However, elderly patrons and others with visual impairments may see the lights, but may not be able to see the steps or their feet as they contact each step.

This loss of visual acuity, especially in a theater auditorium with steps, is accelerated when the auditorium is showing the main feature and can be heightened if the actual picture has low levels of lighting. Additional lighting is usually needed, and the timing of this lighting is especially important.

Additional lighting can come from various sources, but the most beneficial is the use of wall sconces in smaller auditoriums where main aisles are either on a wall or within one row (seven seats or less) of the wall. In larger auditoriums, with multiple aisles and entrances, additional lighting should be at the ceiling level directly over the aisles. These lights will direct their illumination directly on the aisles for their full length, without over-illuminating the remainder of the auditorium.

The timing of the additional lighting is also critical. Most house lights are left on at low levels until the movie trailers begin, when they are either turned off or significantly reduced in intensity. At the end of the main feature, the house lights are not usually brought up until the feature credits have run. It is at these two times when patrons with visual impairments or reduced acuity can easily misstep and fall.

The additional lighting should be tied to the film projectors and timed so that they will stay on during trailers and preliminary announcements, and reduce in intensity once the main feature begins. This will allow latecomers to see aisles and steps as they search for a seat. Likewise, at the end of the main feature, the additional lighting should be brought up as the credits start to run. Most patrons leave the auditorium at the start of the credit run, so bringing up lighting at this time will be beneficial to all patrons as they make their way to exit the auditorium.

Signage

Highly visible signs should be permanently attached to the walls at the entrance to each auditorium where it will not be obstructed by doors or displays. Signs should state there are steps in the auditorium or depict this with a universal type drawing (or have both) to prevent misunderstandings. If possible, signs should also advise patrons to allow time for their eyes to adjust before attempting to use the steps in the aisles.

Highly visible signs should also be placed or posted at or near ticket windows, the main entrance, or at the entrances to the hallways leading to the auditorium doors.

Verbal Warnings by Ticket Takers/Ushers

All employees, especially ticket takers and ushers, should be trained to remind patrons about the steps in the auditoriums, to "watch your step," and to allow time for patrons' eyes to adjust if they are walking in late.

Housekeeping and Maintenance

Provide extra diligence in keeping the aisle steps clear of spills or discarded items to minimize the potential for a slip or trip. If possible, check steps before the end of the main feature to remove or clean up such conditions.

Employee Orientation

Discuss the above strategies with employees during staff meetings. New employees need to be made aware of the special exposures associated with stadium-style seating during the orientation process. It is the responsibility of each theater manager and employee to ensure that falls do not occur in the auditoriums.

EXHIBIT 11.2

Theater Inspection Log: Common Areas

Facility: _____   Date: _____

Time	am or pm	Lobby(s)		Hallway	Restroom(s) Men			Restroom(s) Women			Snack Bar(s)		Parking Lot(s)	
		Main	Aux		1	2	3	1	2	3	1	2	Front	Rear

Comments:

Theater Manager Review:

Instructions:

This log is a tool to help control slip and fall hazards. The person conducting the inspection should initial the box after all the hazards identified during the inspection are eliminated and/or corrected. Restrooms should be surveyed every 15 minutes during time of peak usage to promptly identify and clean up or mark wet floor areas. All other areas should be checked at least hourly.

Special emphasis should be placed on lighting, spills, marking of changes in elevation, condition of carpet, and visibility of signs and marking.

The time completed and the employees' initials should be in the boxes next to the time completed.

The theater manager should review the log daily and initial the box provided.

Reminder:

Wet Floor signs should be used when conditions warrant. Signs should be at least 36" high and positioned to visibly indicate the location of the spill.

EXHIBIT 11.3
Sample Footwear Policy for Trucking Fleets

Falls have attributed to a significant portion of employee injuries over the past several years. This type of injury is often serious and has been responsible for tens of thousands of dollars in costs. Employee injuries also interrupt customer service, cause dispatching headaches, result in excessive driver turnover, and endanger the employee's own financial security.

THE IMPACT OF FOOTWEAR ON INJURIES

Although there may be many contributing factors to any injury, several fall injuries appear to have a common contributing factor—improper footwear. Shoes are made with specific characteristics to provide the best attributes for a particular activity. For example, a sprinter's shoes are very lightweight, have almost no heel, and a sturdy sole for flat-out speed. A linesman's boot has heavy, ankle-high leather uppers to prevent twisted ankles and a steel sole shank to ease pressures on the foot while standing on narrow telephone pole pegs. A cowboy's boot has calf-high uppers for abrasion resistance, a pointed toe and cutback heel for easy insertion into a stirrup.

Many specific attributes also create hazards for other types of activities. A tennis shoe provides good traction but does not provide puncture resistance or toe protection. A cowboy boot's pointed toe doesn't allow a foot to be fully inserted into many ladder rungs, the cut-back heel increases the chances of forward slides, and the leather sole is very slippery when wet.

SAMPLE FOOTWEAR POLICY

We have placed restrictions on the type of shoes our drivers and mechanics are permitted to wear while on company time, not as a punishment to anyone, but to reduce the incidents of falling and the resulting injuries. Boots with a cowboy "look" will be allowed only if they have a squared-off toe, a full, square heel and nonleather soles. We recommend a good slip-resistant sole. (Steel-toed shoes will only be required of certain drivers who typically handle specific commodities). Tennis shoes will also not be allowed, because they provide little puncture resistance, toe protection, or resistance to twisted ankles.

Required footwear will have the following characteristics:

- Oil-resistant sole
- Slip-resistant tread
- Sturdy leather upper (6 or 8")
- Lace tied for proper fit

Sample Purchase Arrangements

Through [insert name of vendor], we have arranged for a (insert percentage) discount for our employees on [number] boot styles, which fit our recommended traits.

Using a new payroll deduction option, you can obtain the boots and then pay for them through this option, or you can accumulate the funds before purchasing the boots. In most cases, this will account for only a few dollars per week. Your account will earn [insert percentage] interest for the time it is held. Upon termination for any reason, any funds left in the account will be returned to you. Any funds still owed to the account for boots purchased will be withheld from your last check.

12 Accident Investigation and Mitigation

Organizational Responsibilities

Component	Potential Responsibility
Accident Response	Line Supervisor
Accident Report/Process Development	Safety, Operations, Security, Legal
Accident Report Completion	Security, Operations
Incident Reporting	Employees, Line Supervisor
Occurrence Analysis and Action	Safety, Others as Indicated
Reporting of Potential Fraud	All Staff
Fraud Investigation	Management, Legal
Selecting a Trial Expert	Safety, Legal
Documentation	All Departments as Indicated
Staff Training	Safety, Operations, Human Resources

What we call failure is not the falling down but the staying down.

Mary Pickford (1893–1979)

12.1 INTRODUCTION

Slip and fall accidents are notoriously poorly investigated. The potential severity of slips and falls is often underestimated, and there is a tendency to attribute these types of losses to carelessness. As a result, corrective action is less likely to be identified or taken.

Although the basic principles of accident investigation can be effectively applied to this type of loss, several aspects of slip and fall accident investigation require special handling. The types of documentation and the approach to documenting physical evidence and information about the relevant management controls are unique to slip and fall accidents. More than many other types of accidents, slips and falls are the frequent subject of fraud.

12.2 PITFALLS OF ACCIDENT REPORTING AND INVESTIGATION

It can often be difficult to prepare slip and fall cases for defense and trial because adequate investigation was not performed. This can compromise your defense and may result in significantly higher awards and legal expenses. Issues commonly found are:

- Witnesses have not been identified, or their statements were inadequately or improperly taken.
- Little information is collected at the time of the event when conditions and recollections are still fresh.
- Pictures of the scene are not taken immediately after the incident.
- No attention is paid to the type and condition of the claimant's footwear.
- Employees have little training or instruction on how to appropriately respond and handle slip and fall incidents, increasing the likelihood of litigation and aggravated injury.
- Employee statements are either not taken or are taken inadequately or improperly.
- Response time does not meet internal standards or is otherwise inadequate.

Procedures and controls should be in place to gather and document key information immediately after the event. Employees should be trained in what their roles will be in the event they must respond to an accident. Employees should be instructed to not make any statements that may imply fault, especially regarding aspects of the accident that they have not witnessed.

12.3 THEORIES OF LIABILITY

Any of several theories of liability may be put forth in a claim in order to place the moniker of liability upon management. Commonly encountered theories are:

- Failure to comply with code—a physical condition, such as a stairway or ramp, that does not comply with local or state building code specifications or those of another authority having jurisdiction.
- Failure to correct—the presence of an unabated hazardous condition of which management had received sufficient prior notice or should have known existed.
- Failure to warn—providing inadequate warning or not advising individuals in advance of entry into a hazardous area or exposure to a hazardous condition.
- Failure to inspect or maintain—an inadequate safety program or nonobservance of a material existing safety program, policy, procedure, or other requirement.

12.4 ACCIDENT INVESTIGATION

Most factual investigation can be scrutinized through the legal process of discovery at a later point in time. Because of this, it is essential that all factual investigation be as accurate as possible. This includes providing specific comments about the following critical areas:

- Claimant and witness information

Accident Investigation and Mitigation

- Information about the actual event
- Condition and design of the building components involved, such as stairs, handrails, and landings
- Adequacy of lighting levels

It is strongly recommended that an established accident investigation format, which is tailored to guide the investigator through a fall occurrence, be used. Some examples are included in this chapter's exhibits.

12.4.1 Claimant and Witness Information

- Claimant and witness names and contact information must be collected.
- The purpose for which the claimant was on the premises is also important. Different classes of guests may be owed different standards of care. A trespasser, for example, might be afforded a lower degree of care than would an "invitee."
- The proximity and vantage point of witnesses to the event can be important in determining what they were able to observe and the relative value of what they allegedly saw.

12.4.2 General Occurrence Information

- Determine whether the occurrence was a slip, a trip (over something), a stumble (excessive traction), or another type of event. This will help to compare the consistency of the claimant's later version of the accident with the reported nature of the event. Thus, if the claimant originally cited a slippery surface, but later claimed the event was a trip (or physical evidence or witness statements indicate such), the validity of the claimant's assertion can be questioned.
- Determine which foot initiated the event and if the claimant fell backward, forward, or sideways. Again, these facts can be compared with other versions of the event and the injury for consistency.
- Establish the nature of the injury and the body parts involved.

Obtaining this type of information as soon as possible after the event can help prevent successful manipulation of the facts later.

Depending on the circumstances, a proper investigation may require a biomechanical analysis. The purpose of a biomechanical analysis is to evaluate the consistency between the type and extent of injury and the fall kinematics and cited conditions. For example, abrasions to the front part of the knee would be expected due to a forward fall, which would be the result of a trip rather than a slip. Some biomechanical analyses are relatively simple, while others require the expertise of a biomechanist. Data for biomechanical analysis, including age, height, weight, pre-existing conditions, incident description, and injury descriptions, are gathered from medical records (Joganich, 2008).

12.4.3 Detailed Occurrence Information

- Establish if the claimant usually wears glasses or contact lenses and whether they were being worn when the accident occurred. Later, you may want to determine the nature of the prescription and possible impact on the event (e.g., is the prescription for constant use, or just for reading or driving?). Finally, it could be important to find out if the prescription is current. This could be a material fact, especially if the claimant had been due or overdue for a new one.
- Find out what, if anything, the claimant was carrying and if it may have played a role in the accident. The dimensions, weight, and ease of carrying (e.g., handles) should be determined.
- Does the claimant normally take any medication or drugs? If so, find out:
  - The type of medication so the effects (e.g., drowsiness or other impairment) can be determined
  - The interval between the last dose and the incident
  - If medication was not taken, but should have been, because it may later be determined to have contributed to the accident
  - If illegal drugs or alcohol were used, so the police can be contacted early on to assist in the investigation
- Determine if the claimant has any physical disabilities, including those which require the use of a walker, cane, or other walking aid. Establishing whether or not the aid was in use, its condition, and where it landed following the incident could be meaningful to the results of the investigation.
- Of prime importance and often ignored is the footwear the claimant was wearing. Photographs of the shoes, including the soles and heels, can prove valuable information. Important details include:
  - Type of shoe and sole
  - Tread design, amount of wear, and overall condition
  - Presence of contaminants
- Determine if weather conditions were a factor.

Online Resource: Obtaining NOAA Weather Data

Securing definitive information on weather conditions at the time of the accident can be essential in confirming or refuting such assertions by claimants. Weather data can be obtained through the NOAA (National Oceanic and Atmospheric Administration) Web site. The cost for weather data from a station is usually $3. Once on the site below, follow these steps:

- Select STATE
- Select CITY OF WEATHER STATION
- Select YEAR AND MONTH
- Select DAY
- Proceed to checkout

http://cdo.ncdc.noaa.gov/qclcd/QCLCD?prior=N

Accident Investigation and Mitigation

12.4.4 Location Information

Identify and describe the following:

- Specific area where the accident occurred
- Type of floor treatment in use, schedule of application, and if it was used in accordance with manufacturer's specifications
- Construction, textures, patterns, or profiles of the flooring material
- Floor maintenance schedule, type of cleaners, and date/time last performed
- Presence or absence of:
  - Foreign substances or contaminants such as snow, ice, or water accumulation, and oils, dusts, or sands
  - Surface irregularities and slope
  - Subtle changes in levels and other unexpected obstructions
- Any apparent deviations from applicable building code requirements

If the accident allegedly involves a slippery surface, it is advisable to conduct a slip resistance test of the area in the direction the claimant was traveling. This should be done prior to cleanup or other changes in the floor and should be made part of the accident investigation report. It is important to conduct testing as soon as possible. If too much time elapses between the event date and the test date, the claimant may contend that floor surface and maintenance conditions had materially changed. The sooner in the factual investigation process that testing can be arranged, the more useful the results will be.

Photograph and diagram the site. The photographs should include a visual indicator pointing to the specific location of the incident, using a pen or other pointing device. Photographic documentation should be taken from three perspectives: the "big picture," specific items of interest, and the claimant's view. Consider that site conditions change over time, and photographs taken after the fact do not necessarily represent site conditions at the time of the incident. Photographing specific elements of interest are normally close-ups, sometimes taken with a scaling device such as a tape measure.

A common error made in fall investigation is the use of supplementary lighting and wide-angle lenses in photographing the scene. Use of such measures alters a primary purpose of capturing the fall conditions—to observe the site from the perspective of the subject of the fall. Excess illumination can make hazardous conditions more apparent. Similarly, wide-angle photography shows more than a human would be able to perceive. While such images may have value as supplementary information, it is essential that photographs represent what the victim was able to observe at the time of the fall. This would mean that photographs should be taken under lighting conditions comparable to those at the time of the fall using camera techniques more representative of human vision (Pauls and Harbuck, 2008).

12.4.5 Stairs or Ramps

Document the following:

- Direction of travel (i.e., ascending or descending)
- Specific step or steps involved
- Tread surface construction and condition
- Type, design, and condition of nonslip tread strips
- Step geometry (stair width, riser height, tread depth, nosings)
- Stairway slope/angle
- Presence or absence of deviations in riser height of more than $3/16$ in.
- Presence or absence of deviations from applicable consensus standards and building code requirements

Any of four types of rulers/tape measures may be useful: a 10-foot, a 25-foot, a carpenter's square, and a measuring wheel. The 10-foot tape measures are easier to hold while taking photographs than the heavier 25-foot alternative. A carpenter's square is best for measuring step treads and risers.

The slope of an inclined surface can be made by calculating the rise over run (see Chapter 1, Physical Evaluation) or by using a digital level, which allows quick and accurate measurement in degrees, percent slope, and rise over run.

While a tape measure is often used to determine smaller elevation changes, a profile gauge is optimal, because it also permits the tracing of the elevation difference profile onto graph paper. The profile gauge is also useful for establishing the height of elevation changes such as door thresholds (see Figure 12.1).

FIGURE 12.1 An example of a profile gauge. (Courtesy of QEP Co. U.K. Ltd. [Vitrex].)

Accident Investigation and Mitigation

12.4.6 HANDRAILS

Identify the following:

- Handrails on one or both sides of the stairway and whether they are accessible from the most remote part of the stairs
- Handrails extending 12 in. beyond the top and bottom steps
- Shape and diameter of handrail and distance from wall (this impacts how "graspable" the handrail is)
- Condition of handrail and how well it is secured
- Whether the claimant was using the handrail
- Apparent deviations from applicable consensus standards and the building code requirements

12.4.7 LANDINGS

Identify the following:

- Landing geometry, including width and length
- Rise over run (angle) of ramp lengths to and from the landing
- Whether the door swings out over the landing and, if so, how far
- Apparent deviations from applicable consensus standards and the building code requirements

12.4.8 LIGHTING

Before taking light readings, develop a grid in the area of interest. It is also important to ensure that lighting conditions are comparable to those prevailing at the time of the incident. Light readings should be compared against accepted standards and design guidelines such as those developed by the IESNA (see Chapter 2, Management Controls).

Evaluate the following:

- Presence, extent, and impact of extreme lighting transitions (dark to light, light to dark)
- Position of lighting source(s) relative to the claimant's direction of travel (e.g., does it cast a shadow over the direction of travel?)
- Overall illumination levels as compared with accepted standards (Illuminating Engineering Society of North America [IESNA] and the National Fire Protection Association [NFPA])
- Normally operating sources of illumination that were unavailable due to maintenance, and the impact
- Automatic or manual-controlled lighting activation
- Obstructions to light sources
- Type, appropriateness, and condition of lighting units

See Exhibit 12.5, Sample Claims Investigation: Parking Garage

> **Additional Information on Slip and Fall Accident Investigation**
>
> ASTM F1694 is the Standard Guide for Composing Walkway Surface Evaluation and Incident Report Forms for Slips, Stumbles, Trips, and Falls.
>
> Under the jurisdiction of Subcommittee F13.50 Walkway Surfaces, this guide provides guidance in recording and evaluating the conditions of walkway surfaces, including components such as ramps and stairs, that may present a hazard or an exposure to slip, stumble, or trip.
>
> http://www.astm.org/cgi-bin/SoftCart.exe/DATABASE.CART/REDLINE_PAGES/F1694.htm?E+mystore

12.4.9 MANAGEMENT/OPERATIONAL CONTROL INFORMATION

The extent and frequency of relevant management programs should be evaluated. Not only should documentation regarding the program as intended be obtained, but documentation (e.g., logs, checklists, work orders) regarding the degree of implementation should be obtained as well. Depending on the circumstances of the event, controls to be evaluated may include a wide range of inspection/housekeeping and maintenance programs, training programs, lighting, construction, and special event planning (also see Chapter 2, Management Controls).

12.5 INCIDENT REPORTING

Why is incident reporting important? An incident can turn into a claim if not handled properly. Using incident reports can provide a larger statistical basis for determining causes of loss, thereby making your conclusions much more accurate. For instance, if 200 slip/fall claims have occurred, adding incident reports may bring your total occurrence rate to 600—any of which could develop into a claim. A larger population of numbers leads to more reliable statistics, a better predictor of where and how future accidents are likely to occur lacking meaningful intervention.

In terms of investigation, each incident report should be treated as if it were an accident report. Because incidents are accidents that almost happened, they deserve to be investigated and acted upon as if they actually had resulted in injury.

12.6 OCCURRENCE ANALYSIS

Occurrence analysis is the process of gathering all incident and accident reports to determine if any common elements exist between them. A systematic tracking system should be developed to aid in this process.

Occurrence analysis should be incorporated into your overall safety program. This type of analysis provides a more efficient and complete method of observing and handling overall loss trends. In conjunction with implementing corrective action for each individual claim, the occurrence analysis may call for forms of corrective action broader in scope, once trends are developed at this level. For example, it may be determined that the cleaning regimen should be improved. In many instances, this should be done not only for the area in question, but as part of a global change in maintenance across all locations and areas of the operation.

Periodic occurrence analysis will also allow for the tracking of incidents in terms of frequency and severity. A graphical method is often helpful, such as plotting slip/fall incidents on a facility map. In many cases, this will provide clear and unequivocal indications of areas that need attention. These measures are also helpful when following up on a prior corrective action in determining whether or not the action implemented was effective.

12.7 CLAIM MITIGATION

In terms of mitigation, management can do several things to help mitigate a loss. Demonstrating respect and empathy for the injured party and showing a sincere concern for their well-being can often reduce the potential for a lawsuit. Although care must be taken regarding the degree of assistance offered to the injured party (to avoid a potential assumption of liability, or becoming exposed by delivering medical care), such measures are generally of more benefit than harm. It is not uncommon to offer "complimentary services/products" (e.g., gift certificates, coupons, or discounts) in order to show good faith and sensitivity. Also helpful is to follow up with the injured party by telephone.

Sixty percent of individuals surveyed who lodged malpractice lawsuits against physicians indicated that they would not have brought legal action if the doctor had been more congenial and cooperative. So, if you too do not mitigate, the chances are that you may have to litigate.

12.8 FRAUD CONTROL INDICATORS

> An investigation by the Ohio Department of Insurance revealed that Sylvia Davido staged slip-and-fall accidents at various businesses in Franklin and Delaware counties. She then filed 13 false insurance claims, using fictitious names, social security numbers, dates of birth, and medical records with different insurance companies. Robert Nicholas assisted Davido by faxing fictitious medical information to the insurance companies and securing drop boxes in the Columbus area.
>
> Davido and Nicolas each pled guilty to two counts of insurance fraud, and Davido also pled guilty to one count of conspiracy to engaging in a pattern of corrupt activities. Davido was sentenced to 3 years in prison and was required to pay $35,000 in restitution. Robert Nicholas was ordered to serve 6 months and make restitution of $35,000. Nicholas was further sentenced to 5 years community control.
>
> **"The Old Slip-and-Fall Scam"**
> *Ohio Department of Insurance, November 16, 2007*
> *http://blogs.dixcdn.com/capitalblog/?p=2803*

In order to combat fraud, it is first necessary to identify indicators of potential fraud. There are four general categories of fraud indicators. Keep in mind that the presence of any single indicator alone does not necessarily mean that fraud was involved. Instead, a composite of multiple factors should be considered relative indicators to suggest further investigation of the circumstances surrounding the event and the background of the claimant.

12.8.1 Fraud—Manner of Claimant

- Overly aggressive and demanding for a quick settlement
- Uncharacteristically familiar with insurance terminology, processes, and procedures
- Handles all contact in person or by phone, avoiding the use of means of communication that can leave documentation (e.g., mail, e-mail, fax)

12.8.2 General Indicators

- Claimant is a transient or an "out-of-towner" on vacation.
- An overly enthusiastic witness is at the scene of the accident.
- The owner's, manager's, or employee's account of what happened suggests a setup.
- The address provided by claimant or witness is a post office box or a hotel.
- There is no supporting evidence of the presence of the reported foreign substance or other causative agent at the site of the accident.
- Statements between the claimant and witnesses are inconsistent.

12.8.3 Medical or Dental Fraud Indicators

- Injuries are of the subjectively diagnosed variety (e.g., headaches, nausea, inability to sleep, soft tissue injury).
- Injuries are inconsistent with the hazard or circumstances of the incident as described by the claimant and/or the evidence.
- Ailments appear to be chronic, persisting for several weeks or more.
- Medical or dental bills are photocopies; third- or fourth-generation photocopies are especially suspicious.

12.8.4 Lost Earnings Fraud Indicators

- Claimant is self-employed.
- Claimant is employed by a small or unheard of business, or the business address is a post office box.
- Claimant started employment shortly before the accident.

12.9 FRAUD CONTROL

Television and print exposés have highlighted the fact that slip and fall fraud is rated as the third easiest and quickest way to earn $50,000 or more. Retail shopping malls, supermarkets, and other high-traffic volume businesses are more prone to fraud, although it could happen anywhere. In addition to a thorough accident investigation process, more business owners and property managers are installing closed-circuit television (CCTV) systems covering strategic areas of the facility. While often done for the purposes of preventing theft, CCTV can serve other useful purposes. A perpetrator will be less prone to staging an event in view of a camera, thus serving as a deterrent. If a claim is made, the recorded event can be used to help determine whether it is fraudulent, and can be used to prosecute the perpetrator.

Good incident reporting, accident investigation, housekeeping, and maintenance all help to control opportunities for fraud. With good management controls in place, you can help deter and defend fraudulent claims.

12.10 ADMISSIBILITY OF EXPERT TESTIMONY

A sound evaluation of facts can be difficult without the application of some scientific, technical, or other specialized knowledge. The most common source of this knowledge is the expert witness.

In a study of one state, expert testimony was proffered in 86% of trials. There was an average of 3.3 experts per trial. In many cases, trial by jury is essentially trial by expert, and experts often offset each other. It is therefore important that an expert be knowledgeable, credible, and make a good appearance. This is why it is important to understand the underlying principles of expert testimony and its admissibility in a court of law.

12.10.1 THE FRYE TEST

http://www.law.ufl.edu/faculty/little/topic8.pdf

Frye v. United States (*293 F.2d 1013, D.C. Cir., 1923*) held that scientific evidence is admissible only after the methods from which the deduction is made was sufficiently established and accepted in the field in question. This became the basis for what is known as the Frye test (or the general acceptance test). The Frye test was widely applied through the 1980s and 1990s in the federal courts. By then, it became the most widely applied test for admissibility of novel scientific evidence. Still widely accepted in state courts, some have also accepted and applied Daubert since its introduction.

12.10.2 THE DAUBERT RULING ET AL.

http://caselaw.lp.findlaw.com/cgi-bin/getcase.pl?court=US&vol=509&invol=579

In 1993, the U.S. Supreme Court handed down a significant decision in *Daubert v. Merrell Dow Pharmaceuticals*, in which the court stated that evidence based on novel scientific knowledge is admissible only after having been established that the evidence was reliable and scientifically valid. While not accepted in every state, it is clearly a known precedent and can impact success in court even when not formally adopted.

In Daubert, the court assigned the trial judge the duty of gatekeeper, charged with preventing "junk science" from entering the courtroom. In this vein, Daubert discussed four reliability factors: testing, peer review, error rates, and acceptability in the relevant scientific community. The general observations the court specified in Daubert included the following types of factors trial judges might consider:

- Whether a theory or technique can be and has been tested
- Whether the theory or technique has been subjected to peer review and publication (understanding that publication alone does not necessarily mean admissibility)
- The known or potential rate of error

In 1997, the court clarified Daubert in *General Electric v. Joiner* (http://caselaw.lp.findlaw.com/scripts/getcase.pl?navby=search&court=US&case=/us/000/96%2D188.html). It added that, in reviewing the decision of a trial court to admit or deny admission of expert testimony, a court of appeals must use an abuse-of-discretion standard. This applied whether the issue was the judge's determination of reliability or the ultimate conclusion reached by the expert.

Guidelines in assessing the admissibility of expert testimony under Daubert can include:

- Is the underlying premise upon which a method rests empirically validated?
- Is there professional literature that describes the purpose to be achieved and the method to realize that purpose?
- Are there professional associations or societies offering continuing education to which members with established credentials belong?
- Is there a rigorous training program so that basic proficiency can be achieved in the discipline under the supervision of persons with established credentials?
- Is there a meaningful certification program that attests to competence and proficiency?
- Has an examination protocol been developed whereby investigations can be reliably carried out and which will yield reasonably consistent results when followed by properly credentialed examiners?

12.10.3 KUMHO TIRE CO. V. CARMICHAEL (97-1709)

http://supct.law.cornell.edu/supct/html/97-1709.ZS.html

The next step in progression in the federal courts was the Kumho ruling, which effectively loosened the requirements set in Daubert.

A tire on the vehicle driven by Patrick Carmichael blew out, and the vehicle overturned. One passenger died and others were injured. A suit was brought against the tire's maker and its distributor claiming that the tire was defective, based largely on the deposition of a tire failure analyst. He intended to testify that, in his expert opinion, a defect in the tire's manufacture or design caused the blowout. That opinion was based upon a visual and tactile inspection of the tire and upon the absence of at least two of four physical symptoms indicating tire abuse.

Kumho Tire moved to exclude this testimony because the methodology failed to satisfy Federal Rule of Evidence 702 (see Section 12.10.5) which says: *"If scientific,*

Accident Investigation and Mitigation

technical, or other specialized knowledge will assist the trier of fact . . . a witness qualified as an expert . . . may testify thereto in the form of an opinion." Granting the motion and entering summary judgment for the defendants, the District Court acknowledged that it should act as a "reliability gatekeeper" under Daubert.

On the plaintiffs' motion for reconsideration, the court agreed that Daubert should be applied flexibly, that its factors were merely illustrative, and that other factors could argue in favor of admissibility. In reversing, the circuit court held that the district court had erred as a matter of law in applying Daubert. Believing that Daubert was limited to the scientific context, the court held that the Daubert factors did not apply to the expert's testimony, which it characterized as skill or experience-based.

12.10.4 Practical Suggestions for Meeting Daubert/Kumho

- The Daubert/Kumho factors focus substantially on the methodology used by the expert, so it is important that experts clearly identify their methodology and be prepared to support how it is used to analyze or resolve the issue in question and that it is generally accepted in the field. The expert should also be able to state how the methodology is repeatable.
- When practical, experts should publish in peer-reviewed journals. It is also important that files are prepared to show numerous supporting peer-reviewed articles and books demonstrating that the methodology is well supported. This can also assist in showing general acceptance of the methodology. The ability to establish that peer-reviewed articles use the same methodology should convince most impartial judges that it is generally accepted.
- Daubert/Kumho also suggests that credentials alone may not suffice. Experts must be more than credible; they should have experience in the area in which they are testifying. The expert must also be prepared to support all methodology and opinions with objective documentation. The expert should be able to support opinions with peer-reviewed literature and published guidelines.
- In federal court, experts must show that the experience is in conformity with the given field of expertise. Merely claiming 20 years of "experience" in the field may not be sufficient.
- It is helpful to establish that the expert's methodology is the same or very similar to that used by opposing experts. This will make it more likely to pass the Daubert tests. If not, the expert should be prepared to provide a detailed explanation as to why a different generally accepted test method was chosen.
- The expert should stress that the method used for the courtroom is the same method with the same thoroughness that they use when working in the relevant industry in a nonlitigation setting.
- The expert's report and deposition should use the same language used in Daubert/Kumho:
  - The opinion is based on knowledge and experience.
  - The method has achieved general acceptance in the field.

- The method has been rigorously tested.
- The method has been subject to peer review and publication.
- Positively state the known or potential rate of error in the method.
• If the trial court rules against admissibility of expert testimony, request a detailed statement of reasons and of the standard used. If the wrong standard was used, this can be grounds for appeal.

12.10.5 Federal Rules of Evidence

http://www.law.cornell.edu/rules/fre/overview.html#article%20vii

These rules govern proceedings in the courts of the United States. Their purpose is to *"secure fairness in administration, elimination of unjustifiable expense and delay, and promotion of growth and development of the law of evidence to the end that the truth may be ascertained and proceedings justly determined."* In 2000, a number of amendments went into effect. The changes are shown below:

> If scientific, technical, or other specialized knowledge will assist the trier of fact to understand the evidence or to determine a fact in issue, a witness qualified as an expert by knowledge, skill, experience, training, or education, may testify thereto in the form of an opinion or otherwise, if (1) the testimony is based upon sufficient facts or data, (2) the testimony is the product of reliable principles and methods, and (3) the witness has applied the principles and methods reliably to the facts of the case.

This revision is intended to limit the use but increase the reliability of party-initiated opinion testimony bearing on scientific and technical issues.

The use of expert testimony has greatly increased since enactment of the Federal Rules of Evidence. While much expert testimony now presented is illuminating and useful, much is not.

When the evidence involves highly technical matters and each side selects experts favorable to its position, the jury is unlikely to be capable of assessing the validity of dramatically opposing testimony.

While the admissibility of such evidence remains subject to the general principles of Rule 403, the revision requires that expert testimony be "reasonably reliable" and "substantially assist" the fact-finder. The court is directed to reject testimony that is based on premises lacking significant support and acceptance within the scientific community, or that is only marginally useful.

> The facts or data in the particular case upon which an expert bases an opinion or inference may be those perceived by or made known to the expert at or before the hearing. If of a type reasonably relied upon by experts in the particular field in forming opinions or inferences upon the subject, the facts or data need not be admissible in evidence in order for the opinion or inference to be admitted. Facts or data that are otherwise inadmissible shall not be disclosed to the jury by the proponent of the opinion or inference unless the court determines that their probative value in assisting the jury to evaluate the expert's opinion substantially outweighs their prejudicial effect.

Accident Investigation and Mitigation

> **For More Information on Admissibility of Expert Testimony**
>
> Using Field Methods—Experiences and Lessons: Defensibility of Field Data
> Barton P. Simmons, Chief, Hazardous Materials Laboratory
> Department of Toxic Substances Control
> http://www.hanford.gov/dqo/training/appendix/pdfs/UsingField Methods.pdf
> Kumho Tire and the Admissibility of Expert Testimony
> Kenneth M. Dennis, Bankruptcy Analyst
> U.S. Trustee Program—San Diego
> http://www.usdoj.gov/ust/eo/public_affairs/articles/docs/testimony-01.htm

Data upon which expert opinions are based may be derived from:

- Firsthand observation of the witness, with opinions based thereon traditionally allowed.
- Presentation at trial, such as a hypothetical question or having the expert attend the trial and hearing the testimony establishing the facts.
- Presentation of data to the expert outside of court and other than by his/her own perception. In this respect the rule is designed to broaden the basis for expert opinions beyond that current in many jurisdictions and to bring the judicial practice into line with the practice of the experts themselves when not in court.

Note that the rule requires that the facts or data *"be of a type reasonably relied upon by experts in the particular field."* The language does not warrant admitting in evidence the opinion of an expert as to the slip resistance of a fall accident based on statements of bystanders.

12.11 DOCUMENTATION

- Document everything. Documenting management controls offers several major advantages. It can provide valuable evidence that reasonable measures have been taken to exercise a prudent and reasonable degree of care. This can help stem weak lawsuits whenever documentation is requested during the discovery phase. It is important that established policies and procedures are observed. Fewer things hurt your case more than the claimant being able to successfully show that you failed to observe your own control procedures.
- Rules, policies, and procedures need to be provided in writing. Dates or version numbers should indicate when updates were made.

- The results of self-inspections need to be documented in the form of checklists or reports that include a means of documenting and initiating corrective action.
- Accident reports and investigations must be clear and detailed. In some cases, months or years can go by before these records are needed—long after the event.
- Other inspection procedures and the results of such inspections, including the spill and wet program, housekeeping, sweep logs, and maintenance/repair records need to be documented to provide evidence that procedures were followed and to provide an accounting of the conditions present at that snapshot in time.
- The results of slip resistance testing should become a routine part of inspection/maintenance and accident investigation records.
- Written specifications and results of staff training and orientation contribute to demonstrating that reasonable efforts are made to ensure that the staff is prepared to perform activities required for your safety program.

12.12 STAFF ISSUES

For effective management controls and accident response and investigation, the responsibility and authority for related activities must rest in the hands of qualified and adequately trained staff. Neglecting to provide the authority can mean unnecessary delays in the correction of a hazardous condition or response to an injured person. At the same time, staff needs to be adequately supervised. By failing to enforce safety policies and procedures, you run the risk that activities will not be performed properly (or at all), thus increasing the potential for lingering hazardous conditions, injury, and liability.

A periodic, independent review of policies and procedures, implementation, and the results of your loss prevention program can help identify improvements that you might not otherwise have identified.

For an extensive discussion of legal issues related to slip and fall accidents, see Charles E. Turnbow's, *Slip and Fall Practice*.

EXHIBIT 12.1
Accident Reporting and Investigation Guide for Slips, Trips, and Falls

Date of Incident:	Time of Incident:	Inside or Outside:
Location/Address:		
Claimant Information		
❑ Name of Claimant, Address, Telephone Number		
❑ For what purpose was the claimant on the premises?		
Witness Information		
❑ Name of Witness, Telephone Number, Address		
❑ How close was witness to claimant?		
❑ What position (right, left, front, behind)?		
❑ Obtain witness statement (what did they observe?)		
Occurrence Information*		
❑ Was the fall a slip or trip?		
❑ Which foot slipped?		
❑ Was fall backward, forward, sideways?		
Describe injury and body parts (knee, ankle, shoulder, upper arm, forearm, hand, lower back, mid-back, cervical, head)		
❑ Specify type of footwear worn by claimant*		
❑ Sole material, heel size and material, condition of sole*		
❑ Does claimant normally wear glasses?*		
❑ Specify type of glasses (bifocal, trifocal, normal)*		
❑ Were glasses worn at the time of the incident?*		
❑ Specify object, size, weight, and positioning of object if claimant was carrying anything.*		
❑ Where was object after the fall?*		
❑ Does claimant normally take any medication or drugs?		
❑ Any medication taken prior to incident?		
❑ Time interval between last dose and slip/fall incident?		
Location Design and Condition Information*		
❑ Specific area where incident occurred		
❑ Describe type of floor surface		
❑ Describe type of floor treatment/brand.		
❑ Foreign substances or water on the floor/stair surface?*		
❑ If so, obtain a sample.		
❑ If known, identify/describe substance.		
❑ Any snow/ice accumulations (if so, describe)?*		
❑ Are any cracks, bulges, potholes, protrusions, or irregularity 1/2" or more present? Describe.*		
❑ If applicable, are changes in levels clearly marked?*		
Stairs (if present)		
❑ Was the claimant ascending or descending the stairs?		
❑ Which step or steps were involved?		

Continued

EXHIBIT 12.1 (*Continued*)
Accident Reporting and Investigation Guide for Slips, Trips, and Falls

❑ Describe tread surface (slip resistant?)*	
❑ Specify width of stairs, riser heights, tread depths, and condition of stairs.*	
Handrails (if present)*	
❑ Are handrails on one or both sides of the stairs?	
❑ Do they begin and end beyond the top and bottom steps?	
❑ Specify shape and diameter	
❑ Specify distance from nose of treads to handrail (at top and bottom steps)	
❑ Was handrail loose at the time of or before the incident?	
❑ Were handrails being used by the claimant before/during/after the fall?	
Landing (if present)	
❑ Specify dimensions	
❑ Specify rise and run between adjacent landings	
❑ Does door swing out into landing? (Specify dimensions)	
Lighting Conditions	
❑ Was claimant moving from a dimly lit area to a well lit area, or vice-versa?	
❑ Was the light source behind or in front of the claimant?	
❑ Specify illumination levels under same conditions	
❑ Are lighting controls manual or automatic?	

* Photograph

Information on Individual Completing Report

Name:	Title:	Telephone Number:
Address:	Date:	

© ESIS, Inc.–Global Risk Control Services. All rights reserved.

EXHIBIT 12.2
Accident Investigation Guide

Ramp Fall Accidents

Tread Surface

Composition	☐ Concrete ☐ Asphalt ☐ Carpet
	☐ Other (specify):
Nonslip	☐ No ☐ Brushed concrete ☐ Friction strips
	☐ Other (specify):
Cross-cleated or grooved	☐ Cross cleated ☐ Grooved
Results of slip resistance testing	
Width of ramp tread surface (edge to edge)	

Slope of Ramp

Inches (vertical)	Inches (horizontal)
Feet (vertical)	Feet (horizontal)

Use of Ramp

Check All that Apply	☐ Pedestrian ☐ Handicapped ☐ Vehicle
Ramp notice, sign, or warning (describe):	

Handrails

Location	☐ One side ☐ Both sides
Orientation of claimant to handrails	☐ On left ☐ On right ☐ N/A
Was claimant using handrail?	☐ Yes ☐ No
Diameter of handrails	
Vertical distance between handrail and ramp surface	
Indication handrails were loose	☐ Yes ☐ No If yes, explain:

Landing Dimensions (Length and Width)

Top:	
Intermediate:	
Bottom:	
Does door open onto landing or ramp?	☐ Yes ☐ No If yes, specify distance:

Type of Fall

Direction:	☐ Backwards ☐ Forwards To side: ☐ Left ☐ Right
Portion of foot that slipped or tripped	☐ Toe ☐ Heel
	☐ Right foot ☐ Left foot

Location of Fall

Where did slip or trip begin?	
Where did person end up?	
What was their body position?	

Continued

EXHIBIT 12.2 (*Continued*)
Accident Investigation Guide

Type of Injury

❏ Hip	❏ Arm (upper)	❏ Mid-back
❏ Knee	❏ Forearm	❏ Cervical (neck)
❏ Ankle	❏ Hand	❏ Head
❏ Shoulder	❏ Lower back	

Eyewear

Does claimant normally wear glasses?	❏ Yes ❏ No
Type of glasses	❏ Bifocal ❏ Trifocal ❏ Normal
	❏ Other (specify):
Were glasses being worn at time?	❏ Yes ❏ No

Footwear

Type	
Sole material	
Heel size and material	
Available for inspection?	❏ Yes ❏ No
Have they been worn since the accident?	❏ Yes ❏ No If yes, explain:
When were they purchased?	
Did they fall off during the accident?	❏ Yes ❏ No If yes, explain:

Lighting

Walking from dimly lit area to well lit area?	❏ Yes ❏ No
Windows nearby the fall	❏ Yes ❏ No
	If yes, specify location in relation to pedestrian path:
Was the light behind or in front of the claimant?	❏ Behind ❏ In front ❏ Balanced
Illumination levels	
Lighting controls manual or automatic?	❏ Manual ❏ Automatic If automatic, describe:

Witnesses

How close were they?	
How many?	
What was their orientation to the accident scene?	❏ Left ❏ Right ❏ In front ❏ Behind

Load Carrying

Method of carrying	❏ Left arm ❏ Right arm ❏ Two arms
Dimensions and weight of object	
Location of object after fall	

Weather

Describe	

EXHIBIT 12.3
Accident Investigation Guide

Slippery Surfaces Accidents

Portion of Foot That Slipped or Tripped

❏ Toe ❏ Heel ❏ Right foot ❏ Left foot

Location of Fall

Where did slip or trip begin?	
Where did person end up?	
What was their body position?	

Type of Flooring

Did claimant walk off a different type of surface?	❏ Yes ❏ No
If yes, specify type and number of steps:	
Type of flooring involved in fall	❏ Resilient ❏ Ceramic ❏ Marble ❏ Terrazzo ❏ Concrete ❏ Asphalt ❏ Carpeting ❏ Other (specify):

Floor Finish

Type	❏ Wax ❏ Sealer ❏ Slip-resistant treatment (specify): ❏ Other (specify):
Sample available for testing	❏ Yes ❏ No
Product test results (if any)	
Manufacturer specifications (use and COF)	
Nonslip additive used	❏ Yes ❏ No If yes, specify:
Floor care:	
• Materials and mixtures	
• Frequency of care	
• Procedures (cleaning, finishing)	
• Date last refinished prior to accident	

Foreign Substances

Did claimant or witness see wetness on shoe sole after fall?	❏ Yes ❏ No If yes, explain:
Did claimant or witness see substance on floor before or after the fall?	❏ Yes ❏ No If yes, explain:
Were claimant's clothes moist, wet, or soiled after fall?	❏ Yes ❏ No If yes, explain:

Floor Mats (if any)

Type and design	
Condition	

Type of Fall

Direction	❏ Backwards ❏ Forwards To side: ❏ Left ❏ Right

Continued

EXHIBIT 12.3 (*Continued*)
Accident Investigation Guide

Type of Injury

❏ Hip	❏ Arm (upper)	❏ Mid-back
❏ Knee	❏ Forearm	❏ Cervical (neck)
❏ Ankle	❏ Hand	❏ Head
❏ Shoulder	❏ Lower back	

Footwear

Type	
Sole material	
Heel size and material	
Available for inspection?	❏ Yes ❏ No
Have they been worn since the accident?	❏ Yes ❏ No If yes, explain:
When were they purchased?	
Did they fall off during the accident?	❏ Yes ❏ No If yes, explain:

Eyewear

Does claimant normally wear glasses?	❏ Yes ❏ No
Type of glasses	❏ Bifocal ❏ Trifocal ❏ Normal ❏ Other (specify):
Were glasses being worn at time?	❏ Yes ❏ No

Lighting

Walking from dimly lit area to well lit area?	❏ Yes ❏ No
Windows nearby the fall	❏ Yes ❏ No If yes, specify location in relation to pedestrian path:
Was the light behind or in front of the claimant?	❏ Behind ❏ In front ❏ Balanced
Illumination levels	
Lighting controls manual or automatic?	❏ Manual ❏ Automatic If automatic, describe:

Witnesses

How close were they?	
How many?	
What was their orientation to the accident scene?	❏ Left ❏ Right ❏ In front ❏ Behind

Load Carrying

Method of carrying	❏ Left arm ❏ Right arm ❏ Two arms
Dimensions and weight of object	
Location of object after fall	

Weather

Describe	

EXHIBIT 12.4
Accident Investigation Guide

Stairway Fall Accidents

Tread Surface

Composition	❑ Concrete ❑ Wood ❑ Carpet ❑ Other (specify):
Results of slip resistance testing	

Stair Dimensions

Stair width	
Riser height	Top to bottom:
	Side to side:
Tread overhang	
Tread depth (top to bottom)	
Nosing features	❑ Friction strips ❑ Lighting ❑ Other (specify):

Handrails

Location	❑ One side ❑ Both sides
Do they extend beyond the top and bottom steps?	❑ Yes ❑ No If yes, explain:
Distance handrail extends above nose of treads at top and bottom	
Shape and diameter of handrails	
Handrail construction	❑ Metal ❑ Wood ❑ Other (specify):
Indication handrails were loose	❑ Yes ❑ No If yes, explain:
Was claimant using handrail?	❑ Yes ❑ No

Landings

Dimensions	Length: Width:
Rise and run between adjacent landings	
Does door open onto landing or stairs?	❑ Yes ❑ No If yes, specify distance and door dimension:
Carpet or mats present	❑ Yes ❑ No If yes, describe:

Type of Fall

Slip or trip	❑ Slip ❑ Trip
Direction	❑ Backwards ❑ Forwards To side: ❑ Left ❑ Right
Portion of foot that slipped or tripped	❑ Toe ❑ Heel ❑ Right foot ❑ Left foot

Type of Injury

❑ Hip	❑ Arm (upper)	❑ Mid-back
❑ Knee	❑ Forearm	❑ Cervical (neck)
❑ Ankle	❑ Hand	❑ Head
❑ Shoulder	❑ Lower back	

Continued

EXHIBIT 12.4 (*Continued*)
Accident Investigation Guide

Footwear

Type	
Sole material	
Heel size and material	
Available for inspection?	❑ Yes ❑ No
Have they been worn since the accident?	❑ Yes ❑ No If yes, explain:
When were they purchased?	
Did they fall off during the accident?	❑ Yes ❑ No If yes, explain:

Eyewear

Does claimant normally wear glasses?	❑ Yes ❑ No
Type of glasses	❑ Bifocal ❑ Trifocal ❑ Normal ❑ Other (specify):
Were glasses being worn at time?	❑ Yes ❑ No

Lighting

Walking from dimly lit area to well lit area?	❑ Yes ❑ No
Windows nearby the fall?	❑ Yes ❑ No If yes, specify location in relation to pedestrian path:
Was the light behind or in front of the claimant?	❑ Behind ❑ In front ❑ Balanced
Illumination levels	
Lighting controls manual or automatic?	❑ Manual ❑ Automatic If automatic, describe:

Witnesses

How close were they?	
How many?	
What was their orientation to the accident scene?	❑ Left ❑ Right ❑ In front ❑ Behind

Load Carrying

Method of carrying	❑ Left arm ❑ Right arm ❑ Two arms
Dimensions and weight of object	
Location of object after fall	

Weather

Describe	

EXHIBIT 12.5
Example Claims Investigation: Parking Garage

DESCRIPTION

This accessible ramp is located on the lower level of a parking garage adjacent to a bank of elevators and across from handicapped parking. The entire landing, ramp, and curb area is of concrete construction, leading to an asphalt vehicle parking area.

THEORY(IES) OF LIABILITY

The claimant has provided three different versions of the event. Information material to each was evaluated. All three were reported by the claimant the day after the event:

- A report was completed by facility security in which the claimant indicated that *"After exiting the elevator, the claimant walked towards the sloping handicapped concrete ramp and slid on ramp and fell. . . ."* This report also documents the security camera images which did show the four individuals in the group, but did not capture the event. The claimant's husband indicated that the claimant *"did not walk in the center of the inclined concrete ramp but approached from the left side which has a steep downward slope. She slid down from there."*
- A police report was filed indicating that *"After exiting the elevator [the claimant] walked toward the sloping handicapped concrete ramp and slid down. . . ."*
- Information from the claim file indicated that a witness stated the claimant might have been on the center of the decline or the edge slope when she slipped. The witness was convinced that the painted area was the cause of the fall because it was *"damp and slippery."*

Based on these descriptions, it appears that the curb paint was not involved in the event, but rather the issues are (1) the slope of the ramp flares and (2) the slip resistance of the enamel.

Ramp

The concrete on the ramp is grooved and painted with battleship gray paint which appears to have sand or grit mixed into it. The slope is 1:10.6, steeper than the 1:12 specified by the ADA. The overall condition of the ramp was satisfactory, with no material cracks, protrusions, or holes noted.

No additional information was available on the gray paint. The parking garage was built by a developer about 5 years ago and then purchased by the city. This parking company manages the operation on behalf of the city. The gray paint was applied by the builder when it was constructed, and it has not been changed by the parking company.

Note: Slip resistance readings in the downhill direction (as indicated by the claim information) indicate dry readings averaging 0.60 (0.65 minus a correction factor of 0.054 for the slope) and wet readings of 0.65 (0.71 minus the same correction factor). These results are consistent with generally accepted standards of slip resistance of 0.50 or higher.

Ramp Flares

The flares of the ramp are painted the same way with the same grooves as the ramp itself. The slope of these flares is 1:3.5, significantly steeper than the 1:10 specified by the ADA. There is also a set of outer flares with a rough brush texture, painted with the same gray paint. The overall condition of ramp flares was good.

Curb

The curb is 8 in. high and painted with industrial grade (Industrial Maintenance Coating) "Safety Yellow" paint.

The property was in compliance with city codes.

The overall condition of the curb was adequate, although there were small portions of paint missing and some chipping of concrete. These do not appear material to the claim and are consistent with the condition expected for this type of facility.

The paint is urethane alkyd, an enamel. No grit is applied to this paint. The label included no information on coefficient of friction or slip resistance, but refers to the reader to the MSDS and Product Sheet on the product. Although recommended for floors (among many other items), a limitation listed is "not for high abuse floor areas."

Note: In terms of slip resistance, averaged test results for the curbing indicated dry readings of 0.69 and wet readings of 0.61, exceeding the generally accepted standard of 0.50.

Lighting

Using a digital light meter, readings were taken at the ramp area. Because this is located at the lowest level and there are no windows or sources for external lighting, it can be expected to be representative of lighting 24/7, especially as this parking facility services not only a large mall, but also a hotel. Readings indicated 30 foot-candles (fc) at the curb, and 40 foot-candles at eye level. These exceed the minimum requirements for lighting set by the Illuminating Engineers Society of North America (IESNA), which specifies a minimum of 5 foot-candles for enclosed garages for safety purposes.

Marking

Marking of the ramp area meets ADA specifications. The curb is clearly marked with safety yellow paint; the white crosswalk markings delineate the end of the ramp at the asphalt parking lot area.

Supplemental Information

- Contact indicated that all maintenance is documented with maintenance sheets, which may be of some use in establishing reasonable and prudent actions on the part of the parking company.
- According to the security report, the claimant was wearing leather-soled stacked-heel shoes. This type of footwear, both from the standpoint of instability and the lack of traction afforded by the leather sole, is likely to be a contributor to any slip that may have occurred.
- In the last week, this facility saw approximately 15,000 members of the public. At peak time, that number reaches 30,000 a week.
- There is extensive security camera monitoring (and taping) throughout the facility. While the claimant and her party were captured on tape, the event was not.

Note: The slip resistance tester (or tribometer) used is the English XL, operated in accordance with the manufacturers' instructions and current recommendations pertaining to its use by the ASTM International Standard F1679, Standard Test Method for Using a Variable Incidence Tribometer (VIT).

Conclusions/Additional Items to Investigate

- **Footwear**. It would be helpful to obtain more details on the type and condition of the footwear worn, most especially the condition of the sole and heel bottoms.
- **Clarify version of events**. There are no apparent issues with the curb. The ramp itself appears adequate from a slip resistance standpoint, but does not quite meet the ADA on the slope ratio. The flares appear fine from a slip resistance standpoint, but are about three times the maximum slope per the ADA.
- **Indemnification**. If the version of events involves the design of the ramp or flares (or the gray paint), this would be inherent to the original construction of the facility. Unless the parking company is indemnifying the city, it would seem that the city and/or the developer and builder/contractor would be most culpable for the event.

Accident Investigation and Mitigation

ATTACHMENTS

- ADA Analysis
- Slip Resistance Test Results Report
- Curb Paint Material Safety Data Sheet
- Curb Paint Technical Information Product Data Sheet

ADA ANALYSIS

4.7 Curb Ramps

4.7.1 Location. Curb ramps complying with 4.7 shall be provided wherever an accessible route crosses a curb.	OK
4.7.2 Slope. Slopes of curb ramps shall comply with 4.8.2. The slope shall be measured as shown below. Transitions from ramps to walks, gutters, or streets shall be flush and free of abrupt changes. Maximum slopes of adjoining gutters, road surface immediately adjacent to the curb ramp, or accessible route shall not exceed 1:20.	

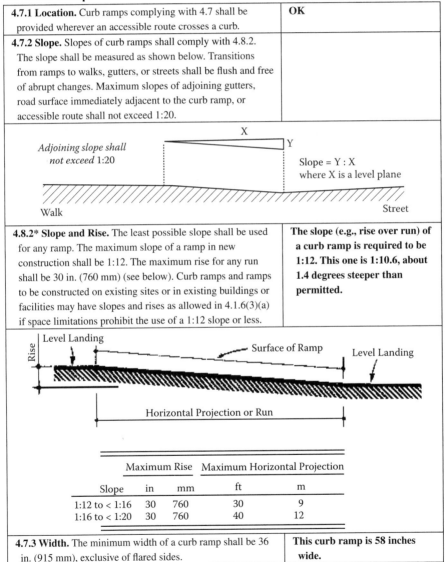

4.8.2* Slope and Rise. The least possible slope shall be used for any ramp. The maximum slope of a ramp in new construction shall be 1:12. The maximum rise for any run shall be 30 in. (760 mm) (see below). Curb ramps and ramps to be constructed on existing sites or in existing buildings or facilities may have slopes and rises as allowed in 4.1.6(3)(a) if space limitations prohibit the use of a 1:12 slope or less.	The slope (e.g., rise over run) of a curb ramp is required to be 1:12. This one is 1:10.6, about 1.4 degrees steeper than permitted.

	Maximum Rise		Maximum Horizontal Projection	
Slope	in	mm	ft	m
1:12 to < 1:16	30	760	30	9
1:16 to < 1:20	30	760	40	12

4.7.3 Width. The minimum width of a curb ramp shall be 36 in. (915 mm), exclusive of flared sides.	This curb ramp is 58 inches wide.

4.7.4 Surface. Surfaces of curb ramps shall comply with 4.5 (see below).	
4.5 Ground and Floor Surfaces.	
4.5.1* General. Ground and floor surfaces along accessible routes and in accessible rooms and spaces including floors, walks, ramps, stairs, and curb ramps, shall be stable, firm, slip-resistant, and shall comply with 4.5.	
4.5.2 Changes in Level. Changes in level up to ¼ in. (6 mm) may be vertical and without edge treatment. Changes in level between 1/4 in and 1/2 in. (6 mm and 13 mm) shall be beveled with a slope no greater than 1:2. Changes in level greater than 1/2 in. (13 mm) shall be accomplished by means of a ramp that complies with 4.7 or 4.8.	Slip resistance on the ramp is 0.60 dry and 0.65 wet. The threshold of safety for slip resistance is generally recognized as 0.50, so these results exceed that threshold.
4.7.5 Sides of Curb Ramps. If a curb ramp is located where pedestrians must walk across the ramp, or where it is not protected by handrails or guardrails, it shall have flared sides; the maximum slope of the flare shall be 1:10 (see below). Curb ramps with returned curbs may be used where pedestrians would not normally walk across the ramp.	The flares of curb ramps are required to be no greater than 1:10. The flares of this curb ramp are 1:3.5, significantly steeper than permitted.
4.7.6 Built-up Curb Ramps. Built-up curb ramps shall be located so that they do not project into vehicular traffic lanes.	N/A
4.7.7 Detectable Warnings. A curb ramp shall have a detectable warning complying with 4.29.2. The detectable warning shall extend the full width and depth of the curb ramp.	OK
4.7.8 Obstructions. Curb ramps shall be located or protected to prevent their obstruction by parked vehicles.	OK
4.7.9 Location at Marked Crossings. Curb ramps at marked crossings shall be wholly contained within the markings, excluding any flared sides.	N/A

4.7.10 Diagonal Curb Ramps. If diagonal (or corner type) curb ramps have returned curbs or other well-defined edges, such edges shall be parallel to the direction of pedestrian flow. The bottom of diagonal curb ramps shall have 48 in. (1220 mm) minimum clear space. If diagonal curb ramps are provided at marked crossings, the 48 in. (1220 mm) clear space shall be within the markings. If diagonal curb ramps have flared sides, they shall also have at least a 24 in. (610 mm) long segment of straight curb located on each side of the curb ramp and within the marked crossing.	N/A

Bibliography

Access Board. Technical Bulletin: Ground and Floor Surfaces, United States Access Board, August 2003. http://www.access-board.gov/Adaag/about/bulletins/surfaces.htm.

Access Board. Update of ADA and ABA standards. October 2007, http://www.access-board.gov/ada-aba/standards-update.htm#ada(accessed April 11, 2008).

Adams, N. Slips and Falls—Some Arguments about Measuring Coefficients of Friction, Productivity Ergonomics and Safety—The Total Package Conference, Gold Coast, 1997, http://www.publicliability.net.au/slip_resistance_testing.htm#SLIPS_AND_FALLS_ñ_SOME_ARGUMENTS_ABOUT_MEASURING_COEFFICIENTSOF_FRICTION.

Adler, S.C., and R. Brungraber. Technical Support for a Slip-Resistance Standard, ASTM STP 649, American Society for Testing Materials Standards, West Conshohocken, PA, 1978.

Adler, S.C., and B.C. Pierman. A History of Walkway Slip-Resistance Research at the National Bureau of Standards, National Bureau of Standards, Gaithersburg, MD, 1979.

Alden Chemical. Why All the Confusion over Surface Types? http://www.aldonchem.com/popup-explain-surfacing-confusion.htm.

American Apparel and Footwear Association (AAFA). Shoe Stats 2005, 2006a.

American Apparel and Footwear Association (AAFA). Trends—An Annual Compilation of Statistical Information on the U.S. Apparel and Footwear Industries, 2006b.

American Association of State Highway Transportation Officials. A Policy on Geometric Design of Highways and Streets, 4th ed. AASHTO, 2001.

American Geriatrics Society. Guideline for the Prevention of Falls in Older Persons, *The Journal of the American Geriatrics Society*, 49:664–672, 2001.

Americans with Disabilities Act (ADA), A4.5 Ground and Floor Surfaces/A4.5.1 General, July 26, 1991, http://www.usdoj.gov/crt/ada/adahom1.htm.

American National Standards Institute (ANSI), http://www.ansi.org.

ANSI Electronic Standards Store, http://webstore.ansi.org/ansidocstore/default.asp.

ANSI Procedures for the Development and Coordination of American National Standards, ANSI, March 2001.

ANSI A117.1 Standard on Accessible and Useable Buildings and Facilities, 2003.

ANSI A1264.2, Standard for the Provision of Slip Resistance on Walking/Working Surfaces, 2001.

ANSI Z41-1999, Personal Protection—Protective Footwear, 1999.

ANSI Z535.1, Safety Color Coding, 2006.

American Society of Mechanical Engineers (ASME), http://www.asme.org.

American Society for Testing Materials Standards (ASTM), http://www.astm.org.

ASTM C21.06, Meeting Minutes, Ceramics, Slip ResistanceSubcommittee, Orlando FL, May 3, 2005.

ASTM, Interlaboratory Test Data for F609, Test for Static Slip Resistance of Footwear Sole, Heel, or Related Materials by Horizontal Pull Slipmeter, ASTM Research Report, West Conshohocken, PA, 1979.

ASTM C1028, Standard Test Method for Determining the Static Coefficient of Friction of Ceramic Tile and Other Like Surfaces by the Horizontal Dynamometer Pull-Meter Method, 2007.

ASTM D2047, Standard Test Method for Static Coefficient of Friction of Polish-Coated Floor Surfaces as Measured by the James Machine, 2004.

ASTM D21, Gray Pages, Book of Standards (V 15.04).

ASTM D3758 Standard Practice for Evaluation of Spray-Buff Products on Test Floors, 2003.

ASTM D4103-90, *Standard Practice for Preparation of Substrate Surfaces for Coefficient of Friction Testing*, 1995.
ASTM D4518-91, *Standard Test Method for Measuring Static Friction of Coating Surfaces*, 1991.
ASTM D5859, Standard Test Method for Determining the Traction of Footwear on Painted Surfaces Using the Variable Incidence Tester, 2000.
ASTM D6205 Standard Practice for Calibration of the James Static Coefficient of Friction Machine, 1998.
ASTM E303, Method for Measuring Surface Frictional Properties using the British Pendulum Tester, 2008.
ASTM F-13 Pedestrian/Walkway Safety and Footwear, 2003.
ASTM F446 Standard Consumer Safety Specification for Grab Bars and Accessories, 1985.
ASTM F462 Standard Consumer Safety Specification for Slip-Resistant Bathing Facilities, 1979.
ASTM F489, Standard Test Method for Using a James Machine, 1996.
ASTM F609, Standard Test Method for Using a Horizontal Pull Slipmeter (HPS), 2005.
ASTM F695 Standard Practice for the Evaluation of Test Data Obtained by Using the Horizontal Pull Slipmeter (HPS) or the James Machine for Measurement of Static Slip Resistance of Footwear Sole, Heel, or Related Materials, 2001.
ASTM F802 Standard Guide for Selection of Certain Walkway Surfaces When Considering Footwear Traction, 2003.
ASTM F1240 Guide for Categorizing Results of Footwear Slip Resistance Measurements on Walkway Surfaces with an Interface of Various Foreign Substances, 2001.
ASTM F1637, Practice for Safe Walking Surfaces, Section 7 (Speed Bumps) and Section 4.2 (Walkway Changes in Level), 1995.
ASTM F1646 Terminology Relating to Safety and Traction for Footwear, 2005.
ASTM F1677, Standard Test Method for Using a Portable Inclineable Articulated Strut Slip Tester (PIAST), 1996.
ASTM F1678 Standard Test Method for Using a Portable Articulated Strut Slip Tester (PAST), 1998.
ASTM F1679 Standard Test Method for Using a Variable Incidence Tribometer (VIT), 2004.
ASTM F1694 Standard Guide for Composing Walkway Surface Evaluation and Incident Report Forms for Slips, Stumbles, Trips and Falls, 1996.
ASTM F2048 Standard Practice for Reporting Slip Resistance Test Results, 2000.
ASTM F2333-04 Standard Test Method for Traction Characteristics of the Athletic Shoe-Sports Surface Interface, 2004.
ASTM F2412-05 Standard Test Methods for Foot Protection, 2005.
ASTM F2413-05 Standard Specification for Performance Requirements for Foot Protection, 2005.
Americans with Disabilities Act (ADA), ADA Accessibility Guidelines for Buildings and Facilities (ADAAG), A4.5 Ground and Floor Surfaces/A4.5.1 General, July 26, 1991, http://www.usdoj.gov/crt/ada/adahom1.htm.
Andres, R.O., and D.B. Chaffin. Ergonomic analysis of slip-resistance measurement devices, *Ergonomics*, 28(7):1065–1079, 1985.
Archea, J., B. Collins, and F. Stahl. Guidelines for stair safety. NBS building science series 120. Washington, DC: National Bureau of Standards, 1979.
Architectural and Transportation Barriers Compliance Board (ATBCB), ATBCB Bulletin Arizona Polymer Flooring, Inc., Advice as to How to Select Surface Particles to Enhance Slip Resistance, http://www.apfepoxy.com/includes/APFslipresistance.pdf, 2002 (accessed February 29, 2008).

Bibliography

The Association of Pool and Spa Professionals (APSP). The Sensible Way to Enjoy Your Aboveground/Onground Swimming Pool, http://www.apsp.org/clientresources/documents/apsp_aboveground_v2.pdf (accessed February 12, 2008).

Barnett, K. Reducing Patient Falls Project, January 2001–March 2002, Mid Yorkshire Hospitals NHS Trust, http://www.premierinc.com/safety/topics/falls/downloads/E-14-falls-project-uk.pdf, 2002 (accessed April 21, 2008).

Bell, J.L., et al. An Evaluation of a Comprehensive Slip, Trip, and Fall (STF) Prevention Program for Hospital Workers, *Proceedings from the International Conference on Slips, Trips, and Falls: From Research to Practice,* Hopkinton, MA: IEA Press, 2007.

Berufsgenossenschaftliches Institut fur Arbeitssicherheit (BIA), http://www.hvbg.de/d/bia/starte.htm.

Bowman, R. The impact of new standards on the specification of tiling systems and products, Commonwealth Scientific & Industrial Research Organisation (CSIRO), Australia, http://www.infotile.com/services/techpapers/95tilex.shtml, 1995.

———. Learning how best to live with multiple slip resistance test results, CSIRO, Australia, http://www.dbce.csiro.au/vb2/345/61/index.htm, revised June 1999.

———. Legal and practical aspects of problems arising from slippery floors, CSIRO, Australia.

———. Should we recognize foreign slip resistance test results? *Tile Today,* August 2001.

———. Slip Resistance and social responsibility, Issue 45, http://tiletoday.com.au, 2005.

———. Slipping to your death: why it's all too easy in Australia and what needs to be done, CSIRO, Australia, April 2000.

———. The Tortus floor friction tester—round robin tests, in Ceramics, Adding the Value: AUSTCERAM 92, M.J. Bannister, Ed., CSIRO Publications, Australia, 1992.

———. Understanding the new slip resistance standards and its implications.

Bowman, R, C. Strautins, M. D. Do, D. Devenish, and G. Quick. comparison of standard footwear for the oil wet ramp slip resistance test, *Contemporary Ergonomics* 2004, 1(2) April 2004, 33–37.

Bowden, F.P., and L.L. Leben. The nature of sliding and the analysis of friction, *Proc. Royal Soc. London,* 169:371–391, 1939.

Boyles, H., ASTM Committee D-21 Position Paper, ASTM Committee D-21, December 7, 1993.

Braun, R. and D. Roemer. Influence of waxes on static and dynamic friction, *Soap, Cosmetics, Chemical Specialties,* December 1974, 60–72 (even).

Braun, R., and R.J. Brungraber. A comparison of two slip-resistance testers, ASTM STP 649, 1978.

Breakey, P., Delhi teen, 16, killed in fall, Delhi News Bureau, http://www.thedailystar.com/local/local_story_004040041.html, 2008.

British Portable Skid Tester (BPST), http://www.munro-group.co.uk/skid.htm.

British Standards Institute (BSI), http://www.bsi-global.com/group.xalter.

BS 5395, Part 1 Stairs, Ladders & Walkways. Code of Practice for the Design, Construction and Maintenance of Straight Stairs and Winders, 2000.

BS 7976-1 Pendulum Testers—The Pendulum Tester Part 1: Specification, 2002.

BS EN 14231 Natural Stone (CEN standard PrEN 14231), 2003.

BS 99/109274 DC Ceramic Tiles and Other Rigid Tiling (CEN standard PrEN 13552 1999), 1999.

BS 00/120903 DC (Resilient Floor Coverings) (CEN standard PrEN 13845 2000).

BS 00/121797 DC (Resilient Floor Coverings) (CEN standard PrEN 13893 2000).

Brungraber, R.J., National Bureau of Standards Technical Note 953, A New Portable Tester for the Evaluation of the Slip-Resistance of Walkway Surfaces, July 1977.

———. Personal communication to Henry Boyles, December 19, 1988.

Brungraber, R.J. and J. Templer. Controlled slip resistance, Progressive Maintenance, March 1991, 112–116.

Brungraber, R.J., et al. Tech news—Bucknell University F-13 workshop to evaluate various slip resistance measuring devices, *Standardization News,* May 1992, 21–24.

Buczek, F.L. and S.A. Banks. High-resolution force plate analysis of utilized slip resistance in human walking, *J. Testing Evaluation,* 24(6), November 1996, 353–358.

Bunte, L., Traffic calming programs and emergency response: a competition of two public goods, 2000, http://www.digitalthreads.com/utarpt/utarpt5.html.

Bureau de normalization du Quebec (BNQ), http://www.criq.qc.ca/bnq.

Canadian General Standards Board (CGSB), http://w3.pwgsc.gc.ca/cgsb.

Canadian Standards Association (CSA) or CSA International, http://www.csa.ca.

    CAN2-75.1-M77, Slip Resistance Standard for Wood Products Commonly Used for Decks (Wet and Dry).

    25.1 NO.30.1-95-CAN/CCSB, Methods of Sampling and Testing Waxes and Polishes Slip Resistance.

Carpet and Rug Institute (CRI). Commercial Carpet Maintenance Manual, 1987.

_____. Floor Covering Maintenance for School Facilities, http://www.carpet-maint.com.

Castle Rock Industries. Outsourcing Floor Care Equipment Apart from Labor, https://uclean.securerain.com/resource/training/outsource_floor_care.html(accessed August 27, 2007.

Centers for Disease Control (CDC). The costs of fall injuries among older adults, CDC, Washington, DC, January 2000.

_____. Help Seniors Live Better, Longer: Prevent Brain Injury, 2008, http://www.cdc.gov/BrainInjuryInSeniors/ (accessed June 30, 2008).

Chandler, H., Foot Protection, Beyond Safety Boots, OSH Canada, 2002.

_____. Preventing Slip and Fall Injuries and Liability, *Cleaning Business Mag.,* December 1998.

Chang, W.R., Preferred Surface Microscopic Geometric Features on Floors as Potential Interventions for Slip and Fall Accidents, *Journal of Safety Research,* 35(1):71–79, 2004.

Chang, W.R. and T.B. Leamon, The effect of surface roughness on measurement of slip resistance, *Experience to Innovation, Volume 3,* Proceedings of the 13th Triennial Congress of the International Ergonomics Association, Finnish Institute of Occupational Health, Helsinki, Finland, 1997, 365–367.

Chang, W.R., M.F. Lesch, and C.C. Chang, The effect of contact area on the friction measured, *Journal of Testing & Evaluation,* 24(6), 377–385, 2007.

Chemical Specialties Manufacturers Association (CSPA), Comparative Determination Slip Resistance of Floor Polishes, CSMA Bulletin No. 245–70, CSMA, Washington, D.C., 1970.

Comite Europeen de Normalisation (CEN)

    CEN/TC 67, PrEN 13552 (Draft) Ceramic Tiles—Determination of Coefficient of Friction.

    CEN/TC 134, PrEN 13893—(Draft) Resilient, Laminate and Textile Floor Coverings—Parameters for the Measurement of Dynamic Coefficient of Friction on Floor Surface)

    CEN/TC 246, PrEN 14231—(Draft) Natural Stones Test Methods—Determination of Slip Resistance by Means of the Pendulum Tester.

    PrEN 13845—(Draft) Resilient Floor Coverings—Polyvinyl Chloride Floor Coverings with Enhanced Slip Resistance—Specification

Commercial Building Entrance Care, Crown Mats and Matting and Mat Tech http://www.mat-tech.ca/eng/products/entrance/index.html (accessed January 8, 2007).

Commonwealth Scientific and Industrial Research Organisation (CSIRO), http://www.dbce.csiro.au.

Consumer Product Safety Commission (CPSC), CPSC Product Hazard Index, Release 73–032, Washington, DC, 1973.

Bibliography

CPSC: 11,000 Escalator-Related Injuries in 2007, *Occupational Health and Safety,* May 13, 2008, http://ohsonline.com/Articles/2008/05/CPSC-11000-EscalatorRelated-Injuries-in-2007.aspx

Cramp, P., and L. Masters. Preliminary Study of the Slipperiness of Flooring, National Bureau of Standards, 1974.

Coulomb, C., Theory of simple archives, memoire de L'academie Royale Services, Vol. 10.

Courtney, Theodore K., et al. Factors influencing restaurant worker perception of floor slipperiness, *Journal of Occupational and Environmental Hygiene,* 3:592–598, 2006.

Curry, D.G., R. Reinke, A. Shah, and J. Kidd. The effect of marking paint on walking surface slip resistance, International Conference on Slips, Trips, and Falls: From Research to Practice, 2007.

Davies, J.C., et al. Bifocal/varifocal spectacles, lighting and missed-step accidents, *Safety Science,* 38:211–226, 2001.

Di Pilla, S. If you want to reduce slips and falls...forget the program, *Food Management Magazine,* June 2006.

_____. Managing traffic to prevent slips and falls, *Food Management Magazine,* November 2006.

_____. Slip-resistant footwear the program doesn't end after the purchase, *Food Management Magazine,* September 2006.

Douglas, Bill, Choosing the best slip-resistant shoe, *Occupational Health and Safety,* April 2004.

Dravitzki, V.K., and S.M. Potter. The use of the Tortus and the pendulum with the 4S rubber for the assessment of slip resistance in the laboratory and the field, J. Testing Evaluation, 25(1):127–134, 1997.

Dunne, J.R., Letter to Judge Ernest F. Hollings, January 19, 1993, http://www.usdoj.gov/crt/foia/tal257.txt.

Derieux, J.B., The coefficient of friction of rubber, *J. Elisha Mitchel Scientific Society,* 50:53–55, 1934.

Ekkebus, C., and W. Kiley. Measurement of safe walkway surfaces, *Soap, Cosmetics, Chemical Specialties,* 40–45, 1973.

English, W., Improved Tribometry on Walking Surfaces, ASTM STP1103, 73–81, 1990.

_____. Pedestrian Slip Resistance: How to Measure It and How To Improve It, Alva: William English Inc., 1996.

English, W., D. Underwood, and K.E. Vidal. Investigation of enhancing footwear traction for ironworkers working at heights, November 24, 1998.

_____. Investigation of means of enhancing footwear traction for ironworkers working at heights, Occupational Safety and Safety Administration (OSHA), Washington, DC, 1998. European Committee for Standardization, http://www.cenorm.be.

ETC of Henderson, Inc., How to buff/strip a floor, http://www.etcpads.com(accessed February 29, 2008).

_____. How to care for floor pads, mops and buckets, http://www.etcpads.com (accessed February 29, 2008).

_____. How to mop a wet floor, http://www.etcpads.com (accessed February 29, 2008).

_____. How to strip floors, http://www.etcpads.com (accessed February 29, 2008).

Evans, D., et al. Falls in acute hospitals, A systematic review. The Joanna Briggs Institute for Evidence Based Nursing and Midwifery, 1998.

Fairfax, R. E., OSHA Standard Interpretation Letter to Mr. Noah L. Chitty of the Tile Council of America, March 21, 2003.

Federal Register, April 1, 1990, Vol. 55, No. 69, p. 13408, 29 CFR, Walking and Working Surfaces and Personal Protective Equipment (Fall Protection Systems); Notice of Proposed Rulemaking—Section 1910.22 General Requirements (Appendix A to Subpart D Compliance Guidelines).

Federal Specification KK-L-115C.

Federal Specification PS-624g Soap, Toilet, Liquid and Paste.

Federal Specification, Grating, Metal, Other than Bar Type (Floor, Except for Naval Vessels), RR-G-1602D, February 29, 1996.

Federal Test Method Standard No. 501a, Method 7121, Floor Covering, Resilient, Non-Textile; Sample and Testing (Obsolete). Federal Test Method Standard, Dynamic Coefficient of Friction, Number 501a, Method 7121, June 15, 1966.

Federal Test Method Standard No. 601, for rubber test heel and leather, Federal Specification KK-L-165 (Obsolete).

Feeney, R. and G.M.B. Webber. Safety Aspects of Handrail Design: A Review (Building Research Establishment Report), IHS BRE, 1994.

Feldman, E. The great, lowly dust mop, *Cleaning Maintenance Manage.,* January 2001.

Fendley, A.E., and M.I. Marpet. Required coefficient of friction versus walking speed: Potential influences of footwear and walkway surfaces, *J. Testing Evaluation,* 24(6): 359–367, November 1996.

Finnish Institute of Occupational Health (FIOH). http://www.occuphealth.fi/e/.

Floor Care, Build up finish to protect VCT, December 17, 2001, http://www.cmmonline.com/ENewsArticle.asp?A ArticleID=2.

Floor Care, Long-lasting shine, long-lasting floors, March 27, 2002, http://www.cmmonline.com/ENewsArticle.asp?ArticleID=28.

Floor Care, Top 10 signs that it's time to strip and refinish a floor, August 29, 2002, http://www.cmmonline.com/ENewsArticle.asp?ArticleID=110.

Flynn, J.E., and D.C. Underwood. Precision and bias testing of the English XL variable incidence tribometer and the Brungraber Mark II portable inclinable articulated strut slip tester, NOIRS 2000, October 18, 2000.

Friedlander, M.M., CTIOA/COF report: Development of safety standards for bathing facilities, Ceramic Tile Institute of American, Inc., Culver City, CA, http://cpj.sagepub.com/cgi/content/abstract/44/4/311

Frohnsdorff, G., and M. Martin. Seeking validation and consensus on slip-resistance measurements and statistics, NIST Building & Fire Research Laboratory, Gaithersburg, MD, 1996.

Fu, D., Health Service Impacts and Costs of Falls in Older Age, Ageing and Life Course, Family and Community Health, WHO, http://www.who.int/entity/ageing/projects/4%20Health%20service%20impacts%20and%20costs%20of%20falls2.pdf (accessed April 28, 2008).

German Institute for Standardization, http://www.en.din.de.

DIN 18032 P2, Sport Halls; Halls for Gymnastics and Games; Floors for Sporting Activities; Requirements, Testing.

DIN 51097, Testing of Floor Coverings; Determination of the Anti-Slip Properties; Wet-Loaded Barefoot Areas; Walking Method; Ramp Test.

DIN 51130, Testing of Floor Coverings; Determination of the Anti-Slip Properties; Workrooms and Fields of Activities with Raised Slip Danger; Walking Method; Ramp Test.

DIN 51131, Testing of Floor Coverings—Determination of the Anti-Slip Properties—Measurement of Sliding Friction Coefficient.

EN 1341 (DIN), Slabs of Natural Stone for External Paving-Requirements and Test Methods.

EN 1342 (DIN), Setts of Natural Stone for External Paving-Requirements and Test Methods.

Glass Association of North America. Glass Floors and Stairs, Glass Informational Bulletin GANA LD 06-1107, November 2007, http://glasswebsite.org/publications/reference/LD%2006-1107%20-%20Glass%20Floors%20and%20Stairs.pdf (accessed February 29, 2008).

Gleason ESP Electronic Slip/Fall Prevention, http://www.gleasonesp.com/espapp/html/index.jsp (accessed February 11, 2008).

Bibliography

Global Engineering Documents (GED), http://global.Ihs.com.
Gotte, T., and A. Heisig. Prevention of slipping accidents, *Die BG,* 9:518–523, 1999.
Goodwin, V.D., Slip and fall accidents—Floor testing, March 1999, http://www.mde.cc/publications/slipfall.html.
Gray, E., R. Jablonsky, and H. Yoshida. Biomechanical measurements: pedestrian coefficient of friction, ASTM STP 1103, 1990.
Greater London Council/Inner London Education Authority, Sheet safety surfaces for floors, methods for evaluation, GLC Bulletin 145, Building Technical File, Number 9, April 1985.
Greene, Lisa, Crocs no longer in style at safety-conscious hospitals, St. Petersburg Times, August 9, 2008, http://tampabay.com/news/article764176.ece (accessed September 15, 2008).
Grieser, B., T. Rhoades, and R. Shah. Slip resistance—field measurements using two modern slipmeters, *Professional Safety,* June 2002, 43–48.
Grönqvist, R. Slips and falls, in *Biomechanics in Ergonomics,* Boca Raton: Taylor & Francis, 351–375, 1999.
Grönqvist, R., M. Hirvonen, and E. Rajamaki. Development of a portable test device for assessing on-site floor slipperiness: an interim report, *Applied Ergonomics,* 32:163–171, 2001.
Grönqvist, R., M. Hirvonen, and A. Tohv. Evaluation of three portable floor slipperiness testers, *Int. J. Industrial Ergonomics,* 25:85–95, 1999.
Guevin, P.R., Status of skid and slip resistance testing relatable to painted floors, P.R. Guevin Associates.
Gushue, David L., et al. Biomechanics for risk managers—analysis of slip, trip & fall injuries, ASSE 2007 Professional Development Conference, Session No. 644, 2007.
Hallas, K., and R. Shaw. Evaluation of the Kirchberg Rolling Slider and SlipAlert Slip Resistance Meters, HSL/2006/65, Health and Safety Laboratory (HSL), 2006.
Hallas, K., et al. Roller coaster slip tests putting slip testing back on the rails! in *Contemporary Ergonomics*, Boca Raton: Taylor & Francis, 514–518, 2005.
Hanson, J., M. Redfern, and M. Mazumdar. Predicting slips and falls considering required and available friction, *Ergonomics*, 2(12):1619–1633, 1999.
Harris, G.W., and S. R. Shaw. Slip resistance of floors: users' opinions, Tortus instrument readings and roughness measurement, *J. Occupational Accidents,* 9:298–298, 1988.
Hazardous polymerization, http://homepage.eircon.net/~safestride/index1.html.
Health and Safety Executive. Assessing the slip resistance of flooring. Technical information sheet, http://www.hse.gov.uk/pubns/web/slips01.pdf. No.4—Surfaces, ATBCB, Washington, DC, July 1993.
Health and Safety Executive, Evaluation of the Kirchberg rolling slider and SlipAlert slip resistance meters, 2006. http://www.hse.gov.uk/research/hsl_pdf/2006/hsl0665.pdf.
Health and Safety Executive, Slips and Trips: Fractured skull from a ìwholly preventableî catering slip, December 2008, http://www.hse.gov.uk/slips/experience/skullfracture.htm.
Health and Safety Executive, Slips and Trips: The importance of floor cleaning, HSE information sheet, Slips and Trips 2, September 2005, http://www.hse.gov.uk/pubns/web/slips02.pdf.
Health and Safety Executive, Slips and Trips: Preventing slips and trips in the workplace, http://www.hse.gov.uk/slips.
Health and Safety Executive. Slippery store entrance in wet weather—supermarket entrance matting, http://www.hse.gov.uk/slips/experience/store-entrance.htm (accessed January 15, 2008).
Health and Safety Executive. Slips and trips: Find out more. http://www.hse.gov.uk/slips/findoutmore.htm.
Health and Safety Executive. Slips and trips: Relevant laws and standards. http://www.hse.gov.uk/slips/law.htm.

Health and Safety Executive. Slips and trips: Summary of guidance for the food industry, Food Sheet No 6, HSE Information Sheet, December 2005, http://www.hse.gov.uk/pubns/fis06.pdf.
Health and Safety Executive. Stop slips in kitchens: Get a grip—How to choose suitable footwear for a kitchen environment, http://www.hse.gov.uk/slips/kitchens/footwearguide.pdf (accessed April 8, 2008).
Health and Safety Laboratory. Falls on Stairways—Literature Review, Re-port Number HLS/2005/10, 2005, http://www.hse.gov.uk/research/hsl_pdf/2005/hsl0510.pdf.
Health and Safety Executive, Role of manufacturers and suppliers of footwear, http://www.hse.gov.uk/slips/manufactfoot.htm (accessed January 14, 2008).
Healey, F., and S. Scobie. Slips, trips and falls in hospital, PSO/3, National Patient Safety Agency, 2007.
Heavyweight Solutions, Inc., Statistics, http://www.theheavyweight.com/Core/Statistics.html.
Hedden, J. Strengthen your safety net: Innovative safety programs give restauranteurs extra protection in the kitchen, 1997, http://www.restaurant.org/rusa/magArticle.cfm?ArticleID=497.
Hendrich, A., et al. Hospital falls: development of a predictive model for clinical practice. *Appl. Nurs. Res.* 8(3):129–139, 1995.
High, Steven. Slip Resistance Testing Research Activities, http://www.highsafety.com/hsl/Resources/Services/SlipTesting/research.html, 2007 (accessed February 27, 2008).
Houston, D. Paving the way for seniors, *Facilities Design Arrangement,* March 2002.
Hughes, R.C., and D.I. James. Slipping determinations: magic and myth, undated paper, RAPRA Technologies, Shawbury, Shrewsbury, Shropshire, United Kingdom.
Hunter, R.B. A method of measuring frictional coefficients of walk-way materials, *Bureau of Standards J. Res.,* 329–347, 1929.
International Code Council (ICC). ICC/ANSI A117.1, Accessible and usable buildings and facilities, http://www.intlcode.org, 1998.
International Conference of Building Officials, http://www.icbo.org.
Interpolymer Corporation, Birth of a floor polish, http://www.interpolymer.com/contentmgr/showdetails.php/id/24 (accessed February 29, 2008).
Illuminating Engineering Society of North America (IESNA). The IESNA Lighting Handbook Reference and Application, Illuminating Engineering Society of North America, 9th ed., 2000, http://www.iesna.org.
Slips, Trips and Same-Level Falls, in *Industrial Safety and Occupational Health Markets,* Richard K. Miller & Associates, 5th ed., 1999, 218–220.
Institut National de Recherche et de Sécurité (INRS, National Research and Safety Institute), http://www.inrs.fr.
Institute for Business and Home Safety (IBHS), Summary of State-Mandated Building Codes, http://www.ibhs.org.
Institute of Transportation Engineers (ITE) Task Force TENC-5TF-01 Staff, Guidelines for the Design and Application of Speed Humps, ITE, Washington, DC, 1993.
International Association of Athletics Federations (IAAF), http://www2.iaaf.org/TheSport/Technical/Tracks/Appendix4.html.
International Organization for Standardization (ISO), ISO 7176-13:1989, Wheelchairs—Part 13: Determination of Coefficient of Friction of Test Surfaces, http://www.iso.ch/iso/en/ISOOnline.frontpage, 1989.
Irvine, C.H., Evaluation of some factors affecting measurements of slip resistance of shoe sole materials on floor surfaces, *J. Testing Evaluation,* 4(2):133–138, 1976.
_____. Evaluation of the effect of contact-time when measuring floor slip resistance, *J. Testing Evaluation,* 14(1), ASTM, 19–22, 1986.
Italian Standards. DM 14 Guigno 1989 n. 236 (slip resistance as dynamic coefficient of friction).

Jablonsky, R.D. Standardization of test methods for measurement of floor slipperiness, ASTM STP 649, 1978.

Jackson, Patricia L., and H. Harvey Cohen. An in-depth investigation of 40 stairway accidents and the stair safety literature, Journal of Safety Research, 26(3), Autumn 1995.

James, D.I. Assessing the slip resistance of flooring materials, ASTM STP 1103, 1990.

———. Slip resistance tests for flooring: two methods compared, *Polymer Testing,* 1985, 5: 403–425.

———. A standard slider for slip measurements, Polymer Testing, 8:9–17, 1989.

Janowitz, Angel. Slip resistance: a subject for standardization, KANBrief, Commission for Occupational Health and Safety and Standardization, March 2006. http://www.kan.de/uploads/tx_kekandocs/2005-3-Rutschsicherheit-e.pdf (accessed January 14, 2008).

Joganich, Tim. Investigating slip, trip and fall mishaps, American Society of Safety Engineers Professional Development Conference, June 11, 2008.

Joganich, Tim, and Len McCuen. Influence of groove count on slip resistance using NTL Test Feet, *Journal of Forensic Science,* 50(5):1141–1146, 2005.

Joh, A., et al. Why walkers slip: Shine is not a reliable cue for slippery ground. *Perception & Psychophysics,* 68:339–352, 2006.

Johnson, Dan. Effect of sand on the slipperiness of ice, Human Factors and Ergonomics Society Annual Meeting Proceedings, Safety, 1763–1766(4), Human Factors and Ergonomics Society, 2005.

Jung, K., and U. Diecken. Slip resistance of floors—Some critical remarks on a device used for workplace measurements, *Die BG,* 7:422–429, 1992.

———. Testing slip resistance of floors with the Schuster machine, *Zbl Arbeitsmedizin,* 42(10), 1992.

Jury Awards $500,000 in Slip and Fall Verdict Against CVS in West Palm Beach, http://www.enewspf.com/index.php?option=com_content&task=view&id=1411&Itemid=88889609 (accessed Febrary 2, 2008).

Kalula, S. Z., A WHO global report on falls among older persons. Prevention of falls in older persons: Africa case study. http://www.who.int/ageing/projects/AFRO.pdf.

Kamins, T.L. Marble Baths: Watch Your Step, Practical Traveler, *New York Times,* January 18, 2004.

Kaufmann, M. Reducing slip, trip and fall accidents on walkways and stairs, ASSE Professional Development Conference Proceedings, Session 716, 2007.

Kim, I.-J., and H. Nagata, Effective roughness levels of the floor surface for the reduction of slips and falls accidents, International Conference on Slips, Trips, and Falls: From Research to Practice, 2007.

Kime, G.A., Slip resistance and the U.K. Slip Resistance Research Group, *Safety Science,* 14:223–229, 1991.

Klein, J. The extremities of safety: Hand and foot protection—Choosing and using safety footwear, *Compliance Mag.,* July 2002.

Klik, M., and A. Faghri. A comparative evaluation of speed humps and deviations, *Transportation Quarterly,* 47(3): 459, July 1993.

Kohr, Robert L. *Accident Prevention for Hotels, Motels, and Restaurants,* Van Nostrand Reinhold, 1991.

———. Using ASTM slip test methods, American Ceramic Society Bull., June 1995, 74(6), 75–78, 1992.

Krafft, L.E. Fix slopping mopping, *Cleaning Maintenance Manage.,* April 2002.

Kulakowski, B.T., F. Buczek, and P. Cavanaugh. Evaluation of performance of three slip resistance testers, *J. Testing Evaluation,* 17(4):234–240, 1989.

Kuzel, M., et al. Comparison of subjective ratings of slipperiness to the measured slip resistance of real-world walking surfaces, International Conference on Slips, Trips, and Falls: From Research to Practice, 2007.

LaReau, R. Operators can build mat market on preventing slips and falls, Textile Rental Services Association (TRSA) Online, http://www.trsa.org, August 1996.

Leach, Ben. Study: Popular prescription drugs raise seniors' fall risk, pressofAtlanticCity.com, July 11, 2008, http://www.reducedrugprices.org/read.asp?news=1982 (accessed July 28, 2008).

Lehtola, Carol J., William J. Becker, and Charles M. Brown. Preventing Injuries from Slips, Trips, and Falls, CIR869, NASD, http://www.cdc.gov/nasd/docs/d000001-d000100/d000006/d000006.html, 2001.

Lesch, M.F., W.R. Chang, and C.C. Chang. Reliability of visual cues in predicting judgments of slipperiness and the coefficient of friction of floor surfaces, International Conference on Slips, Trips, and Falls: From Research to Practice, 2007.

Li, K.W., H.H. Wu, and Y.C. Lin. The effect of shoe sole tread groove depth on the friction coefficient with different sole materials, floors, and contaminants, *Applied Ergonomics,* 37(6):685–808, 2006.

Li, Kai Way, et al. Slips and falls—Employee experience and perception of floor slipperiness: A field survey in fast-food restaurants, *Professional Safety,* September 2006.

Liberty Mutual Insurance Company. Restaurant Floor Cleaning, 1995, http://www.libertymutual.com/research/about/1995_report/tribology-research.html.

Lombardi, F., Slip fall cases cost city a bundle, *New York Daily News,* December 3, 2005.

Mahoney, J. E. Immobility and falls. *Clinics in Geriatric Medicine,* 14(4):699–726, 1998.

Maki, B.E., S.A. Bartlett, and G.R. Fernie. Effect of Stairway pitch on optimal handrail height, *Human Factors,* 27(3): 355–359, 1985.

_____.Influence of stairway handrail height on the ability to generate stabilizing forces and moments, *Human Factors,* 26(6):705–714, 1984.

Malkin, F., and R. Harrison, R. Measurement of the coefficient of friction of floors, *Applied Physics,* 12:517–528, 1979.

Mangan, Benjamin W. Sometimes a shoe is just a shoe? Not in the workplace, *Occupational Health and Safety,* http://www.ohsonline.com/articles/45077/, 2006.

Marletta, W. The effect of the use of differing sanding protocols on readings of slip resistance when using the English XL and Brungraber Mark II tribometers, Doctoral Dissertation, New York University, NY, 1994.

Marlowe, D. Letter to ASTM Technical Committee Chairmen on CEN and ISO, January 12, 2001.

Marpet, M.I., Comparison of walkway safety tribometers, *J. Testing Evaluation,* 24(4):245–254, 1996.

Marpet, M.I., and D.H. Fleisher. Comparison of walkway-safety tribometers: part II, *J. Testing Evaluation,* 25(1):115–126, 1997.

Martin, G., and J. Dimopoulos. Reliability of dry slip resistance test results using the Tortus floor friction tester, Productivity Ergonomics and Safety—The Total Package Conf., Gold Coast. http://www.homemods.info/resource/bibliography/reliability_of_dry_slip_resistance_test_results_using_the_tortus_floor_friction_tester.

Mastrad Quality and Test Systems, http://www.mastrad.com/tortus.htm.

Mat Tech. Commercial Building Entrance Care, Crown Mats and Matting, Mat Tech. http://www.mattech.ca/eng/products/entrance/index.html.

Maynard, W.S. Tribology: preventing slips and falls in the workplace, *Occupational Health and Safety,* 2002.

Maynard, Wayne, and George Brogmus. Reducing slips, trips and falls in stairways, *Occupational Hazards,* October 1, 2007.

McCabe, Paul T. (Ed.), *Contemporary Ergonomics 2004*, CRC Press, Boca Raton, FL, 2004, p. 43.

McCagg, M. Get a charge out of your equipment, *Cleaning Maintenance Manage.,* May 2002.

Medoff, H., D. Fleisher, and S. Di Pilla. Comparison of slip resistance measurements between two tribometers using smooth and grooved Neolite test liner test feet, STP 1424, Metrology of Pedestrian Locomotion and Slip Resistance, ASTM International, 2003.

Miller, J.M., and T.P. Rhoades. Slip Resistance Predictions for Various Metal Step Materials, Shoe Soles and Contaminant Conditions, Society of Automotive Engineers Technical Paper Series, 1987.

Miller, Norman. Two cops injured in icy fall, *The MetroWest Daily News,* December 11, 2007, http://www.metrowestdailynews.com/homepage/x517197292.

Minnesota Snow and Ice Control—Field Handbook for Snowplow Operators, Manual Number 2005-1, Minnesota Local Road Research Board, August 2005. http://www.ttap.mtu.edu/publications/2005/snowicecontrolhandbook.pdf (accessed February 11, 2008).

Mishky, S. Lighting for improved workplace safety, *Occupational Hazards,* August 1, 2001.

Morse, J. M. *Preventing Patient Falls.* Newbury Park, CA: Sage, 1997.

Murphy, T. In search of slip-resistant flooring, *Plant Services,* March 2001.

Myung, R., J. Smith, and T. Leamon. Subjective assessment of floor slipperiness, *Int. J. Industrial Ergonomics,* 11:313–319, 1993.

Nagata, H. Rational index for assessing perceived difficulty while descending stairs with various tread/rise combinations, *Safety Science,* 21(1), November 1995, 37–49.

Nagata, Hisao, and In-Ju Kim. Fall accidents In Japan and the classification of fall-risk factors, Proceedings of the International Conference on Slips, Trips, and Falls 2007—From Research to Practice, August 23–24, 2007, Hopkinton, MA, IEA Press.

NAHB Research Center, Stair Safety—A Review of the Literature and Data Concerning Stair Geometry and Other Characteristics, November 30, 1992, http://www.centerforhealthyhousing.org/stair_safety.pdf.

Nation's Restaurant News, http://www.nrn.com/operations/worksafety.html.

National Building Code (NBC). http://www.iccsafe.org/index.html.

National Bureau of Standards (NBS) Preliminary Study of the Slipperiness of Flooring, NBSIR 74-613, 1974.

National Occupational Health and Safety Commission (NOHSC). http://www.nohsc.gov.au.

National Safety Council (NSC), Falls on Floors, Data Sheet I-495, NSC, Itasca, IL, 1991.

NSC, *Injury Facts,* NSC, Itasca, IL, 2007–2008.

NSC, *International Accident Facts,* 2nd ed., NSC, Itasca, IL, 1999.

National Fire Protection Association (NFPA). http://www.nfpa.org.
    NFPA 101, Life Safety Code, National Fire Protection Association, 2000.
    NFPA 1901, Fire Equipment Apparatus, National Fire Protection Association, 1999.

Newman, M.A. Magic Carpet Ride—Treat your carpet like a BMW and you'll get a lot of miles out of it, *Facilities Design and Management,* 37–38, February 2003.

NSSN: A National Resource for Global Standards, http://www.nssn.org.

NFPA 101, Life Safety Code, National Fire Protection Association, 2006.

Occupational Safety and Health Administration, OSHA 1910 and 1926, http://www.osha.gov.
    Manlifts 1910.68(c)(3)(v).
    OSHA 29 CFR Part 1910.136.
    1926, Subpart M on Fall Protection.
    1926, Subpart R—Steel Erection Regulatory (3).
    Appendix B to Subpart R—Acceptable Test Methods for Testing Slip Resistance of Walking/Working Surfaces (1926.754(c)(3)). Non-Mandatory Guidelines for Complying with 1926.754(c)(3).
    1926, Subpart X on Stairways and Ladders.
    Notice of Proposed Rulemaking—Section 1910.22 General Requirements (Appendix A to Subpart D Compliance Guidelines).

Ohio Bureau of Workers' Compensation, Safety Grant Best Practices, Reducing Slip and Fall Injuries in Restaurants: One Company's Experience.

On-site slip analysis with new portable tester, *Buildingtalk*, June 9, 2006, http://www.buildingtalk.com/news/sat/sat105.html (accessed February 20, 2008).

Parker, M. J., W. J Gillespie, and L. D. Gillespie, Effectiveness of hip protectors for preventing hip fractures in elderly people: systematic review, *BMJ*, 2006, 332:571–574.

Partnership for Health and Accountability, Footwear & Safety, A Safety Bulletin June 20, 2007, http://www.psqh.com/enews/0607feature.html (accessed September 15, 2008).

Pauls, J. Predictable and preventable missteps that are not "slips, trips and falls," but result in serious injuries, International Conference on Slips, Trips, and Falls: From Research to Practice, 2007a.

———. Rating How Well U.S. Model Building Codes Cover Environmental Factors in Fall Prevention, International Conference on Slips, Trips, and Falls: From Research to Practice, 2007b.

Pauls, J.L., and S.C. Harbuck. Ergonomics-based methods of inspecting, assessing and documenting environmental sites of injurious falls, American Society of Safety Engineers Professional Development Conference, June 12, 2008.

Perkins, P., Measurement of slip between the shoe and ground during walking, ASTM STP 649, 1978.

Pinnacol Assurance, Innovative footwear programs pay big returns by reducing slips, trips, and falls, 2006, http://yaktrax.com/PressRelease/FourthQuarter2006Focus.pdf (accessed April 8, 2008).

Playground Safety. Brain Injury Association of America. http://www.biausa.org (accessed February 12, 2008).

Porcelanosa Group Limited. Slip resistance. http://www.ribaproductselector.com/Docs/5/10365/external/COL1710365.pdf?ac=, May 2005 (accessed April 11, 2008).

Powers, C.M., J.R. Brualt, M.A. Stefanou, Y.J. Tsai, J. Flynn, and G.P. Siegmund. Assessment of walkway tribometer readings in evaluating slip resistance: A gait-based approach. *Journal of Forensic Science,* 52(2):400–405, 2007.

Powers, C.M., K. Kulig, J. Flynn, and J.R. Brault. Repeatability and bias of two walkway safety tribometers, *J. Testing Evaluation,* 27(6): 368–374, 1999.

Powers, Christopher M, Maria A. Stefanou, Yi-Ju Tsai, John R. Brualt, and Gunter P. Siegmund. Assessment of walkway tribometer readings in evaluating slip resistance: a gait based approach, Proceedings of the XIX Annual International Occupational Ergonomics and Safety Conference Las Vegas, Nevada, USA, 27–29 June 2005.

Proctor, T.D., Slipping accidents in Great Britain—An update, *Safety Science,* 16:367–377, 1993.

Proctor, T.D., and V. Coleman. Slipping, tripping and falling accidents in Great Britain—present and future. *J. Occupational Accidents,* 9:269–285, 1988.

QI Project. Fall prevention program yields quick results. http://www.premierinc.com/safety/topics/falls/downloads/S-11-case-hty-ne-falls-text-final-02-23-04.doc, 2004 (accessed April 18, 2008).

Quirion, F. Floor cleaning as a preventive measure against slip and fall accidents, RF-366 Technical Guide, IRSST, March 2004.

Quirion F., P. Poirier, and P. Lehane. Optimal cleaning depends on the type and condition of the flooring being cleaned, Proceedings of the International Conference on Slips, Trips and Falls: From Research to Practice, (August 23–24, 2007: Hopkinton, U.S.A.), IEA Press, 178–182, 2007.

Rabinowicz, Ernest. Stick and Slip, *Scientific American,* May 1956, 194(5):109–110, 112, 115–117.

Rapra Technology, http://www.rapra.net.

Redfern, M., and T. Rhoades. Fall prevention in industry using slip resistance testing, *Occupational Ergonomics—Theory and Application,* 27, 1996.

Bibliography

Research Shows Supermarkets Have Room for Improvement at Reducing Slips and Falls, Business Editors, January 30, 2001, http://findarticles.com/p/articles/mi_m0EIN/is_2001_Jan_30/ai_69704873.

Restaurants. RIC Reports Restaurants "Slipping and Tripping" Their Way to Potential Financial Disaster, *Restaurants,* November 15, 2004, http://www.claimsjournal.com/news/national/2004/11/17/47798.htm.

Ritter, Malcom. Pa. researchers explore preventing falls, *Washington Post*, May 17, 2007.

Roger Wilde Ltd. Safety aspects of walking on glass—Slip resistance, http://www.rogerwilde.com/slip_resistance.pdf (accessed February 29, 2008).

Rosen, S. *Slip and Fall Handbook,* Parts I, II, and II, Harlow Press, 1995.

Rowland, F.J., D.J. James, and R.C. Hughes. Slip resistance—An overview: The work of the UK Slip Resistance Group, *Polymer Testing '95,* 17, 1995.

Roys, M, and M. Wright. Proprietary nosing for non-domestic stairs. Information paper 15/03, BRE Centre for Human Interaction, Garston, Watford, UK. http://www.bre.co.uk, 2003.

Rubbersidewalks, Inc. http://www.rubbersidewalks.com/default.aspx (accessed May 12, 2008).

Sacher, A. Slip resistance and the James Machine 0.5 static coefficient of friction—sine qua non, *American Ceramic Society Bull.,* 30–37, 1993.

Sanford, C. Help your customers prevent slip-and-falls, *Cleaning Maintenance Manage.,* July 1999.

SATRA Technology Centre, Ltd. http://www.satra.co.uk.

Schroder, D., and H.J. Garvens. WFK slip resistance, WFK Cleaning Technology Research Institute. http://www.wfk.de/rho/slipresistance_e.html.

Sani-Floor. 2007. http://www.sanifloor.net/home.html (accessed May 12, 2008).

Severn Science. http://www.severnscience.co.uk/pages/tortus.html.

Sherman, R. Role of the expert in a slip and fall lawsuit, *Professional Safety,* 37–40, January 1999.

Shipman, John. Ramp test for floorcoverings, SATRA Spotlight, 13, July 2007.

Shoemaker, Dawn, Do you know your floor types? Cleaning & Maintenance Management Online. 2004.

Shoes for Crews, Testimonials. http://www.shoesforcrews.com/sfc3/index.cfm?changeWebsite=CA_en&route=inserts.testimon/testimonials (accessed January 14, 2008).

Sigler, P. Calibration and Operation of the Sigler Pendulum-Impact Type Slipperiness Tester, June 27, 1949.

Sigler, P., M. Geib, and T. Boone. Measurement of the Slipperiness of Walkway Surfaces, National Bureau of Standards, NBS RP1879, 40, 1948.

Slip Alert. http://www.slipalert.com/index.html (accessed February 20, 2008).

Slip and Fall Accidents No Joke—Inside Edition Finds 13 Million Injuries Per Year, http://www.consumeraffairs.com/news02/slipfall.html (accessed February 11, 2007) 2002.

Slip Slidin' Safety, QSR Web, July 8, 2005, http://www.qsrweb.com/article.php?id=344.

Smith, R.H. Assessing testing bias in two walkway-safety tribometers, *ASTM Journal of Testing and Evaluation,* 31(3):169–177, May 2003.

The Soap and Detergent Association, Chemistry. http://www.sdahq.org.

Sotter, G. *Stop Slip and Fall Accidents,* Soner Eronenine Corporation, Mission Viejo, CA, 2000.

Sotter, George, Comments Accompanying George Sotter's Negative Vote on ASTM C 1028, April 14, 2005.

Southern Building Code Congress International, http://www.sbcci.org.

Standards Australia, http://www.standards.com.au.

    AS/NZS 1141.42:1999, Methods for Sampling and Testing Aggregates—Pendulum Friction Test.

    AS/NZS 2983.4:1988 Methods of Test for Synthetic Sporting Surfaces—Test for Slip Resistance (Method 4: Test for Slip Resistance).

    AS/NZS 3661.1:1993, Slip Resistance of Pedestrian Surfaces.

AS/NZS 3661.2:1994, Slip Resistance of Pedestrian Surfaces—Guide to the Reduction of Slip Hazards.
AS/NZS 3661.2:1994, Slip Resistance of Pedestrian Surfaces—Guide to the Reduction of Slip Hazards.
AS/NZS 3661.2: 1994, Slip Resistance Classification of Pedestrian Surface Materials.
AS/NZS 4586:1999, Slip Resistance Classification of New Pedestrian Surface Materials.
AS/NZS 4663:2002, Slip Resistance Measurement of Existing Pedestrian Surfaces.
Draft DR 99447 CP: 1999, Slip Resistance Measurement of Existing Pedestrian Surfaces.
HB 197:1999, Introductory Guide to the Slip Resistance of Pedestrian Surface Materials.
Stevens, J.A., et al. The costs of fatal and non-fatal falls among older adults, *Injury Prevention* 12:290–295, 2006; doi:10.1136/ip.2005.011015, http://injuryprevention.bmj.com/cgi/content/full/12/5/290.
Strandberg, L. The effect of conditions underfoot on falling and overexertion accidents, *Ergonomics,* 28(1):131–147, 1985.
_____. Ergonomics Applied to Slipping Accidents, in *Ergonomics of Workstation Design,* Kvålseth, T.O., Ed., London: Butterworths, 201–228, 1983.
Swedish National Road and Transport Research Institute, Effects on Cycling of Road Maintenance and Operation, July 26, 1999. http://www.tft.lth.se/kfbkonf/5bergstromnew.PDF.
Swedish Standards Institute (SIS) SS 92 35 15, Floorings—Determination of Slip Resistance.
System for logging premises hazard inspections. Free Patents Online, United States Patent 6078255. http://www.freepatentsonline.com/6078255.html (accessed February 11, 2007).
Templer, J. A. Stair Shape and Human Movement, doctoral dissertation. New York, NY: Columbia University, 1975.
_____. *The Staircase—Studies of Hazards, Falls, and Safer Design,* Cambridge: The MIT Press, 1992.
Templer, J., J. Archea, and H.H. Cohen. Study of factors associated with risk of work-related stairway falls, *Journal of Safety Research,* 16:183–196, 1985.
Textile Rental Services Association (TRSA). http://www.trsa.org.
Tideiksaar, R. *Falls in Older People: Prevention and Management,* 3rd ed., Baltimore: Health Professions Press, 2002.
Torres, Katherine. Stepping into the kitchen: foot protection for food service workers, *Occupational Hazards,* February 15, 2007.
Trebilcock, Bob. Better dock safety and ergonomics, *Modern Materials Handling,* June 1, 2006, http://www.mmh.com/article/CA6341890.html.
Turman, D., Slip and fall prevention: Floor chemical basics. http://Foodservice.com.
_____. Slip resistant footwear basics. http://Foodservice.com.
Turnbow, C.E. *Slip and Fall Practice,* 2nd ed., James Publishing, Costa Mesa, CA, 2001.
U.K. Slip Resistance Group, The Measurement of Floor Slip Resistance: Guidelines Recommended by the UK Slip Resistance Group, 2005.
UK Slip Resistance Group, Frequently Asked Questions, http://www.ukslipresistance.org.uk/faq.php (accessed February 20, 2008).
Underwriters Laboratories (UL). U.L. 410, Slip Resistance of Floor Surface Materials, http://www.ul.com,1996.
Underwriters Laboratories of Canada (ULC). http://www.ulc.ca.
U.S. Bureau of the Census, Life Expectancy at Birth, 1996.
U.S. Bureau of the Census, Projections of the Population by Age and Sex: 1995 to 2050, 1996.
U.S. Department of Justice. Civil Rights Division. *Enforcing the ADA: A Status Report from the Department of Justice, October 1995–March 1996.* Prepared by the Civil Rights Division, 1996.
U.S. Department of Justice. Hundreds of victims of an ADA business can to get money back, U.S. Department of Justice, Press Release, June 28, 1996.

U.S. Federal Highway Administration, *Manual on Uniform Traffic Control Devices for Streets and Highways* (MUTCD), 2001, http://mutcd.fhwa.dot.gov.

U.S. General Services Administration Specification P-F-430C(1), Finish, Floor, Water Emulsion (Obsolete).

U.S. Military Specifications (Navy)
  MIL-D-0016680C (Ships) and MIL-D-18873B, Deck Covering Magnesia Aggregate Mixture.
  MIL-D-17951C (Ships), Deck Covering, Lightweight, Nonslip, Silicon Carbide Particle Coated Fabric, Film, or Composite, and Sealing Compound.
  MIL-D-23003a(SH), Deck Covering Compound, Nonslip, Rollable.
  MIL-D-24483A, Nonslip Flight Deck Compound.
  MIL-D-3134J, Deck Covering Materials, 1988.
  MIL-W-5044C, Walkway Compound, Nonslip, and Walkway Matting, Nonslip.

U.S. Trademark Electronic Search System (TESS). http://www.uspto.gov/main/trademarks.htm.

van Dieën, J.H., M. Pijnappels, and M.F. Bobbert. Age-related intrinsic limitations in preventing a trip and regaining balance after a trip, *Safety Science,* 43(7):437–453, 2005.

Vidal, K. Slips, trips and falls, in *Safety Management Handbook,* Chicago: CGH, Inc., 2000.

Veteran's Administration, National Center for Patient Safety 2004 Falls Toolkit, May 2004, http://www.va.gov/ncps/safetytopics/fallstoolkit/notebook/completebooklet.pdf.

Visual Interpretation of the International Building Code 2006 Stair Building Code, Stairway Manufacturers Association, http://www.stairways.org/pdf/2006%20Stair%20IRC%20SCREEN.pdf, 2006.

Wallask, S., Crocs, OSHA, and you, Hospital Safety Center, August 3, 2007, http://www.hospitalsafetycenter.com/blog/index.cfm/2007/8/3/Crocs-OSHA-and-you (accessed September 15, 2008).

Waltzing to the Vienna Agreement—A refresher course, ISO Bulletin, August 1998.

Webster, T. Workplace falls, Compensation Working Conditions, Spring 2000.

Weller, S.C. Floor covering ills—A prescription for mat rental services, *Textile Rental,* TRSA Online, http://www.trsa.org, September 1998.

WFK Research Institute for Cleaning Technology, http://www.wfk.de/, http://www.wfk.de/rho.

White, M., Crocs a health hazard, hospitals say, *The Gazette,* August 1, 2007, http://www.canada.com/montrealgazette/news/story.html?id=f6db2228-5084-4059-95d2-49f7c689b83a (accessed September 15, 2008).

Williams, W.D., J.A. Smith, and F.J. Draugelis. Topaka: a new device and method for measuring slip resistance of polished floors, *Soap, Cosmetics, Chemical Specialties,* 44:47–48, 52, 60, July 1972.

Winter Parking Lot and Sidewalk Maintenance. Minnesota Pollution Control Agency, July 2006. http://proteus.pca.state.mn.us/publications/roadsalt-clipboardpages.pdf (accessed February 11, 2008).

Wisconsin Transportation Information Center. Using Salt and Sand for Winter Road Maintenance, Wisconsin Transportation Bulletin No. 6, Revised August 2005. http://epdfiles.engr.wisc.edu/pdf_web_files/tic/bulletins/Bltn_006_SaltNSand.pdf.

World Health Organization. Table 1: Summary of evidence to support interventions to reduce falls and injuries, Health Evidence Network, November 1, 2004, http://www.euro.who.int/HEN/Syntheses/Fallsrisk/20040318_6 (accessed April 18, 2008).

Yoshida, S. A Global Report on Falls Prevention Epidemiology of Falls, Ageing and Life Course Family and Community, World Health Organization, 2006.

Index

A

AAFS, *see* American Academy of Forensic Sciences
ABA, *see* Architectural Barriers Act
Access covers, physical evaluation of, 10
Accident investigation and mitigation, 381–411
    accident investigation, 382–388
        claimant and witness information, 383
        detailed occurrence information, 384
        general occurrence information, 383
        handrails, 387
        landings, 387
        lighting, 387–388
        location information, 385
        management/operational control information, 388
        stairs or ramps, 385–386
    accident investigation guide
        ramp fall accidents, 399–400
        slippery surfaces accidents, 401–402
        stairway fall accidents, 403–404
    accident reporting and investigation guide for slips, trips, and falls, 397–398
    claim mitigation, 389
    claims investigation (parking garage), example, 405–411
        ADA analysis, 409–411
        additional items to investigate, 408
        attachments, 409
        curb, 407
        description, 405
        lighting, 407
        marking, 408
        ramp, 406
        ramp flares, 406
        supplemental information, 408
        theory of liability, 405–406
    documentation, 395–396
    expert testimony, admissibility of, 391–395
        Daubert ruling et al, 391–392
        federal rules of evidence, 394–395
        Frye test, 391
        Kumho Tire Co. v. Carmichael (97-1709), 392–393
        practical suggestions for meeting Daubert/Kumho, 393–394
    fraud control, 390–391
    fraud control indicators, 389–390
        fraud (manner of claimant), 390
        general indicators, 390
        lost earnings fraud indicators, 390
        medical or dental fraud indicators, 390
    incident reporting, 388
    occurrence analysis, 388–389
    pitfalls of accident reporting and investigation, 381–382
    staff issues, 396
    theories of liability, 382
Acidic floor cleaners, 205
ADA, *see* Americans with Disabilities Act
Alkaline salts, 206
American Academy of Forensic Sciences (AAFS), 184
American National Standards Institute (ANSI), 171–173, 255
    ANSI A1264.2-2006, 172
    ANSI A1264.3-2007, 173
    ANSI A137.1-1988, 173
    curb cutout design specifications, 9
    Electronic Standards Store, 255
Americans with Disabilities Act (ADA), 2, 164–166
    Accessibility Guidelines for Buildings and Facilities, 2, 25
    analysis, claims investigation (parking garage), 409–411
    compliance, accessibility, 57
    curb ramps, 15
American Society of Testing and Materials (ASTM), 228, *see also* ASTM International
    footwear standards, 305
    interlaboratory studies, 239
    meeting of test method, 228
    sticktion, 228
ANSI, *see* American National Standards Institute
Architectural Barriers Act (ABA), 228
Architectural and Transportation Barriers Compliance Board (ATBCB), 164, 228
Assistive devices, 349–352
    bed alarm systems, 350–351
    bed side rails, 351
    hip protectors, 351–352
    identification bracelets, 351
    restraints, 349–350
ASTM, *see* American Society of Testing and Materials
ASTM International, 167–170
    ASTM F695, 169

429

ASTM F802, 170
ASTM F1240, 169
ASTM F1637, 169–170
ASTM F1646, 170
ASTM F1694, 170
ASTM F2048, 170
technical committee ASTM F-13, 168–170
ATBCB, *see* Architectural and Transportation Barriers Compliance Board
Australian Commonwealth Engineering Standards Association, 262
Australia/New Zealand standards, 262–265
  Australian standards, 263–265
  standards process, 263

B

Balancing a committee, overseas standard development, 239
BCA, *see* Building Code of Australia
Bed alarm systems, 350–351
Berufsgenossenschaftliches Institut fur Arbeitssicherheit (BIA), 272
BIA, *see* Berufsgenossenschaftliches Institut fur Arbeitssicherheit
Bicycle racks, 12, 13
BLS, *see* Bureau of Labor Statistics
BPNE, *see* British Pendulum Number Equivalent
Braked-wheel/skiddometer instrument, 255
British Pendulum Number Equivalent (BPNE), 252
British standards, 260–262
  Committee B/208, 261
  Committee B/539, 262
  Committee B/545, 261
  Committee B/556, 261
  Committee PRI/60, 262
Building Code of Australia (BCA), 264
Building codes, state-mandated, 64–66
Building inspection and maintenance programs, 71–72
Bureau of Labor Statistics (BLS), 301, 311

C

Carpet, physical evaluation, 56–57
Carpet and Rug Institute (CRI), 48
CCTV, *see* Closed-circuit television
CEN Rubber, 144
Ceramic Tile Institute (CTI), 229
Ceramic Tile Institute of America (CTIOA), 177–178
Certified Floor Safety Technician (CFST), 217
CFST, *see* Certified Floor Safety Technician
CGSTF, *see* Contact Group on Slips, Trips, and Falls
Closed-circuit television (CCTV), 391

Coefficient of friction (COF), 106, 122, 161
  footwear, 208
  tribometers, 122
COF, *see* Coefficient of friction
Commonwealth Scientific and Industrial Research Organisation (CSIRO), 270–271
Consumer Product Safety Commission (CPSC), 365
Consumer Specialty Products Association (CSPA), 181–182
Contact Group on Slips, Trips, and Falls (CGSTF), 183
CPSC, *see* Consumer Product Safety Commission
Creep, 106
CRI, *see* Carpet and Rug Institute
Crocs footwear, 355–356
CSIRO, *see* Commonwealth Scientific and Industrial Research Organisation
CSPA, *see* Consumer Specialty Products Association
CTI, *see* Ceramic Tile Institute
CTIOA, *see* Ceramic Tile Institute of America
Curbing, 6–8
  construction and design, 6–8
  marking, 8
  physical evaluation of, 6–8

D

Daubert ruling et al, 391–392
DCOF, *see* Dynamic coefficient of friction
Degreasers, 206
Digitized dragsleds, overseas standards, 247–251
  BOT-3000, 251
  issues, 249–251
  Tortus round robin, 249
  variations, 247
DIN, *see* German Institute for Standardization
Dock plates (trucking), 373–374
Doorstep, well-camouflaged, 14
Doorstops, physical evaluation of, 13
Drainage grates, physical evaluation of, 10–12
Dynamic coefficient of friction (DCOF), 105, 167, 196

E

Elevators, physical evaluation, 52–53
Emulsification, 205
Engineering precept, management controls, 68
Escalators, physical evaluation, 54–55
European standards, 255–258, *see also* Overseas standards, international standards
Expectation, definition of, 1

Index

Expert testimony, admissibility of, 391–395
    Daubert ruling et al, 391–392
    federal rules of evidence, 394–395
    Frye test, 391
    Kumho Tire Co. v. Carmichael (97-1709), 392–393
    practical suggestions for meeting Daubert/Kumho, 393–394

F

FIA, *see* Footwear Industries of America
FIDO, 255
Finish coat, structural steel, 187
Finnish Institute of Occupational Health (FIOH), 272
FIOH, *see* Finnish Institute of Occupational Health
Flooring and floor maintenance, 195–236
    assessment of floor treatment/cleaning products and methods, 216
    carpet maintenance, 216
    floor cleaning, 204–215
        acidic cleaners, 205
        basic types of floor cleaners, 205
        case study, 207, 211
        categories of floor cleaners, 206
        emulsification, 205
        floor buffing/polishing, 208–209
        floor care equipment maintenance, 209–212
        floor finish indicators and maintenance issues, 212
        floor stripping, 207–208
        high-pH floor cleaners, 205
        issues and methods, 206–212
        neutral cleaners, 205
        outsourcing of floor care, 212–215
        outsourcing of labor and equipment, 213
        owning equipment and outsourcing of labor, 213–214
        owning equipment and using in-house labor, 214–215
        oxide reduction, 205
        products, 205–206
        saponification, 205
        surface tension, 205
        use of wrong product, 204
        wet mopping floors, 208
    floor finishes and their properties, 200–204
        conventional floor finishes, 201
        etching, 203
        particle embedding, 202
        relationship of shine to slip, 200–201
        "slip-resistant" floor treatments, 202–204
        surface grooving and texturing, 202–203
    floor maintenance certification, 217
        Certified Floor Safety Technician, 217
        IICRC Hard Surface Floor Maintenance Specialist, 217
        Rochester Midland Corporation, 217
    floor treatment study, 217–218
    identifying types of flooring and their properties, 196–200
        extent of exposure, 196
        nonresilient flooring, 198–199
        other types of flooring, 199–200
        resilient flooring, 197
    impact of wear, 204
    self-washing floors, 322–323
    slip-resistant treatment study (2000), 219–230
        American Society of Testing and Materials, 228
        appearance notes, 224
        Ceramic Tile Institute, 229
        comparison of results (ceramic), 224–225
        comparison of results (marble), 226
        consensus standards, 228–229
        credentials and claims, 227
        ESIS risk control services, 219
        federal laws and standards, 227–228
        goals, 220
        independent laboratory testing, 229–230
        maintenance notes, 223–224
        other specifications, 222
        overview, 220–221
        packing and shipping, 223
        protocol, 221–222
        surface preparation, 222–223
        test notes, 223
        treated tiles testing, 223
        Underwriters Laboratories, 229
    statistical analysis, 235–236
    test protocol, 231–234
        testing process, 233–234
        test preparation, 231
        wet testing, 232
    threshold of safety, 195–196
Floor mats, 43–52
    case study, 43–44, 48–49, 49–51
    food service operations, 324–326
        case study, 326
        cleaning requirements, 330
        olefin fiber mats, 326
        rubber mats, 324–326
        wiper/scraper mats, 326
    general precautions, 44–45
    mat cleaning guidelines, 51–52
    mat design, 46
    mat design and arrangement, 46–47
    mat size, 48
    mat storage, 47–48

minimizing contaminants, 45–46
protection of hard flooring surfaces, 51
Food service operations, 311–333
    exposure overview, 312–313
    floor mats, 324–326
        case study, 326
        cleaning requirements, 330
        olefin fiber mats, 326
        rubber mats, 324–326
        wiper/scraper mats, 326
    floor surfaces/housekeeping, 315–324
        case study, 319–320
        chemistry of fat and flooring, 320–322
        common cleaning scenario, 316–319
        keep floors clean, 316–323
        keep floors dry, 315–316
        keep walkways clear, 323
        mop cleaning, 319
        over-the-spill cleanup kit, 324
        perceptions of food service workers, 323
        polymerization, 319
        self-washing floors, 322–323
        spill cleanup, 323–324
    footwear, 326–328
    kitchen slip/fall hazard assessment, 331–333
    multiple intervention study, 328–329
    pedestrian flow and slips, trips, and falls, 313–315
        case study, 314–315
        customer falls, 311, 314
        making undesirable path safer, 313–314
        redirect traffic onto preferred, safer paths, 314
Footwear, 295–309
    advertising, 300
    ASTM F08 sports equipment and facilities, 305
    consensus standards, 305–306
    Crocs, 355–356
    ESD, 301
    federal specification (USPS no. 89C), 305
    footwear design for slip resistance, 296–298
        general guidelines for shoe design and selection, 298
        outsole tread patterns, 297
        sole compounds, 297
    German standard (DIN 4843-100), 309
    industry conditions, 295–296
    international footwear standards for slip resistance, 306–309
    ISO standards, 306–309
    labeling, 298–300
        labeling for slip resistance testing, 299–300
        labeling for usage, 298–299
    maintenance, 302–304
        keeping clean, 302
        replacement, 303–304
        wear and inspection, 302–303
    other protective features, 301–302
    other selection guidelines, 300
    potential impact, 296
    programs, 304–305
        enforcement, 305
        mandate or recommend, 304
        purchase options, 304–305
        specifications, 304
    shoe program, mandated, 296
Footwear Industries of America (FIA), 182–183
Four S (Standard Shoe Sole Simulating) Rubber, 143
Fraud control indicators (accident investigation), 389–390
    fraud (manner of claimant), 390
    general indicators, 390
    lost earnings fraud indicators, 390
    medical or dental fraud indicators, 390
Frye test, 391

G

GED, *see* Global Engineering Documents
German Institute for Standardization (DIN), 258
German standards, 258–260
    DIN 18032 P2 DIN V 18032-2 floors for sporting activities, 259
    DIN 51097 testing of floor coverings; wet-loaded barefoot areas, 259
    DIN 51130 testing of floor coverings; workrooms and fields of activities with raised slip danger, 259–260
    draft standard DIN 51131 testing of floor coverings; measurement of sliding friction coefficient, 260
Global Engineering Documents (GED), 255
Glycol ethers, 206
Graspability (handrail), 35–36
Guest room bathrooms, 363–364

H

Handrails, physical evaluation of, 33–39
    case study, 38–39
    graspability, 35–36
    supports, 36–38
Hazard mapping (slip, trip, and fall), 73
Healthcare operations, 335–360
    calculating fall rates, 344
        number of falls per bed, 344
        number of patients at risk rate, 344
        number of patients who fell rate, 344
    causes of patient falls, 338–344
        categorizing causes of patient falls, 338–341

Index

medical conditions, 341
medication, 342–343
Morse Fall Scale, 338–340
other factors, 343
personal risk factors of patient falls, 341–344
physical conditions, 341–342
specialty units, 343–344
Tideiksaar classification method, 340–341
guidelines for fall prevention program, 357–360
impact of age, 336
interventions/controls, 344–353
assistive devices, 349–352
bathrooms, 347
bed alarm systems, 350–351
beds/bedside, 345–346
bed side rails, 351
coordination and strength training, 352
flooring, 345
footwear, 347
hallways, 347
hip protectors, 351–352
identification bracelets, 351
lighting, 345
other policies and procedures, 348
restraints, 349–350
known parameters of patient falls, 336–338
activity at time of fall, 337–338
length of stay, 338
location of falls, 337
time of day of falls, 337
type of units, 336–337
reducing employee falls, 353–356
all employees, 354
case study, 354
Crocs footwear, 355–356
dietary, 354
housekeeping, 355
laundry, 355
High-pH floor cleaners, 205
High-pressure sodium lights, 88
High-risk industries, profiles of, 361–379
all occupancies, 361–362
exterior controls, 361
interior controls, 362
hospitality (lodging), 362–367
employees, 363
guest room bathrooms, 363–364
inside hazards, 362–363
outside hazards, 362
playgrounds, 365–367
swimming pool/whirlpool areas, 364–367
mercantile (retail), 367–368
stadium-style seating (theater), 375–376
additional lighting and timing of lighting, 375–376
design and use of handrails, 375
employee orientation, 376
housekeeping and maintenance, 376
lighting and identification, 375
signage, 376
verbal warnings by ticket takers/ushers, 376
theater inspection log (common areas), 377
theaters, 369–371
trucking, 371–374
dock plates and dock boards, 373–374
falls from cabs, 372
falls from loading docks, 373–374
trucking fleets, sample footwear policy for, 378–379
impact of footwear on injuries, 378
sample footwear policy, 378
sample purchase arrangements, 379
Horizontal pull slipmeter (HPS), 121, 126
Hospitality (lodging), 362–367
employees, 363
guest room bathrooms, 363–364
inside hazards, 362–363
outside hazards, 362
playgrounds, 365–367
swimming pool/whirlpool areas, 364–367
HPS, *see* Horizontal pull slipmeter
HSE, *see* U.K. Health and Safety Executive
Hydrodynamic squeeze film, 111
Hydroxides, 206

I

IAAF, *see* International Association of Athletics Federations
IEA, *see* International Ergonomics Association
IESNA, *see* Illuminating Engineering Society of North America
IICRC Hard Surface Floor Maintenance Specialist, 217
Illuminating Engineering Society of North America (IESNA), 87, 90, 407
INRS, *see* National Research and Safety Institute
Institute of Transportation Engineers (ITE), 23
International Association of Athletics Federations (IAAF), 272
International Ergonomics Association (IEA), 183
International Organization for Standardization (ISO), 266–268
concerns, 267
flooring committees, 267–268
footwear standards, 306
ISO TC 216 footwear, 309
ISO TC 94 personal safety, 306
standards process, 266–267

ISO, *see* International Organization for Standardization
Italian standards, 265–266
ITE, *see* Institute of Transportation Engineers

J

James Machine, 121–126
  correlation to actual usage, 126
  Jablonsky low-friction model, 124
  operation, 121–122
  operational issues, 123–125
  standards, 125–126
  subsequent versions, 122–123

K

Kitchen slip/fall hazard assessment, 331–333
Kumho Tire Co. v. Carmichael (97-1709), 392–393

L

Landings, 387
Level walkway surfaces, 2–16
  access covers, 10
  architectural designs, 12
  bicycle racks, 12, 13
  case study, 4–5, 14–16
  curb construction and design, 6–8
  curb cutouts, 8–10
  curbing, 6–8
  curb marking, 8
  doorstops and other small trip hazards, 13
  drainage grates, 10–12
  other walkway impediments, 10–16
  planters, trash receptacles, and similar furnishings, 12
  pole farm, 12
  posts, 12
  rubberized sidewalks, 5–6
  sidewalks, 3–6
  temporary fixes, 13
Lighting, 87–91
  high-pressure sodium, 88
  IESNA lighting categories, 90
  lighting levels (categories), 89
  lighting levels (safety only), 88–89
  lighting transitions, 89–90
  light sources, 87–88
  maintenance, 90–91
  mercury vapor, 87
  metal halide, 88
Limonene, 206
Liquid dispersion (surface roughness), 111
Loss analysis, 94–95
  developing solutions, 95
  gathering of data, 94–95
  tracking exposure, 94
Lost earnings fraud indicators, 390

M

Management controls, 67–104
  building inspection and maintenance programs, 71–72
  construction, renovation, and special event planning, 93–94
  contractual risk transfer, 91–93
    fundamental guidelines, 93
    general administrative measures, 92
    specific control measures, 92–93
  engineering precept, 68
  facility and parking lot hourly inspection log, 96
  integrate controls, 68–70
    administrative controls, 70
    elimination, 69
    engineering controls, 69
    hierarchy of controls, 69
    personal protective equipment, 70
    program component design, 69
  lighting, 87–91
    high-pressure sodium, 88
    lighting levels (categories), 89
    lighting levels (safety only), 88–89
    lighting transitions, 89–90
    light sources, 87–88
    maintenance, 90–91
    mercury vapor, 87
    metal halide, 88
  loss analysis, 94–95
    developing solutions, 95
    gathering of data, 94–95
    tracking exposure, 94
  no "silver bullet", 68
  pedestrian traffic flow, behavioral safety and, 70–71
    adapt or adopt, 70–71
    natural paths/observation, 70
  sample slip resistance test results report, 104
  sample snow removal report, 97
  sample walkway surface design audit, 98–100
  sample walkway surfaces inspection, 101–103
  slip/fall hazard self-inspection programs, 72–74
  snow removal, recommended practices for, 78–86
    application of anti-icing, deicing and sand, 81–86
    guidelines for removal, 80–81
    objectives, 78–79

Index

personnel and responsibilities, 79
priorities for removal, 79–80
sample snow removal report, 97
snow storage, 81
spill and wet program, 74–78
electronic inspection system, 75–78
sweep/aisle walk logs, 74
Manual on Uniform Traffic Control Devices for Streets and Highways (MUTCD), 22, 23
MBPN, see Mean British Pendulum Number
Mean British Pendulum Number (MBPN), 243
Metasilicates, 206
MFS, see Morse Fall Scale
Military specifications (Navy), 166–167
    MIL-D-0016680c (ships) and MIL-D-18873B, 167
    MIL-D-17951C (ships), 167
    MIL-D-23003A(SH), 167
    MIL-D-24483A, 167
    MIL-D-3134J, 167
    MIl-W-5044C, 167
Missteps, definition of, 2
Morse Fall Scale (MFS), 338
MUTCD, see Manual on Uniform Traffic Control Devices for Streets and Highways

N

National Bureau of Standards (NBS), 131
    pendulum testers, 243
    Standard Static COF Tester, 131
National Floor Safety Institute (NFSI), 179–181
National Health Service (NHS), 335
National Institute for Standards and Testing (NIST), 131
National Playground Safety Institute (NPSI), 366
National Reporting and Learning System (NRLS), 337
National Research and Safety Institute (INRS), 271
National Safety Council (NSC), 183–184
National standards bodies
    overseas, see Overseas standards
    United States, see U.S. standards and guidelines
NBR, see Nitrile-butadiene rubber
NBS, see National Bureau of Standards
NEISS survey of stair accidents, 27
Neolite patent, 156–160
Neutral floor cleaners, 205, 206
NFPA International, 170–171
    NFPA 101/5000, 171
    NFPA 1901, 171
    requirements, 239
NFSI, see National Floor Safety Institute
NHS, see National Health Service
NIST, see National Institute for Standards and Testing
Nitrile-butadiene rubber (NBR), 297
Nonresilient flooring, 198–199
NPSI, see National Playground Safety Institute
NRLS, see National Reporting and Learning System
NSC, see National Safety Council
NSSN (National Resource for Global Standards), 255

O

Occupational Safety and Health Administration (OSHA), 161–163
    appendix B to 1926 subpart R: steel erection regulatory (3) [withdrawn], 163
    fire brigades 1910.156(e)(2)(ii), 162–163
    manlifts 1910.68(c)(3)(v), 162
    report on slip resistance for structural steel, 185–194
        availability of paints to meet slip resistance benchmark, 191–194
        5214 Federal Register/Vol. 66, No. 12/ Thursday, January 18, 2001/Rules and Regulations (pp. 5214–5218), 185
        "finish coat," 187–190
        hazard, 185–187
        test methods, 190–191
    safety signage, 57
    Section 1910.22 general requirements, 161–162
    Steel Erection Standard, 139
Official Vinyl Composition Tile (OVCT), 125
Olefin fiber mats, 326
OSHA, see Occupational Safety and Health Administration
OVCT, see Official Vinyl Composition Tile
Overseas standards, 237–293
    approaches to development standards, 291–293
        balance, 291–292
        canvass method, 292–293
        due process, 291
        full consensus method, 291–292
        industry method, 293
        not infallible, 292
        transparency, 291
    balancing a committee, 239
    braked-wheel/skiddometer instrument, 255
    digitized dragsleds, 247–251
        BOT-3000, 251
        issues, 249–251
        Tortus round robin, 249
        variations, 247
    DIN slipperiness classification, 242
    FIDO, 255

international standards, 255–268
   American National Standards Institute, 255
   ANSI Electronic Standards Store, 255
   Australian Commonwealth Engineering Standards Association, 262
   Australia/New Zealand standards, 262–265
   British standards, 260–262
   DIN oil-wet ramp test, 264
   European standards, 255–258
   German standards, 258–260
   Global Engineering Documents, 255
   International Organization for Standardization, 266–268
   Italian standards, 265–266
   NSSN (National Resource for Global Standards), 255
   Swedish standards, 262
national standards bodies, 273–290
   Afghanistan, 273
   Albania, 273
   Algeria, 273
   Angola, 273
   Antigua and Barbuda, 273
   Argentina, 273
   Armenia, 273
   Australia, 273
   Austria, 274
   Azerbaijan, 274
   Bahrain, 274
   Bangladesh, 274
   Barbados, 274
   Belarus, 274
   Belgium, 274
   Benin, 274
   Bhutan, 274
   Bolivia, 275
   Bosnia and Herzegovina, 275
   Botswana, 275
   Brazil, 275
   Brunei Darussalam, 275
   Bulgaria, 275
   Burkina Faso, 275
   Burundi, 275
   Camaroon, 275
   Cambodia, 276
   Canada, 276
   Chile, 276
   China, 276
   Colombia, 276
   Congo, The Democratic Republic of, 276
   Costa Rica, 276
   Cote-d'Ivoire, 276
   Croatia, 276
   Cuba, 277
   Cyprus, 277
   Czech Republic, 277
   Denmark, 277
   Dominica, 277
   Dominican Republic, 277
   Ecuador, 277
   Egypt, 277
   El Salvador, 277
   Eritrea, 278
   Estonia, 278
   Ethiopia, 278
   Fiji, 278
   Finland, 278
   France, 278
   Gabon, 278
   Georgia, 278
   Germany, 278
   Ghana, 279
   Greece, 279
   Grenada, 279
   Guatemala, 279
   Guyana, 279
   Honduras, 279
   Hong Kong, China, 279
   Hungary, 279
   Iceland, 279
   India, 280
   Indonesia, 280
   Iran, Islamic Republic of, 280
   Iraq, 280
   Ireland, 280
   Israel, 280
   Italy, 280
   Jamaica, 280
   Japan, 280
   Jordan, 281
   Kazakhstan, 281
   Kenya, 281
   Korea, Democratic People's Republic, 281
   Korea, Republic of, 281
   Kuwait, 281
   Kyrgyzstan, 281
   Lao People's Democratic Rep, 281
   Latvia, 281
   Lebanon, 282
   Lesotho, 282
   Libyan Arab Jamahiriya, 282
   Lithuania, 282
   Luxembourg, 282
   Macau, China, 282
   Macedonia, 282
   Macedonia, the former Yugoslav Republic of, 282
   Madagascar, 282
   Malawi, 283
   Malaysia, 283
   Malta, 283
   Mauritius, 283

Index   437

Mexico, 283
Moldova, Republic of, 283
Mongolia, 283
Montenegro, 283
Morocco, 283
Mozambique, 284
Myanmar, 284
Namibia, 284
Nepal, 284
Netherlands, 284
New Zealand, 284
Nicaragua, 284
Nigeria, 284
Norway, 284
Oman, 285
Pakistan, 285
Palestine, 285
Panama, 285
Papua New Guinea, 285
Paraguay, 285
Peru, 285
Philippines, 285
Poland, 285
Portugal, 286
Qatar, 286
Romania, 286
Russian Federation, 286
Rwanda, 286
Saint Vincent and the Grenadines, 286
Santa Lucia, 286
Saudi Arabia, 286
Senegal, 286
Serbia, 287
Seychelies, 287
Singapore, 287
Slovakia, 287
Slovenia, 287
South Africa, 287
Spain, 287
Sri Lanka, 287
Sudan, 287
Suriname, 288
Swaziland, 288
Sweden, 288
Switzerland, 288
Syrian Arab Republic, 288
Tajikistan, 288
Tanzania, United Republic of, 288
Thailand, 288
Togo, 288
Trinidad and Tobago, 289
Tunisia, 289
Turkey, 289
Turkmenistan, 289
Uganda, 289
Ukraine, 289
United Arab Emirates, 289
United Kingdom, 289
Uruguay, 289
Uzbekistan, 290
Venezuela, 290
Viet Nam, 290
Yemen, 290
Zambia, 290
Zimbabwe, 290
other dragsleds, 251–252
overseas organizations involved in slip resistance, 268–272
    Berufsgenossenschaftliches Institut fur Arbeitssicherheit, 272
    Commonwealth Scientific and Industrial Research Organisation, 270–271
    Finnish Institute of Occupational Health, 272
    Health and Safety Executive, 269
    INRS National Research and Safety Institute, 271
    International Association of Athletics Federations, 272
    SATRA Footwear Technology Centre, 269–270
    SlipAlert, 268
    U.K. Slip Resistance Group, 268–269
overseas standard development, 238–239
pendulum testers, 243–247
    ASTM E-303, 243–245
    issues, 245–247
    operation, 243
    pendulum numbers, 243
    U.K. Pendulum Test, 243
portable friction tester, 254–255
ramp tests, 239–242
    operational issues, 240–242
    slipperiness classifications, 242
roller-coaster tests, 252–254
    operational issues, 253
    results, 253–254
    SlipAlert operation, 252–253
slip and fall statistics overseas, 237–238
Over-the-spill cleanup kit, 324

P

Parking areas, 19–24
    speed bumps, 22–24
    tire stops, 19–22
PAST, *see* Portable Articulated Strut Tester
P&B statement, 134
Pedestrian traffic flow, behavioral safety and, 70–71
    adapt or adopt, 70–71
    natural paths/observation, 70
Pendulum numbers, 243

Pendulum testers, overseas standards, 243–247
  ASTM E-303, 243–245
  issues, 245–247
  operation, 243
  pendulum numbers, 243
Personal protective equipment (PPE), 70, 301
PFT, see Portable Friction Tester
Physical evaluation, 1–66
  accessibility, 57
  bathrooms, 52
  carpet, 56–57
  changes in levels, 24–26
    air steps, 24–25
    design, 25–26
  distractions, 2
  elevators, 52–53
  escalators, 54–55
  expectation, 1–2
    example, 1
    missteps, 1–2
  floor mats/entrances and exits, 43–52
    case study, 43–44, 48–49, 49–51
    general precautions, 44–45
    mat cleaning guidelines, 51–52
    mat design, 46
    mat design and arrangement, 46–47
    mat size, 48
    mat storage, 47–48
    minimizing contaminants, 45–46
    protection of hard flooring surfaces, 51
  guards (or guardrails), 39
  handrails, 33–39
    case study, 38–39
    graspability, 35–36
    supports, 36–38
  level walkway surfaces, 2–16
    access covers, 10
    architectural designs, 12
    bicycle racks, 12, 13
    case study, 4–5, 14–16
    curb construction and design, 6–8
    curb cutouts, 8–10
    curbing, 6–8
    curb marking, 8
    doorstops and other small trip hazards, 13
    drainage grates, 10–12
    other walkway impediments, 10–16
    planters, trash receptacles, and similar furnishings, 12
    pole farm, 12
    posts, 12
    rubberized sidewalks, 5–6
    sidewalks, 3–6
    temporary fixes, 13
  parking areas, 19–24
    speed bumps, 22–24
    tire stops, 19–22
  ramps, 40–43
    design, 40–42
    landings, 43
  reference standards, 59–61
    American National Standards Institute, 59
    ANSI A1264.1, 59–60
    ANSI A1264.2, 60
    ASME International, 61
    ASTM F1637 Practice for Safe Walking Surfaces, 59
    ASTM International, 59
    ICC/ANSI A117.1, 60
    model building codes, 60–61
    NFPA International, 59
    NFPA 101 Life Safety Code (2006), 59
    Occupational Safety and Health Administration, 60
  signage, 57–58
  slip and fall self-assessment checklist, 62–63
  stairs, 26–33
    components and measurements, 28
    good design and condition, 29
    NEISS survey of stair accidents, 27
    stair design, 29–30
    stair landings, 32–33
    stair missteps (overstep and heel scuffs), 26–27
    stair nosings, 30–31
    stair statistics, 27–28
    stair visibility, 31–32
  state-mandated building codes, 64–66
  water, level walkway surfaces and, 16–19
    case study, 18–19
    water accumulation, 17–18
PIAST, see Portable inclinable articulated strut tester
Planters, physical evaluation of, 12
Playgrounds
  CPSC statistics, 266
  top 12 problems found at, 366
Pole farm, 12
Polymerization, definition of, 319
Polyvinyl chloride (PVC), 297
Portable Articulated Strut Tester (PAST), 131
Portable Friction Tester (PFT), 254
Portable inclinable articulated strut tester (PIAST), 121
Posts, physical evaluation of, 12
PPE, see Personal protective equipment
Precision and Bias (P&B) statement, 134
PVC, see Polyvinyl chloride

R

Ramps, 40–43
  design, 40–42
  landings, 43

Index

RAPRA, *see* Rubber and Plastics Research Association
Resilient Floor Covering Institute (RFCI), 182
Restaurant Insurance Corp. (RIC), 311
RFCI, *see* Resilient Floor Covering Institute
Risk transfer, contractual, 91–93
    fundamental guidelines, 93
    general administrative measures, 92
    specific control measures, 92–93
    subcontracted maintenance, 91
RMA, *see* Rubber Manufacturers' Association
Rochester Midland Corporation, 217
Roller-coaster tests, overseas standards, 252–254
    operational issues, 253
    results, 253–254
    SlipAlert operation, 252–253
Rubberized sidewalks, physical evaluation of, 5–6
Rubber Manufacturers' Association (RMA), 143
Rubber and Plastics Research Association (RAPRA), 143, 268
Rubbers, 143–144
    Four S, 143–144
    neoprene, 143
    TRRL or TRL rubber, 144

S

Saponification, 205
SATRA Footwear Technology Centre, 269–270
SBR, *see* Styrene-butadiene rubber
SCOF, *see* Static coefficient of friction
Self-inspection programs, slip/fall hazard, 72–74
Self-washing floors, 322–323
Sidewalks
    comparison of rubber, concrete, and asphalt, 6
    physical evaluation of, 3–6
Signage, OSHA guidelines, 57
SJI, *see* Steel Joist Institute
SlipAlert, 268
Slip and fall self-assessment checklist, 62–63
Slip index, 127
Slip resistance, principles of, 105–119
    causation between incident and injury, 108–109
    classes of tribometers, 117–118
        articulated strut, 118
        horizontal pull (dragsled), 117
        pendulum, 117
    creep, 106
    human perception of slipperiness, 116
    Hunter Machine, 118–119
    mechanics of walking, 107–108
    principles of friction, 105–106
    slip resistance defined, 106–107
    slip resistance factors, 107
    slip resistance scale, 109–110
    stick-slip phenomenon, 106
    surface roughness, 110–115
        height or sharpness, 111
        hydrodynamic squeeze film, 111
        liquid dispersion, 111
        measuring roughness, 111–112
        roughness measurement standards, 114–115
        roughness thresholds, 112–114
    wet surfaces, 115–116
        hydroplaning, 115
        sticktion, 115–116
Slip resistance, test methods
    not recognized in U.S. (comparison chart), 152–155
    U.S. (comparison chart), 151
Slip, trip, and fall (STF), 335
    accident severity (healthcare industry), 335
    hazard mapping, 73
    hazard self-inspection programs, 72–74
    prevention program, 348
    self-assessment checklist, 62–63
Slop mopping, 206
Snow removal, recommended practices for, 78–86
    amount of sand and slip resistance, 86
    anti-icing guidelines, 84
    application of anti-icing, deicing and sand, 81–86
        after the storm, 86
        before the storm, 83–84
        during the storm, 84–86
        temperature and ice melting, 82–83
    efficacy of sand on icy surfaces, 85
    guidelines for removal, 80–81
    melting characteristics, 82
    melt times for salt, 82
    objectives, 78–79
    personnel and responsibilities, 79
        contracted snow removal, 79
        custodians, 79
        facility manager, 79
        grounds maintenance staff, 79
    priorities for removal, 79–80
    sample snow removal report, 97
    snow storage, 81
    variables affecting application rate, 83
Spill and wet program, 74–78
    electronic inspection system, 75–78
    sweep/aisle walk logs, 74
Stadium-style seating (theater), 375–376
    additional lighting and timing of lighting, 375–376
    design and use of handrails, 375
    employee orientation, 376
    housekeeping and maintenance, 376

lighting and identification, 375
signage, 376
verbal warnings by ticket takers/ushers, 376
Stairs, 26–33
  components and measurements, 28
  good design and condition, 29
  NEISS survey of stair accidents, 27
  stair design, 29–30
  stair landings, 32–33
  stair missteps (overstep and heel scuffs), 26–27
  stair nosings, 30–31
  stair statistics, 27–28
  stair visibility, 31–32
Standard Pedestrian Hard Rubber, 143–144
Standards and guidelines, see Overseas standards; U.S. standards and guidelines
State-mandated building codes, 64–66
Static coefficient of friction (SCOF), 105, 167, 196
  classic physics definition of, 106
  equipment measuring, 107
  slip resistance scale, 109
Steel Joist Institute (SJI), 192
STF, see Slip, trip, and fall
Stick-slip phenomenon, 106
Sticktion, 228
Styrene-butadiene rubber (SBR), 142, 297
Surface roughness, 110–115
  height or sharpness, 111
  hydrodynamic squeeze film, 111
  liquid dispersion, 111
  measuring roughness, 111–112
  roughness measurement standards, 114–115
  roughness thresholds, 112–114
Surfactants, 206
Swedish standards, 262
Swimming pool/whirlpool areas (lodging), 364–367
  decking, 364
  pool, 364
  rules, 365
  signs, 265

T

TCNA, see Tile Council of North America
TESS, see U.S. Trademark Electronic Search System
Tile Council of North America (TCNA), 178–179
Trademark Electronic Search System, 142
Transport and Roads Research Laboratory (TRRL), 144
Trash receptacles, physical evaluation of, 12

Tribometers, classes of, 117–118, see also U.S. tribometers
  articulated strut, 118
  horizontal pull (dragsled), 117
  pendulum, 117
Trip hazards, physical evaluation of, 13
TRRL, see Transport and Roads Research Laboratory
Trucking, 371–374
  dock plates and dock boards, 373–374
  falls from cabs, 372
  falls from loading docks, 373–374
  fleets, sample footwear policy for, 378–379
  impact of footwear on injuries, 378
  sample footwear policy, 378
  sample purchase arrangements, 379

U

U.K. Health and Safety Executive (HSE), 112, 268
  DIN test methods and, 242
  floor cleaning information sheet, 354
  footwear, 308
  roughness thresholds, 11
  workplace slips and trips prevention, 269, 312
U.K. Health and Safety at Work etc Act, 261
U.K. Pendulum Test, 243
U.K. Slip Resistance Group, 268–269
UL, see Underwriters Laboratories
Underwriters Laboratories, 173–176, 229
U.S. slip resistance test methods (comparison chart), 151
U.S. standards and guidelines, 161–194
  access board recommendations, 166
  American National Standards Institute, 171–173
    ANSI A1264.2-2006, 172
    ANSI A1264.3-2007, 173
    ANSI A137.1-1988, 173
  Americans with Disabilities Act, 164–166
  ASTM International, 167–170
    ASTM F695, 169
    ASTM F802, 170
    ASTM F1240, 169
    ASTM F1637, 169–170
    ASTM F1646, 170
    ASTM F1694, 170
    ASTM F2048, 170
    technical committee ASTM F-13, 168–170
  federal specifications, 166
  military specifications (Navy), 166–167
    MIL-D-0016680c (ships) and MIL-D-18873B, 167
    MIL-D-17951C (ships), 167
    MIL-D-23003A(SH), 167

Index

MIL-D-24483A, 167
MIL-D-3134J, 167
MIl-W-5044C, 167
model building codes, 176
NFPA International, 170–171
  NFPA 101/5000, 171
  NFPA 1901, 171
obsolete standards, 176–177
  ASTM D-21 gray pages, 177
  ASTM D4518-91, 177
  Federal Test Method Standard 501A, method 7121, 176
  U.S. General Services Administration specification PF-430C(1), 176–177
Occupational Safety and Health Administration, 161–163
  appendix B to 1926 subpart R: steel erection regulatory (3) [withdrawn], 163
  fire brigades 1910.156(e)(2)(ii), 162–163
  manlifts 1910.68(c)(3)(v), 162
  Section 1910.22 general requirements, 161–162
OSHA report on slip resistance for structural steel, 185–194
  availability of paints to meet slip resistance benchmark, 191–194
  5214 Federal Register/Vol. 66, No. 12/ Thursday, January 18, 2001/Rules and Regulations (pp. 5214–5218), 185
  "finish coat," 187–190
  hazard, 185–187
  test methods, 190–191
Underwriters Laboratories, 173–176
U.S.-based industry associations and organizations involved with slip resistance, 177–184
  American Academy of Forensic Sciences, 184
  Ceramic Tile Institute of America, 177–178
  Consumer Specialty Products Association, 181–182
  Contact Group on Slips, Trips, and Falls, 183
  Footwear Industries of America, 182–183
  International Ergonomics Association, 183
  National Floor Safety Institute, 179–181
  National Safety Council, 183–184
  Resilient Floor Covering Institute, 182
  Tile Council of North America, 178–179
U.S. Trademark Electronic Search System (TESS), 142
U.S. tribometers, 121–160
  ASTM "gold" standard, progress on, 146–148
    benefits, 148

    field testing with walkway tribometers, 147
    thresholds, 147–148
    validation and calibration of walkway tribometers, 146
  Brungraber Mark I Articulated Strut Slip Tester, 131–134
    availability and standards, 132–133
    F15 consumer products, 133–134
    NBS Standard Static COF Tester, 131
    operation, 132
  Brungraber Mark II Portable Inclinable Articulated Strut Slip Tester, 134–137
  comparison chart
    slip resistance test methods not recognized in U.S., 152–155
    U.S. slip resistance test methods, 151
  English XL variable incidence tribometer, 137–141
    D01 paint and related coatings, materials, and applications, 141
    WK11411, 141
    Wood-Plastic Composite deck boards, 140
  equipment calibration and maintenance, 145–146
  groundbreaking research, 148–150
  horizontal dynamometer pull-meter, 128–131
    apparatus issues, 130–131
    scope, 129
    significance and use, 129–130
    slip-resistant products, 131
  horizontal pull slipmeter, 126–128
  James Machine, 121–126
    correlation to actual usage, 126
    Jablonsky low-friction model, 124
    operation, 121–122
    operational issues, 123–125
    standards, 125–126
    subsequent versions, 122–123
  Neolite patent, 156–160
  operator qualifications of competency, 145
  slip index, 127
  test pad material, 141–144
    actual footwear bottoms, 144
    leather, 141–142
    Neolite® test liner, 142–143
    rubbers and other footwear materials, 143–144
  uses, 144–145
    accident investigation, 145
    claims defense/documentation, 145
    little or no prior history, 144–145
    problem identification (prevention), 144

V

Variable-angle tribometers (VAT), 141

Variable Incidence Tribometer (VIT), 121, 137, 145
VAT, *see* Variable-angle tribometers
VIT, *see* Variable Incidence Tribometer

W

Walking, mechanics of, 107–108
Walkway surface design audit, sample, 98–100

Wood-Plastic Composite (WPC) deck boards, 140
WPC deck boards, see Wood-Plastic Composite deck boards

Z

Zinc-rich primers, 192